Stainless Steels: An Introduction and Their Recent Developments

Edited By

Joseph Ki Leuk Lai

Department of Physics and Materials Science. City University of Hong Kong

Kin Ho Lo

Department of Electromechanical Engineering. The University of Macau

Chan Hung Shek

Department of Physics and Materials Science. City University of Hong Kong

eBooks End User License Agreement

CONTENTS

FOREWORD

I have no hesitation in recommending this text. It is best described as a considerable work of scholarship. It makes a substantial contribution to the literature and will be of real value to both industry and academy.

Industrialists concerned with the production of stainless steels will find the logical development in this e-Book of real value. Similarly, those concerned with the use of stainless steels in the production of components are catered for in considerable depth. Above all, those interested in the research aspects will find a text of substantial importance. It would certainly make an excellent support text for a post-graduate course on stainless steels. The emphasis on recent developments is to be appreciated.

The text begins with a general introduction (Chapter 1) which briefly covers the history of stainless steels and classifies them according to their microstructures. Chapters 2-7 then consider each of the classifications, ferritic, austenitic, martensitic, duplex, precipitation-hardening and high nitrogen stainless steels. The various problems that occur with particular classes of stainless steels such as embrittlement, sensitisation intergranular corrosion, etc., are outlined. These chapters (2-7) also cover the control of the various transformations (martensitic, spinodal precipitation) together with the control of properties.

Chapter 8 is concerned with the phases which precipitate in stainless steels. Some of the problems encountered with stainless steels are then covered in much more detail in Chapter 9 with the emphasis again on recent advances and discoveries. Here particularly useful reviews are given to hydrogen embrittlement, sensitisation, metal dusting, pitting, stress corrosion cracking, fatigue and creep.

The final four chapters deal with novel and new applications of stainless steels (Chapter 10), improvement of bulk and surface properties of stainless steels (Chapter 11), coloration of stainless steels (Chapter 12) and powder metallurgy of stainless steels (Chapter 13),

Each chapter is associated with a very extensive reference list. These reference lists includes the really significant historical references but the emphasis is, as it should be, on recent papers which has helped to develop the field of stainless steels and will continue to do so.

I feel this text will make a significant contribution to the field for many years and hope the authors will find the time to update it occasionally.

Brian Ralph
School of Engineering and Design,
Brunel University,
W. London

PREFACE

The discovery of stainless steels is generally thought to originate in Europe about a century ago, although many people find this claim debatable. Even within Europe, there is controversy over the 'birthplace' of stainless steel. While there is debate on who should be credited with the discovery of stainlessness, it is incontrovertible that stainless steels have contributed significantly to the wellbeing of humanity. The use of stainless steels ranges from low-end products such as water buckets to very high-end ones, such as the spacecraft propellers and casings. It is no exaggeration that stainless steels are omnipresent in our daily lives. Although stainless steels are such time-honoured materials, there is still a lot of room for improvement. Research and development of these materials are actively pursued at various institutions worldwide. A search with the keywords 'stainless steel' in popular databases such as Science Citation Index-Expanded or Scorpus would reveal thousands of recent publications. This is a vividly testimony to the perennial interest on stainless steels as a key research topic by various research in the world.

Many excellent monographs on stainless steels have been published over the years. The epitomes are *An Introduction to Stainless steels* by Lula and the *Handbook of Stainless steels* by Peckner and Bernstein. These classic texts give overviews on the various classes of stainless steels and were published in the 1970's and 1980's, respectively. Since then, many exciting developments on stainless steels have been made. These include advances in the fabrication of ultra-high nitrogen stainless steel, the discovery of new phases, techniques for refining grain sizes down to the nanometre scale, to name just a few. Phenomena that were not adequately explicated in the 1980's, such as hydrogen-enhanced localised plasticity (HELP), have been satisfactorily explained recently. In addition, stainless steels have been employed in many new and novel applications that were unheard of when these early texts were published. Examples of these applications are the use of stainless steels for temperature sensing, signature authentication, and targeted drug delivery.

Besides the afore-mentioned novel applications, the interest in employing stainless steels in many conventional applications has been rekindled lately. The construction industry is an illustrative example. Although it was long ago when stainless steels was suggested for use as concrete reinforcing bars, ordinary carbon steel still dominate because of their lower initial costs. However, it has been conclusively proved in several recent cost-cycle analyses that stainless steels are more economical than carbon steel in the long run. Nevertheless, the widespread use of stainless steels as concrete reinforcing bars is still hindered by the relative lack of knowledge and experience of these materials for this application. In 2006, an extensive study was undertaken by researchers in the Nordic Innovation Centre in Norway with the aim to 'overcome the knowledge gap for application of stainless steels reinforcements in concrete structures'. This study stressed that although stainless steels belonged to the class of traditional materials, there were still many 'knowledge gaps' that needed to be bridged.

Surprisingly, there are few monographs reporting recent developments on stainless steels in the past 15 years. Texts such as V. G. Gavriljuk's *High Nitrogen Steel* and Mudali and Raj's *High Nitrogen Steel and Stainless steels* are recent examples. Nevertheless, these recent monographs were specifically devoted to the high nitrogen varieties, and as a consequence the numerous new discoveries and developments in the various classes of conventional stainless steels were not included.

This e-Book is intended to fill this gap and report the latest developments on stainless steels since the publications of the above-mentioned texts. Discoveries and developments in various aspects of stainless steels within the past 15 years are introduced. Besides recent developments, the historical backgrounds of some of the topics covered in this e-Book are included (e.g. the timeline of developments made on 475°C embrittlement). The e-Book is divided into two parts: the first part gives an overview of the many classes of stainless steel, while the second part describes recent innovations and novel applications. Chapters 7, 8, 9 and 11 draw heavily from our review paper in *Materials Science and Engineering R Reports* (Vol.65 (3-4), 2009, pp.39-104). We are grateful to Elsevier for granting us the permission for adopting some of the contents of the paper for use in this e-Book.

The Chief Editor (J. K. L. Lai) set the tone for this e-Book and the Associate Editor K. H. Lo acted as the lead author for most of the chapters. During the writing of the first part of this e-Book, the authors made very frequent references to the time-honoured, outstanding handbook on stainless steels by Peckner and Bernstein (*Handbook of Stainless Steel*. Peckner D and Bernstein IM (Eds). McGraw-Hill New York 1977). We therefore would like to express our heartfelt gratitude to the authors and editors of this handbook for producing such an exceptional reference on stainless steel. We also would like to thank the authors of the references that we have quoted in this e-Book. We sincerely hope that those colleagues whose works we have cited and quoted in this e-Book will point out to us any misquotations and misinterpretations, such that we can make corrections for the future editions of this e-Book. Wherever possible, we have tried to acknowledge the contributors whose results we have quoted. However, due to the sheer amount of publications that we have referenced, we are afraid that we might have inadvertently failed to acknowledge some of our colleagues. We will be very appreciative if they can inform us of any such missing acknowledgements, so that we can credit them with their efforts in the future editions.

The editors are greatly indebted to Prof. Brian Ralph for his invaluable comments on the text. The Editors are also indebted to the many anonymous reviewers of our e-Book manuscripts. The Editors would like to record a special thank you to Salma Safaraz, our main contact in Bentham. Without the assistance of Salma, the production of this e-Book would not have been so smooth. Last but not the least, the Editors would like to acknowledge the supports they received from City University of Hong Kong and the University of Macau.

Joseph Ki Leuk Lai
City University of Hong Kong

Kin Ho Lo
The University of Macau

Chan Hung Shek
City University of Hong Kong

CHAPTER 1

A General Introduction to Stainless Steels

Abstract: This chapter gives an overview on the various types of stainless steels and their general classification based on microstructures. This chapter also goes through briefly the history of the discovery of 'stainlessness' and the status quo of stainless steels production in several major economies. Motivations for writing this book and some of the new applications that utilise stainless steels (which are covered in chapter 10) are given at the end of this chapter.

Keywords: Stainless steel, carbon steel, high nitrogen stainless steel, Pt-alloyed stainless steel, J phase, G phase, stainlessness, retention of stiffness, retention of strength, Fe-Cr system, severe plastic deformation, stacking fault energy.

1. AN OVERVIEW OF STAINLESS STEELS AND THEIR GENERAL CLASSIFICATION

The importance of stainless steels to our society is vividly demonstrated by the plenitude of applications that rely on their use. These applications range from the low-end, like cooking utensils and furniture, to the very sophisticated, such as space vehicles [1]. The ubiquity of stainless steels in our daily life makes it impossible to enumerate all their applications.

The word 'steel' means that the material is iron-based, while the adjective 'stainless' implies absence of staining, rusting or corroding in environments where normal steels are susceptible (for instance, in relatively pure, dry air). In order for steels to be stainless, at least about 11wt% of chromium must be alloyed to the base material. At this Cr level, an adherent, self-healing chromium oxide can form on or at the steel surface in relatively benign environments. However, to stave off pitting and rusting in more hostile environments (say, in moist atmospheres or polluted environments) or in the presence of elements like carbon, higher Cr contents and other alloying element (Mo, Ni, *e.g.*) must be added. In addition to being corrosion-resistant, stainless steels also do not discolour in a normal atmospheric environment. The crown of the Chrysler building, which was made of an austenitic stainless steels (tradename Nirosta), stills shines nowadays, although the building was completed in the year 1930 and is located near the seaside. Stainless steels also outperform ordinary steels on high-temperature mechanical properties. Stainless steels are much better in terms of fire resistance and retention of strength and stiffness at elevated temperatures compared with carbon steels (Figs. **1** and **2**) [2-5].

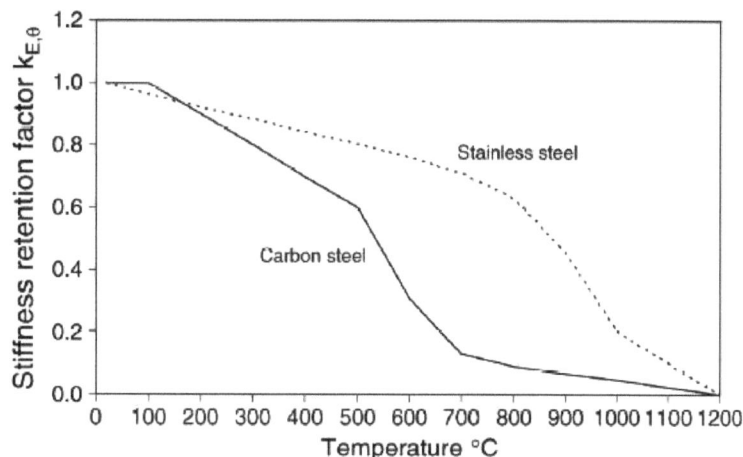

Figure 1: Stainless steels performs better than carbon steel does in terms of retention of stiffness [Gardner L, Baddoo NR. Fire testing and design of stainless steels structures. Journal of Constructional Steel Research 2006; 62(6): 532-543. With permission for reproduction from Elsevier].

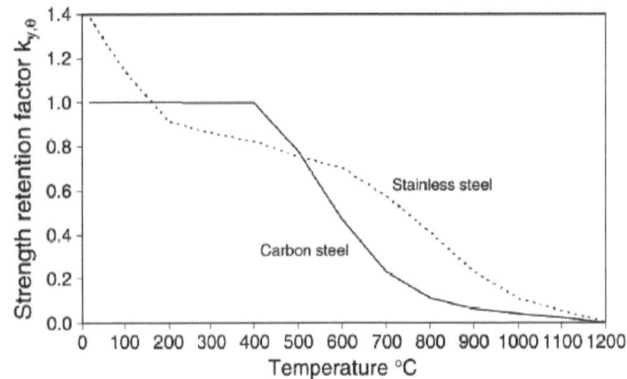

Figure 2: Stainless steels retains strength better than carbon steel does at high temperatures [Gardner L, Baddoo NR. Fire testing and design of stainless steels structures. Journal of Constructional Steel Research 2006; 62(6): 532-543. With permission for reproduction from Elsevier].

As early as about the beginning of the 1820's, the Frenchman Pierre Berthier and the Englishmen Stoddard and Farraday had noted Cr-bearing iron was more resistant to attack by certain acids. Nevertheless, these early materials were not stainless steels because of their low Cr contents. The pernicious effect of carbon on corrosion resistance was recognised in 1875 by Brustlein of France. In 1904, Leon Guilllet, another Frenchman, undertook extensive studies on the metallography, constitution and mechanical properties of a range of iron-chromium alloys, which nowadays are known as 410, 420, 442, 446 and 440C. He also studied an iron-chromium-nickel alloy whose composition resembled the present-day 300 series. P. Monnartz and W. Borchers, in Germany, were arguably the first ones to recognise that passivity (and so stainlessness) was associated with at least about 12% Cr. The discovery of stainlessness in steels struck almost simultaneously in the USA, the UK and Germany, leading to recognitions and patents. Harry Brearley of England is often referred to as the initiator of the age of stainlessness. He made and studied a genuine stainless steels of 12.8% Cr and 0.24% C in 1913. In 1915 and 1916, Brearley obtained patents in Canada and the USA for martensitic alloys with 9 to 16% Cr and less than 0.7% C. In Germany, Edward Maurer and Benno Strauss, during the period of 1909 to 1914, brought the austenitic stainless steels to industrial prominence. Maurer and Strauss worked on a range of austenitic stainless steels containing 15-40%Cr, <20% Ni and <1% C. In the USA, Dantsizen realised that alloys similar to those that were being studied by Brearley were stainless if the carbon content was kept to low levels. The ferritic stainless alloys were used as lead-in wires for light bulbs and for turbine blades. In 1926, Monypenny published a ground-breaking book entitled 'Stainless Iron and Steel'. Bain and co-workers were recognised for their contributions to the development of the constitution diagrams of iron-chromium-carbon alloys. The importance of the work by Bain on the gamma-loop, *i.e.*, the range of composition and temperature over which austenite is stable, and the effects of alloying elements on the retention of austenite at room temperature cannot be overstressed. Contributors since the 1930's are more difficult to identify, because of the sheer number of research efforts on almost all aspects of stainless steels.

While the Fe-Cr system forms the basis, modern stainless steels, besides Cr, also contain a myriad of other alloying elements whose presence enhances specific properties. For example, pitting resistance is enhanced when Mo is present. Ni may be used if it is desired to have some austenite in the microstructure. Cu is used to form the strengthening precipitates in precipitation-hardening stainless steels.

The superferritics and superaustenitics contain very generous amounts of alloying elements. When Cr and Ni are present in substantial amounts, the resulting alloy is called heat-resisting alloy. Although a jumble of alloying elements may be present in stainless steels, their total content is usually kept below the iron content, for the sake of maintaining the resulting alloy to be a steel.

Three main types of microstructures exist in stainless steels, *i.e.*, ferritic, austenitic and martensitic. These microstructures may be obtained by adjusting steel chemistry. Out of these three main microstructures, stainless steels may be categorised into several main classes [1, 6]. These are: **1**. ferritic stainless steels, **2**. austenitic stainless steels, **3**. martensitic stainless steels, **4**. duplex stainless steels **5**. precipitation hardening stainless steels and 6. Mn-N substituted austenitic stainless steels. The different classes of stainless steels possess different properties. For instance, while fully austenitic stainless steels are non-magnetic at room temperature, their martensitic and ferritic counterparts are ferromagnetic. The different properties of the various stainless steels have been studied extensively for a very long period and thus are very well documented in the literature. An early handbook on stainless steels, which contains a series of informative articles, is a very good source on this topic [7].

2. STATUS QUO OF GLOBAL PRODUCTION AND CONSUMPTION OF STAINLESS STEELS

The importance of stainless steels may be appreciated by looking at the tonnages of their production and consumption in some of the major economies (Table **1**).

Table 1: Production and consumption of stainless steels in some of the major economies (production*/consumption)** [6]

	Year						
	1990	**1991**	**1992**	**1993**	**1994**	**1995**	**1996**
China	240/185	260/325	250/470	319/649	325/631	380/717	270/941
France	797/380	772/400	814/408	768/340	919/440	980/518	972/413
Germany	1146/900	1154/935	1163/975	1194/915	1419/960	1490/1093	1280/1061
Italy	574/524	606/570	645/609	720/638	854/900	1017/1031	862/988
Japan	3130/1836	3357/1985	3148/1760	3213/1750	3449/1837	3925/2045	3891/2105
UK	388/300	374/255	388/308	433/302	531/330	548/310	557/291
USA	1851/1537	1708/1479	1808/1612	1774/1742	1835/1960	2055/2015	1870/2088

* Liquid metal.

** Figures in 1000 tonnes.

Overall, the figures show an upward trend on a yearly basis in the 1990s. While Table **1** shows the trend in the last century, Table **2** gives more up-to-date figures of major continents since the year 2000. Table **2** clearly shows that the tonnages of production of stainless steels and heat resisting steels are constantly on the rise.

Table 2: Tonnages of stainless and heat resisting crude steel produced in major continents from 2001 to 2007 (in 000 metric tonnes) [8].

	2001	**2002**	**2003**	**2004**	**2005**	**2006**	**2007**
Western Europe/Africa	8210	8628	9043	9422	8823	9972	9700
Central and Eastern Europe	285	279	322	318	310	363	400
The Americas	2289	2735	2830	2933	2688	2951	2850
Asia	8403	9048	10645	11897	12498	15074	16850
World	19187	20690	22840	24570	24319	28359	29800

Table **2** shows that Asia, whose economy is fast expanding, accounts for a large portion of the stainless steels produced worldwide. The tonnage produced in China, which has overtaken Japan as the world's leading producer of stainless steels [8] and reached a per capita consumption over 4.6 kilogrammes in 2006 [9], is particularly high [10].

The above being said, stainless steels still lag far behind carbon steels in tonnages of crude production. It has been suggested recently that stainless steels production is less than 1% of carbon steels, although yearly production of flat stainless steels has been growing at an average of 6% for more than 50 years [11].

3. MOTIVATION FOR WRITING THIS e-BOOK

Although they constitute an 'old' alloy system (the 'birth' of stainless steels dates back to about the early 1900s) and very extensive research effort has been undertaken on understanding them, stainless steels have never been sidelined in materials science. The fact that a search in popular databases (such as ScienceDirect® and the Science Citation Index Expanded®) with the keywords 'stainless steel' easily turns up thousands of recent articles is a cogent testimony to this statement. The sheer volume of recent publications on stainless steels vividly shows that there are still unresolved problems and 'uncharted territories' in this time-honoured alloy system. The following two paragraphs present very briefly some of the results obtained by recent workers.

Improvements and refinements to existing stainless steels have been made (and are still being made). New steels have been designed for specific applications. For instance, a new Pt-alloyed stainless steels that possesses fluoroscopic radiopacity (radiopaque) has been produced for use in stents (known as Platinum-Enhanced Stainless steels Stent (PRESS)) [12, 13]. Besides steel composition, new developments on heat treatment, fabrication methods, grain refining techniques, identification of new phases (*e.g..*, the J phase) have all been made quite recently.

Some of the recent papers have wrapped up the unfinished works undertaken by early workers. For instance, while the G phase was discovered and characterised a long time ago, a model for its formation had not been formally formulated until recently by Mateo *et al.* [14]. Some of the problems that seemed to be well settled long before have been re-examined in a number of recent publications and new views have been put forward. For example, deformation-induced martensite in metastable austenitic stainless steels is generally believed to be possible *via* the formation of the ε martensite. However, this is thought to be unlikely by Hedstrom *et al.* [15] in their recent work. Another example is the effect of nitrogen on the stacking fault energy (SFE). While a number of early works concluded that N monotonically suppressed the SFE in austenitic stainless steels as its content increased, recent studies, however, have unambiguously proved that the effect of N on SFE is rather complicated and far from being monotonic [16, 17]. Needless to say, recent workers have been able to make new discoveries with an arsenal of sophisticated technologies that were unavailable to early workers. For instance, using 3DXRD, Hedstrom *et al.* [18] have been able to study in-situ the $\gamma \rightarrow \varepsilon$ transformation of individual grains in a polycrystalline AISI301 steel and shown that this transformation occurs in a very localised manner. New stainless steels offering very superior properties have been made [19]. A variety of methods (using very severe plastic deformation) for fabricating stainless steels of grain sizes in the nanometre scale have also been demonstrated recently. Transmission electron microscopes of higher resolution have enabled workers to better characterise the microstructural evolution during very severe plastic deformations (refer to the series of recent works by Belyakov *et al.* [20-23], for example). Several recent publications seem to have settled controversies, such as the main mechanism responsible for causing hydrogen embrittlement. Several recent works have convincingly demonstrated that HELP (hydrogen-enhanced localised plasticity) is the underlying mechanism for causing hydrogen embrittlement in austenitic stainless steels [24].

A large chunk of recent research, whose results shall be presented later on in this e-Book, has been devoted to high nitrogen steels and their stainless siblings. Even though high-N stainless steels are certainly not new inventions, systematic and large scale research on them seems to gain ascendance only since the 1990s, as evidenced by the publication of some frequently cited monographs [25, 26] and a series of international conferences that are dedicated specifically to high-N steels (refer to the prefaces in the afore-mentioned monographs for details of these conferences). In fact, even in 1998, which was not a long time ago, studies aimed at formulating the guidelines on alloy design for high-N duplex stainless steels were still encountered in the literature [27]. A few new approaches for introducing high levels of N into stainless steels were only published within the last decade.

REFERENCES

[1]   Lula RA. Stainless Steel. USA: American Society for Metals; 1986.

[2]   Gardner L. The use of stainless steels in structures. Prog Struc Engin Mat 2005; 7(2): 45-55.

[3]   Gardner L, Ng KT. Temperature development in structural stainless steels sections exposed to fire. Fire Safety J 2006; 41(3): 185-203.

[4]   Gardner L, Baddoo NR. Fire testing and design of stainless steels structures. J Constr Steel Res 2006; 62(6): 532-43.

[5]   Sakumoto Y, Nakazato T, Matsuzaki A. High-temperature properties of stainless steels for building structures. J Struct Eng 1996; 122(4): 399-406.

[6]   Beddoes J, Parr JG. Introduction to Stainless Steels (3rd ed). Materials Park, Ohio. ASM International. 1999.

[7]   Peckner D, Bernstein IM, EDS. Handbook of Stainless Steels. McGraw-Hill Inc. 1977.

[8]   International Iron and Steel Institute. http://www.worldstainless.org (accessed Mar 14, 2008).

[9]   People's Daily Online. http://english.peopledaily.com.cn. (accessed Feb 13, 2010)

[10]   Anonymous. World stainless steels production rises by 30%. Adv Mater Process 2007; 165(2): p.20.

[11]   Charles J. Duplex stainless steels – a review after DSS'07 held in Grado. Steel Res Intl 2008; 79(6): 455-75.

[12]   Craig CH, Friend CM, Edwards MR, *et al.* Mechanical properties and microstructure of platinum enhanced radiopaque stainless steels (PERSS) alloys. J Alloy Compd 2003; 361(1-2): 187-99.

[13]   Craig CH, Radisch HR, Trozera TA, *et al.* In: ASTM STP 1438 (eds G.L.Winters and M.J.Nutt), ASTM International, West Conshohocken, PA, 2003, pp. 28-38.

[14]   Mateo A, Llanes L, Anglada M, *et al.* Characterization of the intermetallic G-phase in an AISI 329 duplex stainless steel. J Mater Sci 1997; 32(17): 4533-40.

[15]   Hedstrom P, Lienert U, Almer J, *et al.* Stepwise transformation behavior of the strain-induced martensitic transformation in a metastable stainless steel. Scr Mater 2007; 56(3): 213-6.

[16]   Gavriljuk VG, Berns H, Escher C, *et al.* Grain boundary strengthening in austenitic nitrogen steels. Mater Sci Eng A 1999; 271(1-2): 14-21.

[17]   Yakubtsov IA, Ariapour A, Perovic DD. Effect of nitrogen on stacking fault energy of FCC iron-based alloys. Acta Mater 1999; 47(4): 1271-79.

[18]   Hedstrom P, Lienert U, Almer J, *et al.* Elastic strain evolution and epsilon-martensite formation in individual austenite grains during in situ loading of a metastable stainless steel. Mater Lett 2008; 62(2): 338-40.

[19]   Yamamoto Y, Brady MP, Lu ZP, *et al.* Creep-resistant, Al2O3-forming austenitic stainless steels. Science 2007; 316(5823): 433-6.

[20]   Belyakov A, Miura H, Sakai T. Dynamic recrystallization under warm deformation of a 304 type austenitic stainless steel. Mater Sci Eng A 1998; 255(1-2): 139-47.

[21]   BelyakovA, Sakai T, Miura H. Fine-grained structure formation in austenitic stainless steels under multiple deformation at 0.5 Tm. Mater Trans JIM 2000; 41(4): 476-84.

[22]   Belyakov A, Sakai T, Miura H, *et al.* Substructures and internal stresses developed under warm severe deformation of austenitic stainless steel. Scr Mater 2000; 42(4): 319-25.

[23]   Belyakov A, Sakai T, Miura H, *et al.* Strain-induced submicrocrystalline grains developed in austenitic stainless steels under severe warm deformation. Philos Mag Lett 2000; 80(11): 711-18.

[24]   Teus SM, Shivanyuk VN, Shanina BD, *et al.* Effect of hydrogen on electronic structure of fcc iron in relation to hydrogen embrittlement of austenitic steels. Physic Status Solid 2007; 204(12): 4249-58.

[25]   Gavriljuk VG, Berns H. High Nitrogen Steels Structure, Properties, Manufacture and Applications. Berlin. Springer-Verlag. 1999.

[26]   Mudali UK, Raj B. EDS. High Nitrogen Steels and Stainless Steels-Manufacturing Properties and Applications. Pangbourne, UK. Alpha Science International Ltd. 2004.

[27]   Weber L, Uggowitzer PJ. Partitioning of chromium and molybdenum in super duplex stainless steels with respect to nitrogen and nickel content. Mater Sci Eng A 1998; 242(1-2): 222-9.

CHAPTER 2

Ferritic Stainless Steels

Abstract: This chapter is on ferritic stainless steels. A brief introduction to the compositions and uses of the various grades of steels in the ferritic class is given. New techniques for grain refinement of the ferritic class are presented. The main problems plaguing the ferritic class, *i.e.*, 475°C-embrittlement, formation of sigma phase and sensitisation, are discussed, with reference to the constitution diagram of the Fe-Cr system. The various theories proposed for the causes of these problems, together with the old and new methods for their alleviation/elimination, are presented in detail.

Keywords: Ferritic stainless steel, sigma phase, sensitisation, 475°C-embrittlement, spinodal decomposition, G phase, constitution diagram, grain refinement, high-temperature embrittlement, intergranular carbide, Cr-depletion theory.

1. INTRODUCTION

Commercial ferritic stainless steels may be classed as Fe-Cr alloys containing about 12 to 30wt% Cr. High Cr ferritic stainless steels are used in a wide range of high-temperature applications, such as boiler tubes and turbine blades [1]. In terms of usage, ferritic stainless steels lag behind their austenitic counterparts for the following reasons: lack of ductility, poor weldability, susceptibility to embrittlement (due to 475°C-embrittlement, e.g), notch sensitivity and poor formability (relative to their austenitic counterparts). Furthermore, Fe-Cr alloys normally do not passivate in a reducing acid environment [2] and high-Cr ferrite is infamous for its brittleness and ductile-to-brittle cleavage transition.

However, ferritic stainless steels, under certain circumstances (*e.g.*, in Cl-containing environments), may have a higher corrosion resistance than austenitic stainless steels when their pitting resistance equivalent number (PERN) exceeds 35 [3]. Superferritic stainless steels, which contain even higher amounts of Cr and some Mo, possess excellent pitting resistance. Ferritic stainless steels are also quite machinable and it is now possible to achieve very good machinability without using Pb [4]. In addition, ferritic stainless steels have a high thermal conductivity, a lower thermal expansion compared with their austenitic stainless steels counterparts, and they are very immune to stress corrosion cracking.

Some of the typical ferritic stainless steels, together with their properties, are listed in Table **1** The low Cr grades (<14%) are primarily used as structural materials and they are not that corrosion-resistant. They are weldable as they form some austenite at high temperatures, which hinders excessive grain growth in the heat-affected zones. AISI430 may be viewed as the workhorse of the ferritic class that contains low Cr levels. It is used for pots, pans and automotive parts. Lately, AISI439 has been utilised as automotive trims because of its higher Mo content. AISI405 is used in steam turbines and AISI409 as exhaust equipment of automobiles. Grades containing about 17%Cr (AISI430, AISI434) are used as cookware and in decorative applications. In countries where deicing salts are used on roads in the winter, it may be desirable to use AISI434 in automobiles as this steel contains some Mo. Some of these grades may contain high-temperature austenite which may subsequently transform to brittle martensite upon cooling to room temperature and so they are not that weldable. The high-Cr grades (>20%) have good oxidation and corrosion resistance and may be used in high-temperature applications. Grades 18/2 and E-Brite26-1 are of much lower carbon contents and contain Mo for enhanced pitting corrosion resistance.

In general, steels having low Cr contents and relatively high levels of C are considered non-weldable. AISI409 contains a low level of C and is stabilised with Ti. Ti is a ferritising element and because of the low C level, AISI409 is ferritic at almost all temperatures. Some of the steels (AISI436, AISI439, e.g) also contain stabilising element like Nb, Ti and Ta. These steels are more adaptable to welding. The E-Brite26-1, AL29-4-2, AISI446 and AISI444 contain very high levels of Cr and may be regarded as the superferritics. These are very corrosion-resistant steels in chloride solutions. Nonetheless, they are lacking in toughness. The E-Brite26-1 and AL29-4-2 steels are of ultra-high purity.

Joseph Ki Leuk Lai, Kin Ho Lo and Chan Hung Shek

Table 1: Compositions and mechanical properties of some commonly used ferritic stainless steels (in wt%)

Designation	N	C	Si	Mn	Ni	Cr	Al	Mo	Se	Nb+Ta	Ti
AISI405	---	0.08	1.0	1.0	---	11.5-14.5	0.1-0.3	---	---	---	---
AISI409	---	0.08	1.0	1.0	0.5	10.5-11.75	---	---	---	---	5×C min
AISI430	---	0.12	1.0	1.0	---	16.0-18.0	---	---	---	---	---
AISI430FSe	---	0.12	1.0	1.25	---	16.0-18.0	---	---	0.15min	---	---
AISI434	---	0.12	1.0	1.0	---	16.0-18.0	---	0.75-1.25	---	---	---
AISI436	---	0.12	1.0	1.0	---	16.0-18.0	---	0.75-1.25	---	5×C min	---
AISI439	---	0.07	0.60	1.0	0.5	17.75-18.75	0.15	---	---	---	5×C (1.0max)
AISI442	---	0.2	1.0	1.0	---	18.0-23.0	---	---	---	---	---
AISI444	0.035	0.025	1.0	1.0	1.0	17.5-19.5	---	1.75-2.50	---	*	*
AISI446	0.25	0.2	1.0	1.5	---	23.0-27.0	---	---	---	---	---
E-Brite26-1	0.001	0.002	---	---	0.1	26.0	---	1.0	---	0.1Nb	---
AL29-4-2	0.005	0.005	0.20	0.30	2.0-2.50	28.0-30.0	---	3.50-4.20	---	---	---

(*(Nb+Ti)=[0.20+4(C+N)] / 0.80)

2. MICROSTRUCTURE OF FERRITIC STAINLESS STEEL

At room temperature, ferritic stainless steels are of the bcc crystal structure and have limited solubility of carbon. Hence, most of the carbon forms the $(Cr, Fe)_7C_3$ and $(Cr, Fe)_{23}C_6$ carbides. Fig. **1** is the phase diagram of Fe-Cr alloys.

From the constitution diagram, it may be seen that for Cr contents lower than about 13wt%, austenite may form at elevated temperatures. Therefore, Fe-Cr alloys containing less than about 13wt% Cr may possess a duplex microstructure consisting of ferrite and austenite if they are not cooled rapidly enough from high temperatures. Beyond 13wt % Cr, their microstructure may be fully ferritic up to the melting point. Because ferritic stainless steels normally do not form austenite on heating (>850°C), they are not amenable to strengthening *via* a martensitic transformation. Nevertheless, the extent of the austenite region in the phase diagram depends strongly on steel chemistry. The austenite region may be expanded when austenite-forming elements are present (*e.g.*, carbon, nitrogen and nickel). In this respect, carbon and nitrogen are particularly effective (Fig. **2**).

Therefore, it is still possible for some Fe-Cr alloys (the 405, 429 grades, for instance) to contain some austenite at high temperatures and then retain the duplex microstructure at room temperature [5]. The duplex microstructure may bring about a certain degree of strengthening and grain refinement. As a matter of fact, the composition of the 1.4003 ferritic stainless steel is designed to contain some austenite on cooling from high temperatures, with the aim of improving weldability and as-welded toughness by lessening grain growth in the heat-affect zone [6]. The austenite usually forms at the ferrite grain boundaries and as a Widmanstatten structure. If the austenite transforms to martensite upon cooling to room temperature, then tempering is necessary to transform the martensite to ferrite and carbides. Fast cooling is needed to suppress austenite formation in the heat-affected zones of welds.

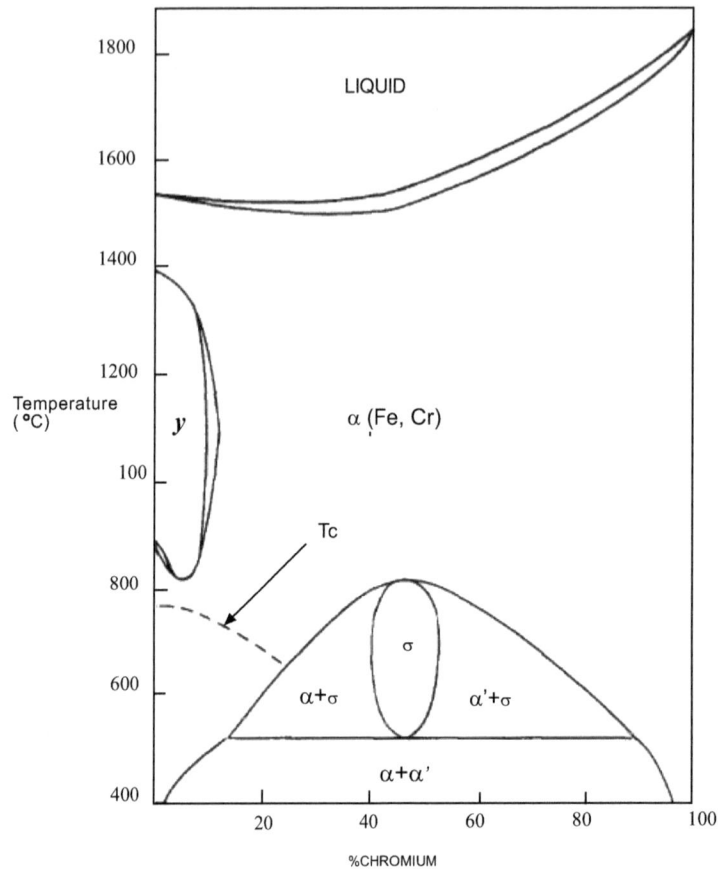

Figure 1: Phase diagram of Fe-Cr alloys [Zwieten ACTMV, Bulloch JH. Some considerations on the toughness properties of ferritic stainless steels – a brief review. Intl J Press Vessel Pip 1993; 56(1): 1-31. With permission for reproduction from Elsevier].

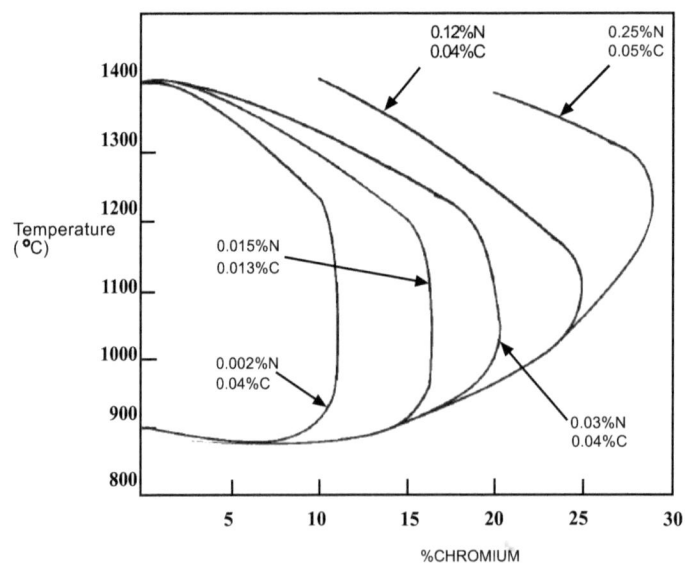

Figure 2: Expansion of the austenite region by addition of C and N [Zwieten ACTMV, Bulloch JH. Some considerations on the toughness properties of ferritic stainless steels – a brief review. Intl J Press Vessel Pip 1993; 56(1): 1-31. With permission for reproduction from Elsevier].

3. GRAIN REFINEMENT OF THE FERRITIC CLASS

A finer grain size leads to higher yield and tensile strengths, and better impact properties. The rate of grain growth of ferritic stainless steels is higher than that of their austenitic counterparts. Also, the temperature at which grain growth starts is lower for ferritic stainless steels (~600°C) than for austenitic stainless steels (~900°C). Excessive grain growth detracts from the weldability of the ferritic class. The presence of Nb(CN) and Ti(CN) may slow down grain growth and raise the lower grain-growth temperature. Welding with a laser [7] or a hybrid method using plasma plus gas tungsten arc [8] may enhance the weldability of the ferritic class.

Cold-working plus a subsequent annealing may refine the grain size somewhat. But the rate of work hardening is low and ductility drops sharply in ferritic stainless steels during plastic deformation. Hence, the ferritic class is not frequently used in the cold-rolled condition. For this reason, grain refinement of ferritic stainless steels is difficult because of the lack of austenitic transformation. Recent researchers have been able to attain a fine-grained microstructure in ferritic stainless steels through oxide-dispersion and heavy deformation [9, 10]. In this method, powders of iron, chromium and yttria (Y_2O_3) are mixed and milled mechanically. Grains reaching sizes of 10 to 20nm surrounded by amorphous grain boundary layers could be achieved. Either the yttirum-and-oxygen-enriched amorphous layer [9] or the fine Y_2O_3 and $YCrO_3$ oxides [10] may prevent grain growth at elevated temperatures. Recently, it has also been found in a high purity 15%Cr ferritic stainless steel that co-doping of the stabilising element niobium (Nb) with boron (B) may lead to a finer grain structure [11]. When B is doped to Nb-stabilised ferritic stainless steels, the niobium carbides will be smaller in size and more numerous in density. A lot of fine boron nitrides also form in the steels. These small and numerous precipitates exert a pinning effect on grain boundaries, thereby preventing grain growth. The segregation of B to grain boundaries plays a role, too. However, co-doping B with another oft-used stabilising element Ti will lead to a coarser grain structure because of the formation of coarse carbides [11].

Increases in strength and hardness in ferritic stainless steels are mainly associated with 475°C-embrittlement (Fig. **3**), formation of the sigma phase, and formation of carbides at high temperatures (high-temperature embrittlement or sensitisation). Attendant to strengthening and hardness increase are losses of ductility, impact toughness and corrosion resistance. Consequently, these strengthening mechanisms are often considered undesirable. In fact, the ferritic class is almost always used in the annealed state and these embrittling phenomenon are the main hindrances to their use. For the high-Cr grades containing intermediate to high levels of impurities, the formation of carbides at high temperatures (>1000°) is especially problematic, inasmuch as welding becomes very difficult [12].

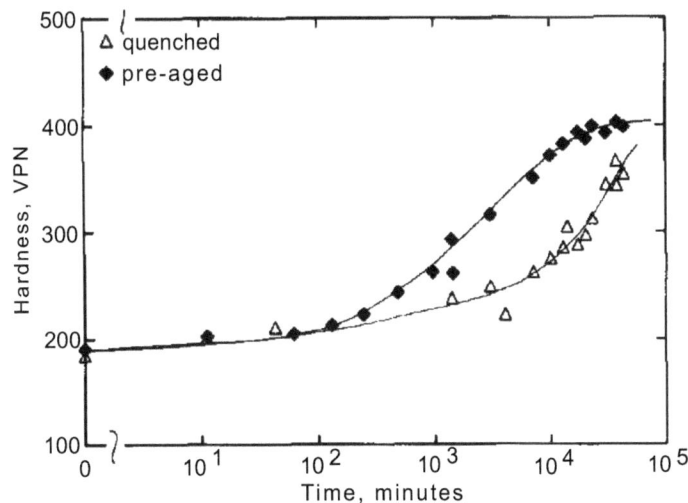

Figure 3: Increase in Vicker hardness of a FeCr alloy containing 38wt%Cr upon exposure to 475°C [Cortie MB, Pollak H. Embrittlement and aging at 475°C in an experimental ferritic stainless steel containing 38 wt% chromium. Mater Sci Eng A 1995; 199(2): 153-63. With permission for reproduction from Elsevier].

4. 475°C EMBRITTLEMENT

This embrittling phenomenon occurs when Fe-Cr alloys are heated between 300°C and 550°C for prolonged periods (the ferrite phase of duplex stainless steels has been found to get embrittled at temperatures as low as 250°C [13]). It is called 475°C embrittlement because the hardness increase and rate of embrittling are the most rapid at this temperature.

4.1. Property Changes During 475°C Embrittlement

Accompanying the hardness increase are decreases in ductility and impact toughness when 475°C embrittlement occurs (Fig. **4**). It has to be noted that the loss of toughness occurs much faster than the increase in hardness. Also, the hardness may increase just a little after extended annealing because of overageing.

In general, yield and tensile strengths increase during 475°C embrittlement. For fully ferritic stainless steels, the changes in hardness and tensile strength might only be noticeable after ageing for hours at 475°C. Nevertheless, the degradation in toughness may occur in a short time (Fig. **4**). For notched specimens and the ferrite phase of duplex stainless steels, embrittlement may become appreciable in even much shorter times. The fast drop in toughness perhaps may be regarded as the most obvious indication of 475°C embrittlement. In ferritic stainless steels whose α_{Cr} forms through nucleation and growth, Grobner attributed the drop in toughness to the immobilisation of dislocations by α_{Cr} particles [14]. In a 38wt% ferritic stainless steel that underwent spinodal decomposition, Cortie and Pollak found that toughness dropped before spinodal decomposition became discernible and they attributed the initial drop in toughness to extremely small Cr-C or Cr-N clusters that could reduce dislocation mobility at high strain rates [15]. Dissociation of dislocation is also believed to be responsible for easier twinning occurrence.

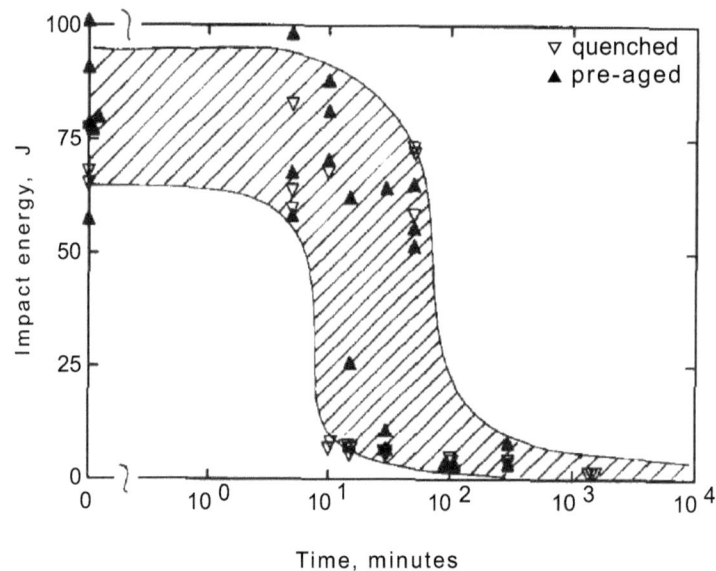

Figure 4: Change in V-impact toughness of a FeCr alloy containing 38wt%Cr upon exposure to 475°C [Cortie MB, Pollak H. Embrittlement and aging at 475°C in an experimental ferritic stainless steel containing 38 wt% chromium. Mater Sci Eng A 1995; 199(2): 153-63. With permission for reproduction from Elsevier].

Besides loss of ductility and impact toughness, corrosion resistance is also compromised [16]. As regards corrosion, Carrasco *et al.* [17] have shown that in MA956, 475°C embrittlement lowers pitting corrosion resistance, but not general corrosion. The ferrite phase of duplex stainless steels also suffers from this problem [18]. Fig. **5** shows that the spinodally-decomposed ferrite phase of a duplex stainless steel became more susceptible to corrosion attack relative to the unaged state. Fig. **6** shows that as annealing at 475°C proceeds, the steel will become progressively more brittle (transgranular fracture). The cleavage planes are mainly {110}, but {102}, {314} and {131} are also involved [16].

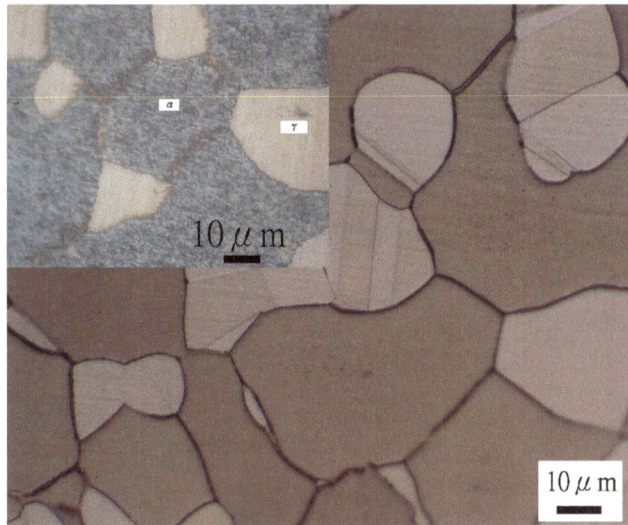

Figure 5: Optical micrographs showing the microstructure of the duplex stainless steel 7MoPLUS in the as-solution-treated state (1100°C, 1h) and the spinodally-decomposed state (500°C, 2000h) [Lo KH, Lai JKL. Microstructural characterisation and change in a.c. magnetic susceptibility of duplex stainless steel during spinodal decomposition. J Nucl Mater 2010; 401(1-3): 143-8. With permission for reproduction from Elsevier].

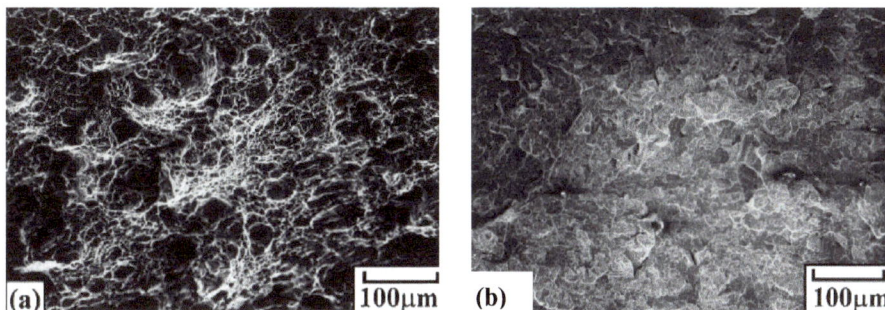

Figure 6: Fractography of the ferritic phase in the duplex stainless steel 2205 after exposure to 475°C for 2h **(a)** and 64h **(b)** [Weng KL, Chen HR, Yang JR. The low-temperature aging embrittlement in a 2205 duplex stainless steel. Mater Sci Eng A 2004; 379(1-2): 119-32. With permission for reproduction from Elsevier].

Presently, it is widely accepted that 475°C embrittlement is caused by the formation of a coherent precipitate because of the presence of a miscibility gap in the Fe-Cr system below 550°C [19] (Fig. **1**). In a system consisting of two regular solutions (A and B) having positive heats of mixing, there exists a critical temperature above which the two solutions are completely miscible. However, the system will be a mixture of two solutions that are respectively enriched in A and B below the critical temperature. The Fe-Cr alloy system may be considered to be such a system of strictly regular solutions, according to the heats and entropies of mixing obtained by Kubaschewski and Chart [20].

Besides temperature, composition also affects the way α_{Cr} forms. Depending on Cr content, the coherent α_{Cr} may either form through spinodal decomposition [35, 36] or the classical nucleation and growth process [14, 21]. Alloys having less than about 18wt%Cr are unlikely to decompose spinodally [14].

When embrittlement is associated with spinodal decomposition, the original ferrite (α) decomposes into two ferrites that are respectively enriched with Cr and Fe (designated as α_{Cr} and α_{Fe}). The Cr-content of α_{Cr} may reach 93at% [22]. Consequently, when α_{Cr} forms, the adjacent α_{Fe} becomes severely depleted with Cr. The severe depletion of Cr in α_{Fe} renders it susceptible to corrosion attack (See inset of Fig. **5**). α_{Cr} is paramagnetic and so as it forms, the overall magnetism of the steel will decrease (as indicated by the amount of ferromagnetic ferrite content in Fig. **7**). However, other magnetic properties like coercive force and the Curie temperature seem to be nearly unaffected [23].

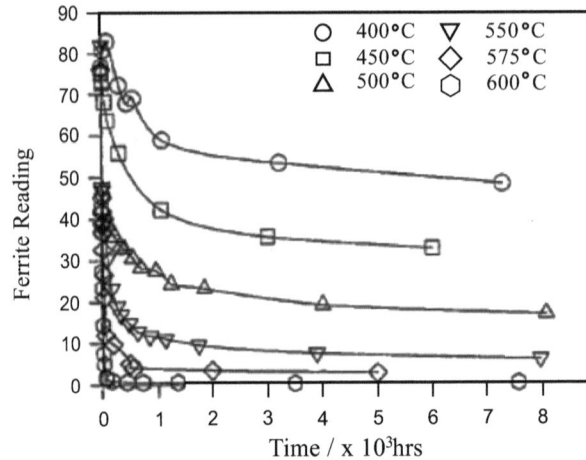

Figure 7: The decrease in ferromagnetic ferrite content with ageing time and duration in the spinodal range [Shek CH, Wong KW, Lai JKL. Review of temperature indicators and the use of duplex, stainless steels for life assessment. Mater Sci Eng R Rep 1997; 19(5-6): 153-200. With permission for reproduction from Elsevier].

4.2. Microstructural Changes During 475°C Embrittlement

The earliest discoverers of the coherent precipitate were Demo [24] and Fishers *et al.* [25]. In a steel containing 28.5%Cr that had been aged at 475°C for 1 to 3 years, they extracted non-magnetic particles having high Cr contents (61 to 83%). These particles were found to be of the bcc crystal structure and have a lattice parameter of about 2.877Å. The morphology of the iron-rich precipitate, designated as α_{Fe}, may either be spherical [25] or plate-like [26, 27].

Because the Cr-rich and Fe-rich ferrites are of nearly the same lattice parameters (α_{Cr} : 2.885Å and α_{Fe}: 2.866Å) [28] and crystal structure, detection of 475°C-embrittlement may be difficult (if not impossible) with conventional techniques like optical metallography and X-ray diffractometry (XRD) [29]. This is why this embrittling phenomenon remained elusive for a long time. Early investigators wrongly related 475°C embrittlement to the early stage of sigma phase formation [30] and impurity/grain boundary precipitation [31]. For instance, Heger [32] hypothesised that 475°C embrittlement was associated with a coherent transition phase when a Fe-Cr alloy transformed to the sigma phase upon heating (α phase $\rightarrow$ coherent transition phase $\rightarrow$ σ phase). Nevertheless, this hypothesis was incompatible with the results obtained by Houdremont [33]. Between 600°C and 900°C, some of the Fe-Cr alloys used by Houdremont had sigma phase precipitation without hardening, whereas significant hardening occurred between 400°C and 550°C. The proposition that impurity precipitation led to 475°C embrittlement was refuted because even very pure alloys would be embrittled [14]. Atomic ordering, *i.e.*, formation of ordered superstructures around the compositions Fe_3Cr, $FeCr$ and $FeCr_3$, was also thought to cause 475°C embrittlement [34]. However, experiments utilising a sensitive neutron diffraction technique failed to come up with concrete evidence to support this hypothesis [35].

While XRD and optical metallography cannot reveal spinodal decomposition, its occurrence may be clearly revealed when an embrittled steel is examined under the transmission electron microscope (TEM). Under the TEM, mottled contrast associated with the segregation of Cr and Fe atoms may be clearly observed (Fig. **8**). Besides substitutional alloying elements, the interstitial element N has also been found to give rise to the mottled contrast [36].

In Fig. **8**, it may be seen that a lot G phase precipitates had formed during spinodal decomposition. Invariably, the precipitation of the G phase lags behind the occurrence of spinodal decomposition. The precipitation of the G phase also contributes to hardening. But views on its role in causing embrittlement seem to be at variance. While Chung considered that the contribution of G phase was only secondary [13], Bonnet *et al.* [37] argued that it might be effective in causing embrittlement below 350°C. Recently, Mateo *et al.* [38] have put forth a model for the precipitation of the G phase. In this model, G-phase forming elements are ejected into domains between the Cr-rich and Fe-rich ferrites as spinodal decomposition proceeds. Later on when the composition of the interdomain region reaches the requisite level for transformation, the G phase forms with very slight adjustments of atoms. The model propounded by Matero *et al.* is consistent with the experimental observation that the G phase only forms at the later stage of spinodal decomposition.

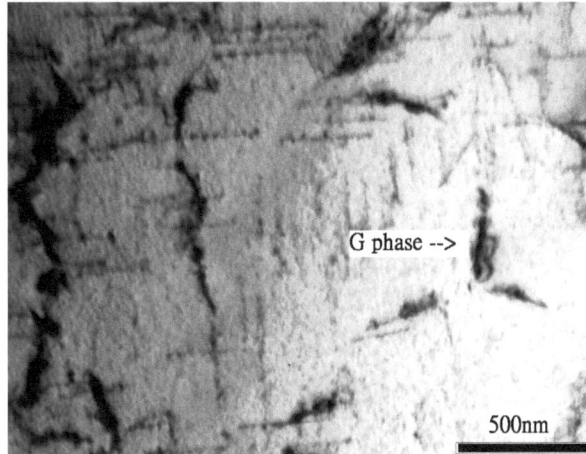

Figure 8: Mottled contrast caused by composition fluctuations associated with spinodal decomposition in the ferrite phase (annealed at 500°C for 1744h) [Lo KH, Lai JKL. Microstructural characterisation and change in a.c. magnetic susceptibility of duplex stainless steel during spinodal decomposition. J Nucl Mater 2010; 401(1-3): 143-8. With permission for reproduction from Elsevier].

Another microstructural feature accompanying spinodal decomposition is the crisscrossing of dislocations (Fig. **9**) [29]. This crisscrossing arrangement of dislocations also contributes to hardening because dislocations pin down one another [28]. Twinning also becomes more prevalent as 475°C embrittlement proceeds, insofar as the immobilisation of dislocation makes slip more difficult to occur.

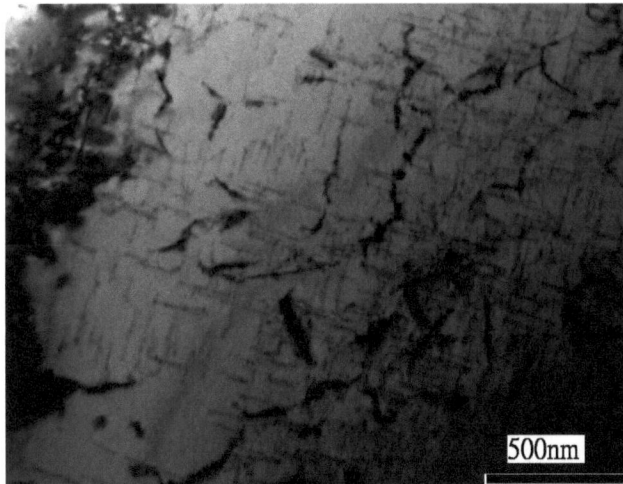

Figure 9: Crisscrossing of dislocations and the G phase upon spinodal decomposition [Lo KH, Lai JKL. Microstructural characterisation and change in a.c. magnetic susceptibility of duplex stainless steel during spinodal decomposition. J Nucl Mater 2010; 401(1-3): 143-8. With permission for reproduction from Elsevier].

The kinetics of spinodal decomposition is often described by using an Arrhenius-type equation as follows:

$$\frac{t_2}{t_1} = \exp[Q / R(1 / T_2 - 1 / T_1)]$$

T_1 and T_2: annealing temperature.

Q: activation energy of annealing.

R: universal gas constant.

t_1 and t_2: annealing times for reaching equivalent material toughness at T_1 and T_2, respectively.

Therefore, as the temperature drops in the spinodal range, the rate of spinodal decomposition drops precipitously [13]. Below 340˚C, spinodal decomposition proceeds extremely sluggishly. Fig. **10**. shows a TEM micrograph of the ferrite phase of a duplex stainless steel after ageing at 350°C for 7000h. It may be seen that the characteristic mottled contrast of spinodal decomposition is absent.

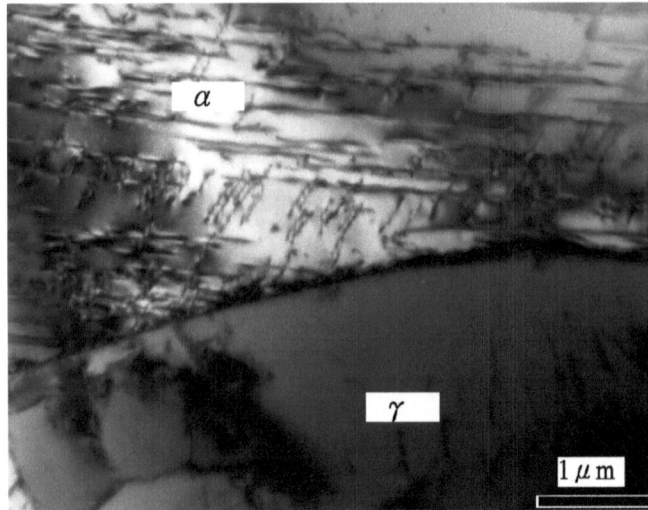

Figure 10: TEM micrograph of a sample annealed at 350˚C for 7000h. The mottled contrast is absent in the ferrite phase [Lo KH, Lai JKL. Microstructural characterisation and change in a.c. magnetic susceptibility of duplex stainless steel during spinodal decomposition. J Nucl Mater 2010; 401(1-3): 143-8. With permission for reproduction from Elsevier].

In fact, Grobner stated that for a 18wt% Cr ferritic stainless steel, embrittlement might only become noticeable between 600000 and 2000000h at 316°C [14]. Because of the extremely sluggish rate of decomposition, accelerated annealing at higher temperatures is frequently used by researchers in order to study the behaviour of embrittlement in the low end of the spinodal range. In so doing, it is tacitly assumed that the mechanism of embrittlement remains unchanged. Nevertheless, the value of Q has been found to change with temperature in the spinodal range [37], implying different mechanisms do exist.

Reduced ductility, toughness and corrosion resistance associated with 475°C-embrittlement may be restored by heating embrittled materials above 550°C for a long enough period of time. Newell [39] reported that 1 hour at 593°C was sufficient to anneal out the embrittlement, while it would take 1000 hours at 538°C.

The kinetics of 475°C embrittlement is altered by alloying [24]. Cr, Ti, Cb, Si, Al, Mo and P have been found to hasten embrittlement, whereas Mn has the opposite effect. Ni, on the other hand, has been found to either hasten or slacken 475°C embrittlement, depending on its content. Plastic deformation is conducive to the embrittling process. Alloying elements like Ni, Cu and Cr tend to aggravate 475°C embrittlement [26]. Mo is also believed to encourage 475°C embrittlement [14], but in the range used by Solomon and Levinson, Mo did not show much effect [26]. Golovin *et al.* [40] found that reducing C and N would hasten spinodal decomposition, whereas Courtnall and Pickering [41] got opposite results. More recent work by Cortie and Pollak has confirmed the retarding effect of interstitials [15].

5. SIGMA PHASE PRECIPITATION

According to the Fe-Cr constitution diagram (Fig. **1**), the intermetallic sigma phase may form in alloys containing as little as 20wt% Cr. Hall and Algie [42] presented a detailed review on the sigma phase as early as 1927. About 80 years since the publication of this review, research on the sigma phase formed in stainless steels (not just restricted to ferritic stainless steels) is still extensive. The sigma phase forms upon heating above 600°C. In fully ferritic stainless steels, it takes relatively long times (hundreds of hours) for the sigma phase to precipitate out. Nonetheless, the sigma phase has been found to occur in the heat-affected zones in the AISI444 steel containing Ti and Nb, which are known to encourage sigma formation [43]. However, in duplex stainless steels or ferritic stainless steels containing small amounts of austenite, this intermetallic compound may form very quickly. More on the sigma phase will be presented in the chapter on precipitates of stainless steels (Chapter **8**).

Besides the sigma phase, other embrittling intermetallic compounds form in the ferritic class, too. These are the chi phase and the Laves phase that form in roughly the same temperature regime where the sigma phase forms. The time-temperature-precipitation (TTP) diagram for a duplex stainless steel in Chapter **5** provides insight into the phases that may form in FSS.

6. HIGH-TEMPERATURE EMBRITTLEMENT (SENSITISATION)

When heated above 950°C and cooled to room temperature, high-Cr ferritic stainless steels containing high interstitial contents lose their ductility [44] and corrosion resistance [45]. Insofar as the temperature involved in welding and casting processes easily exceeds 950°C, this embrittling phenomenon is a big obstacle to the use of ferritic stainless steels, especially the air-melted grades.

It is interesting to note that the effects of thermal annealing on austenitic stainless steels and ferritic stainless steels are opposite. For the ferritic grades, their corrosion resistance is degraded if they are heated to above 950°C, followed by not-so-fast cooling. If the ferritic stainless steels are then subjected to heat treatment between 700°C and 850°C, their corrosion resistance will be restored (the reason to be discussed below). Quite the opposite, when austenitic stainless steels are annealed between 700°C and 850°C, they lose their corrosion resistance because of the precipitation of carbides. However, if they are re-heated to above 950°C, their corrosion resistance is recovered owing to the dissolution of carbides.

6.1. Early Theories Explaining Sensitisation of Ferritic Stainless Steels

Like 475°C-embrittlement, various theories were put forth to explain the embrittlement and loss of corrosion resistance upon heating above 950°C. These included the formation of coherent clusters [46] and the formation of martensite [47]. In the former theory, it was postulated that during cooling from above 950°C, carbon atoms that are dissolved in the supersaturated ferritic matrix would form coherent clusters (instead of carbides). These coherent clusters could harden the matrix. Subsequent annealing between 700°C and 800°C would encourage carbon atoms to form carbides, thereby removing the coherent clusters and relieving the embrittlement. In the martensitic mechanism, it was suggested that regions containing high carbon contents would transform to austenite upon heating. These austenitic regions would then undergo martensitic transformation upon cooling, making the ferritic stainless steel brittle. On re-heating between 700°C and 800°C, the brittle martensite would decompose into ferrite and carbides, leading to removal of the embrittlement.

The formation of austenite was also used to explain the loss of corrosion resistance. Houdremont and Tofaute [48] hypothesised that high-temperature, carbon-rich austenite decomposed into iron carbides at the ferrite/austenite interfaces, which would detract from corrosion resistance. The recovery of corrosion resistance upon heating at about 750°C was attributed by these authors to the conversion of iron carbides to chromium carbides, which could resist chemical dissolution and so the stainless steel would become resistant to intergranular attack. Hochmann [49] also attributed intergranular attack to austenite formation. However, Hochmann thought that the intergranular attack occurred at the austenite grain boundary because of its low chromium and high carbon content [49]. Nonetheless, the necessity of austenite formation was disputed by Lula *et al.* [50] on the basis that ferritic stainless steels subjected to heat-treating operations that could avoid austenite formation following high-temperature exposure were still vulnerable to intergranular corrosion. These authors suggested that it was the stress in the vicinity of carbides and nitrides that led to corrosion of their adjacent matrix. Hence, the recovery of corrosion resistance on re-heating between 650°C and 815°C had to do with the elimination of the stress.

6.2. Role of Intergranular Carbides on Corrosion Resistance and Ductility

6.2.1. Role of Intergranular Carbides on Intergranular Corrosion Resistance: the Cr-depletion Theory

Although several theories were put forward to explain the loss of corrosion resistance of ferritic stainless steels following high-temperature exposure, the most widely accepted theory nowadays is the Cr-depletion theory [24]. In this theory, it is proposed that the preferential formation of precipitates like carbides, nitrides and carbonitrides along grain boundaries leads to depletion of chromium in the neighbourhoods of these precipitates. So, if a ferritic stainless steel is heated above 950°C but not cooled down fast enough, intergranular corrosion occurs because of the precipitation of carbides (the steel is said to be sensitised). When a sensitised steel is reheated between 700°C and 950°C, the Cr-depleted regions are replenished with Cr atoms because Cr diffusion is expedited. Therefore, corrosion resistance of the once Cr-depleted regions is restored and the material is resistant to intergranular attack, even though grain boundary carbides are still present. It is to be noted that precipitation between 700°C and 950°C might occur. But as the precipitates form, Cr atoms diffuse to their vicinity almost simultaneously and so the adjacent regions of the precipitates are never short of Cr, thereby making them corrosion-resistant. It is interesting to note that a 16wt% Cr ferritic stainless steel containing 0.040wt%C and 0.032wt%N was sensitised after solution-treatment at 1160°C, but its pitting potential was high [51]. However, after exposure to 800°C for 10 minutes, the pitting potential became very low, but intergranular corrosion resistance was partially recovered [51]. Therefore, there is no direct relation between degree of sensitisation and pitting potential.

To sum up, except that a ferritic stainless steel is of very low interstitial levels, it must be cooled down fast enough after high-temperature exposure above 950°C (above this temperature, carbides and nitrides are dissolved in solid solution). In this way, carbon and nitrogen will be held in solution. If the steel is re-heated between 500°C and 950°C, then carbides and nitrides would form to relieve supersaturation of the matrix. Between 700°C and 950°C, the precipitation is accompanied by rapid replenishment of Cr in the neighbourhoods of the precipitates and no intergranular corrosion would occur, albeit grain boundary precipitates are present. However, in the range of 500°C to 700°C, the diffusion of Cr atoms is not efficient and so Cr-depleted regions would form in the vicinity of the precipitates. Therefore, the steel must be held in this temperature range for long enough times if corrosion resistance is to be restored. It is important to realise that when the levels of C and N are high, embrittlement may still occur even if cooling is fast, because clustered C and N atoms may form in the supersaturated ferritic matrix [52].

6.2.2. Role of Intergranular Carbides on Embrittlement

As regards intergranular precipitation of carbides on embrittlement, different views have been held out. Demo [45] looked at the effects of heat treatment on the ductility of ferritic stainless steels having different levels of interstitials. For an AISI446 stainless steel having high interstitial contents after solution-treatment at 1100°C and subsequent quenching in water, its ductility was degraded. Nonetheless, the same steel that was slowly cooled did not lose ductility. Furthermore, the ductility of a steel having a low interstitial level was unaffected. Fractographic analyses revealed that the slowly-cooled steel showed intergranular cleavage with localised deformation and grain elongation. Additionally, ductility of the water-quenched steel (brittle) was recovered after a healing annealing at 850°C, although intergranular carbides were still present. Therefore, Demo [45] concluded that intergranular carbide/nitride precipitation did not adversely affect ductility. Instead, he suggested that it was precipitates that form in the grain interiors (on dislocations) of the water-quenched steel that were responsible for the ductility loss. TEM examination done by Demo [45] showed no in-grain precipitates in the ductile air-cooled steel, whereas they were omnipresent in the brittle water-quenched steel. Hence, he concluded that the water-quenched steel was brittle because movement of dislocations in it was constrained. Demo [45] reasoned that the air-cooled steel did not contain in-grain precipitates on dislocations because carbon and nitrogen atoms had more time to diffuse to grain boundaries (which are of higher energy than dislocations) to form precipitate there.

However, other authors (*e.g.*, Semchyshen *et al.* [53]) were convinced that it was the precipitation of grain boundary carbides/nitrides that were to blame for ductility loss after high-temperature exposure. These authors believed that grain-boundary precipitates would lower the effective surface energy of a crack and thus would promote cleavage failure and brittleness.

Although different views exist on the role of intergranular carbide precipitation, both groups of authors agreed that the loss of ductility of ferritic stainless steels after high-temperature exposure hinges on levels of interstitial elements (C, N and O).It has to be noted that the total (C+N) level that can be tolerated without compromising weldability of ferritic stainless steels decreases with increasing Cr content, but the tolerable level increases somewhat for intergranular corrosion resistance as the Cr content goes up.

Besides leading to embrittlement and corrosion attack, interstitials also affect strongly the notch sensitivity of ferritic stainless steels. Ferritic stainless steels in the annealed state are notch-sensitive. This notch-sensitivity gradually decreases as the temperature increases. Therefore, in severe forming operations, it is advisable to pre-heat the ferritic stainless steels (especially those with high interstitial contents) to minimise notch sensitivity.

7. METHODS FOR ACHIEVING LOW INTERSTITIAL LEVELS

Good ferritic stainless steels should possess adequate corrosion resistance and ductility in the as-welded or as-cast conditions, and so the interstitial levels should be kept at low levels. Currently, three methods are commonly used to this end.

The first method is the production of 'clean' steels using techniques like oxygen-argon melting, argon-oxygen desulphurisation-vacuum oxy-desulphurisation (AOD-VOD) processes [54, 55] and electron beam refining [56]. Among the three methods, the cleanness of steels produced by the last method is the highest. In a 25%Cr steel, a total level of (C+N+O) lower than 50ppm has been possible [56]. In this method, a high surface-to-volume ratio of molten metal is exposed to a high vacuum (10^{-1} Pa or less [56]) for prolonged periods of time. As the molten metal flows down the hearths, localised regions of intense heat (>3000K) in the molten metal are achieved by using electron-beam heat sources. The intense heat volatilises impurity elements, making the steel clean. In fact, even refractory elements like niobium, tungsten and tantalum can be melted by using this method [56].

The second method for lowering C and N is by alloying with a strong carbide forming element like titanium, niobium, tantalum and zirconium. This method is known as stabilisation and the afore-mentioned elements are called stabilising elements. These stabilising elements must be added in sufficient levels in order to tie up most of C and N such that stabilisation is attained. For instance, Steigerwald has proposed that the minimum Ti content must be five times the total content of carbon plus nitrogen, *i.e.*, $\%Ti = 5 \times (\%C + \%N)$ [57]. Pardo *et al.* [58], however, have found that this formula needs modification, with more emphasis on the importance of nitrogen (although this study is on an austenitic stainless steel, it may also be used as a reference for the ferritic class). Processes like AOD-VOD can be expensive and require sophisticated equipment, and so stabilisation is quite often used.

Sometimes, the addition of stabilising elements may backfire. For instance, although Nb is effective in lowering the ductile-to-brittle transition temperature of Fe-Cr alloys subjected to high-temperature exposure, it detracts from impact resistance of steels in the annealed condition [53]. Too much Ti is also harmful to the impact behaviour of annealed ferritic stainless steel [53] and TiN may reduce surface quality [59]. Also, the addition of stabilising elements should be chosen carefully according to the environment in which the steel is anticipated to work. For instance, Ti-stabilised ferritic stainless steels are prone to intergranular attack in a highly oxidising environment, but Nb-stabilised steels are not [60]. However, Ti may tie up sulphur and oxygen, whereas Nb does not seem to have a high affinity to these impurities. Although stabilising elements must be added in sufficient amounts in order to be effective, too much of them may actually decrease toughness [61, 62].

The last method for attaining good ductility in the as-welded state is by alloying with low amounts of copper, aluminium and vanadium, whose atomic radii differ from that of the ferrite matrix by less than 15% [63]. This process is known as ductilisation. Good as-weld ductility may be achieved even at relatively high interstitial levels. For example, in the presence of ductilising elements, a Fe-Cr alloy containing 35wt%Cr was found to possess good as-welded ductility at an interstitial level of 250ppm, while the interstitial level must be kept to below 15ppm if no ductilising elements were used.

Although enhanced toughness, corrosion resistance, ductility and weldability are obtained by lowering the total interstitial contents, strengths are reduced. Additionally, lack of C and N makes grain refinement difficult because it is hard to precipitate Ti(CN) and Nb(CN) to hinder grain-boundary movement.

REFERENCES

[1]   Fujita T. Current progress in advanced high Cr ferritic steels for high-temperature applications. ISIJ Int 1992; 32(2): 175-81.

[2]   Truman JE. Stainless steels. In: F.B.Pickering, Ed. Constitution and Properties of Steels. VCH. Weinheim. 1992. pp. 592-582.

[3]   Ujiro T, Yosioka K, Kawasaki T, *et al.* In: Development of high-alloy stainless steels with corrosion resistance to seawater environment. Proceedings of international conference on stainless steels. Kawasaki F., Ed. 1991, Tokyo, Japan. ISIJ Int. pp. 23-27.

[4]   Oikawa K, Mitsui H, Ebata T, *et al.* A new Pb-free machinable ferritic stainless steel. ISIJ Int 2002; 42(7): 806-7.

[5]   Hu J, Song H, Yu M, *et al.* Thermo-Calc calculation and experimental study of microstructure of SUS410S and SUS430 ferrite stainless steels at high temperatures. J Iron Steel Inst Int 2007; 14(Suppl 1): 183-8.

[6]   Grobler C. Ph.D thesis. Weldability studies in 12% and 14% chromium steels. University of Pretoria. South Africa 1987.

[7]   Kaul R, Ganesh P, Tripathi P, *et al.* Comparison of laser and gas tungsten arc weldments of stabilized 17wt% Cr ferritic stainless steel. Mater Manuf Process 2003; 18(4): 563-80.

[8]   Taban E, Kaluc E, Dhooge A. Hybrid (plasma + gas tungsten arc) weldability of modified 12% Cr ferritic stainless steel. Mater Des 2009; 30(10): 4236-42.

[9]   Kimura Y, Takaki S, Suejima S, *et al.* Ultra grian refining and decomposition of oxide during super-heavy deformation in oxide dispersion ferritic stainless steel powder. ISIJ Int 1999; 39(2): 176-82.

[10] Kimura Y, Suejima S, Goto H, *et al.* Microstructure and mechanical properties in ultra fin-grained oxide-dispersion ferritic stainless steels. ISIJ Int 2000; 40(Suppl): S174-8.

[11] Kashif E, Asakura K, Koseki T, *et al.* Effects of boron, niobium and titanium on grain growth in ultra high purity 18%Cr ferritic stainless steel. ISIJ Int 2004; 44(9): 1568-75.

[12] Toit MD, Rooyen GTV, Smith D. Heat-affected zone sensitization and stress corrosion cracking in 12% chromium type 1.4003 ferritic stainless steel. Corrosion 2007; 63(5): 395-404.

[13] Chung HM. Aging and life prediction of cast duplex stainless steel components. Int J Press Vessel Pip 1992; 50(1-3): 179-213.

[14] Grobner PJ. The 885°F(475°C) embrittlement of ferritic stainless steels. Metall Trans A 1973; 4(1): 251-60.

[15] Cortie MB, Pollak H. Embrittlement and aging at 475°C in an experimental ferritic stainless steel containing 38 wt% chromium. Mater Sci Eng A 1995; 199(2): 153-63.

[16] Terada M, Hupalo MF, Costa I, *et al.* Effect of alpha prime due to 475°C aging on fracture behavior and corrosion resistance of DIN 1.4575 and MA956 high performance ferritic stainless steels. J Mater Sci 2008; 43(2): 425-33.

[17] Carrasco JLG, Escudero ML, Martin FJ, *et al.* Influence of the 475°C hardening of MA956 on its corrosion behaviour at room temperature. Corros Sci 2001; 43(6): 1081-94.

[18] Tavares SSM, Pedrosa PDS, Teodosio JR, *et al.* Magnetic properties of the UNS S39205 duplex stainless steel. J Alloy CompD 2003; 351(1-2): 283-8.

[19]　Chandra D, Schwartz LH. Mossbauer effect study of the 475°C decomposition of Fe-Cr. Metall Trans 1971; 2(2): 511-19.

[20]　Kubaschewski O, Chart TG. Calculation of metallurgical equilibrium diagrams from thermochemical data. J Inst Metal 1964-65: 93; 329-38.

[21]　Nys TD, Gielen PM. Spinodal decomposition in the Fe-Cr system. Metall Trans 1971; 2(5): 1423-28.

[22]　Kuwano H, Imamasu H. Determination of the chromium concentration of phase decomposition products in an aged duplex stainless steel. Hyper Interact 2006; 168(1-3): 1009-15.

[23]　Souza, JA, Abreu HFG, Nascimento M, *et al.* Effects of low-temperature aging on AISI 444 steel. J Mater Eng Perform 2005; 14: 367-72.

[24]　Demo JJ (1977). Structure and constitution of wrought ferritic stainless steels. In: Handbook of Stainless Steels. Peckner D, Bernstein (Eds). McGraw-Hill, USA.

[25]　Fisher RM, Dulis EJ, Carroll KG. Identification of the precipitate accompanying 885°F embrittlement in chromium steels. J Metal May 1953: 690-95.

[26]　Solomon HD, Levinson LM. Mossbauer effect study of 475°C embrittlement of duplex and ferritic stainless steel. Acta Metall 1978; 26(3): 429-42.

[27]　Lagneborg R. Metallography of the 475°C Embrittlement in an Iron-30%. Chromium Alloy. Trans Am Soc Metal 1967; 60: 67-78.

[28]　Weng KL, Chen HR, Yang JR. The low-temperature aging embrittlement in a 2205 duplex stainless steel. Mater Sci Eng A 2004; 379(1-2): 119-32.

[29]　Lo KH, Lai JKL. Microstructural characterisation and change in a.c. magnetic susceptibility of duplex stainless steel during spinodal decomposition. J Nucl Mater 2010; 401(1-3): 143-8.

[30]　Link HS, Marshall PW. The formation of sigma phase in 13%-16% chromium steel. Trans Am Soc Metal 1952; 44: 549-59.

[31]　Riedrich G, Loib F. Embrittlement of high-chromium steels in temperature range of 570-1110°F. Arch Eisenhuettenwes 1941-1942; 15(7): 175-82.

[32]　Heger JJ. 885°F embrittlement of the ferritic chromium-iron alloys. Metal Prog August, 1951: 55-61.

[33]　Houdremont E. Handbuch der Sonderstahlkunder. Springer Verlag. Berlin. 1943.

[34]　Masumoto H, Sato H, Sugihara M. Sci Rep Res Inst Tohoku Univ 1953; 5: 203-08.

[35]　Williams RO. Further studies of the iron-chromium system. Transact Metall Soc AIME 1958; 212: 497-502.

[36]　Kobayashi S, Nakai K, Ohmori Y. Decomposition processes of δ-ferrite during continuous heating in a 25Cr–7Ni–0.14N stainless steel. ISIJ Int 2000; 40(8): 802-8.

[37]　Bonnet S, Bourgoin J, Champredonde J, *et al.* Relationship between evolution of various cast duplex stainless-steels and metallurgical and aging parameters-outline of current EDF programs. Mater Sci Techno 1990; 6(3): 221-9.

[38]　Mateo A, Llanes L, Anglada M, *et al.* Characterization of the intermetallic G-phase in an AISI 329 duplex stainless steel. J Mater Sci 1997; 32(17): 4533-40.

[39]　Newell ND. Properties and characteristics of 27% chromium-iron. Metal Prog May 1946: 977-1028.

[40]　Golovin IS, Sarrack VI, Suvorova SO. Influence of carbon and nitrogen on solid solution decay and 475°C-embrittlement of high-chromium ferritic steels. Metall Trans A 1992; 23(9): 2567-79.

[41]　Courtnall M, Pickering FB. The effect of alloying on 485°C embrittlement. Metal Sci 1976; 10(2): 273-276.

[42]　Hall EO, Algie SH. The sigma phase. Metall Rev 1966; 11: 61-88.

[43]　Silva CC, Farias JP, Miranda HC, *et al.* Microstructural characterization of the HAZ in AISI444 ferritic stainless steel welds. Mater Charact 2008; 59(5): 528-33.

[44]　Thielsch H. Weld embrittlement in chromium stainless steels. Weld J 1950; 29: 126s-32s.

[45]　Demo JJ. Mechanism of high temperature embrittlement and loss of corrosion resistance in AISI type 446 stainless steel. Corrosion 1971; 27: 531-44.

[46]　Thielsch H. Physical and welding metallurgy of chromium stainless steels. Weld J 1951; 30: 209s-50s.

[47]　Pruger TA. Flow to get better welding results with 17% chromium steel. Steel Horiz 1951; 13: 10-2.

[48]　Houndremont E, Tofaute W. Resistance in intergranular corrosion of ferritic and martensitic chromium steels. Stahl Eisen 1952; 72: 539-545.

[49]　Hochmann J. Properties of vacuum-melted steels containing 25% chromium. Revue de Metallurgie 1951; 48: 734-58.

[50]　Lula RA, Lena JA, Kiefer GC. Intergranular corrosion of ferritic stainless steels. Trans Amer Soc Metal 1954; 46: 197-230.

[51] Paroni ASM, Falleiros NA, Magnabosco R. Sensitization and pitting corrosion resistance of ferritic stainless steel aged at 800°C. Corros 2006; 62(11): 1039-46.

[52] Zwieten ACTMS, Bulloch JH. Some considerations on the toughness properties of ferritic stainless steels-a brief review. Int J Press Vessel Pip 1996; 56(1): 1-31.

[53] Semchysen M, Bond PA, Dundas HJ. Effects of composition on ductility and toughness of ferritic stainless steel. In: Proceedings of Symposium on Toward Improved Ductility and Toughness. Kyoto, Japan. pp. 239-253. Climax Molydenum Co., Greenwich Connecticut. 1971.

[54] Dowling NJE, Kim H, Kim JN, *et al.* Corrosion and toughness of experimental and commercial super ferritic stainless steels. Corros 1999; 55(8): 743-55.

[55] Beddoes J, Parr JG. Introduction to Stainless Steels (3rd ed). ASM Intrnational. 1999.

[56] Nakao R, Fukumoto S, Fuji M, *et al.* Evaporation of alloying elements and behavior of degassing reactions of high chromium steel in electron-beam melting. ISIJ Int 1992; 32(5): 685-692.

[57] Steigerwald R. Metals Handbook. vol.13, Metals Park (OH), ASM Int 1990, p.123.

[58] Pardo A, Merino MC, Coy AE, *et al.* Influence of Ti, C and N concentration on the intergranular corrosion behaviour of AISI 316Ti and 321 stainless steels. Acta Mater 2007; 55(7): 2239-51.

[59] Fritz JD, Franson IA. Sensitization and stabilization of type 409 ferritic stainless steel. Mater Perform 36(8); 1997: 57-61.

[60] Cowling RD, Hintermann HE. The corrosion of titanium carbide. J Electrochem Soc 1970; 117(11): 1447-1449.

[61] Abo H, Nakazawa T, Takemura S, *et al.* The role of carbon and nitrogen on the toughness and intergranular corrosion of ferritic stainless steels. In: Stainless Steel 77, R.Q.Barr ED. Climax Molybdenum Company, London. UK. Sep, 1977; pp. 35-47.

[62] Steigerwald RF, Dundas HJ, Redmond JD, *et al.* The physical metallurgy of Fe-Cr-Mo ferritic stainless steels. In: Stainless Steel 77, R.Q.Barr (ed). Sep, 1977; pp. 57-72.

[63] Sipos DJ, Steigerwald RF, Whitcomb NE. Ductile corrosion-resistant ferrous alloys containing chromium. US Patent no. 3672876. 27 Jun, 1972.

CHAPTER 3

Austenitic Stainless Steels

Abstract This chapter gives a brief introduction to the compositions and uses of the various grades of steels in the austenitic class. The phenomena of deformation-induced transformation (DIM) and sensitisation are then covered. The underlying mechanisms causing DIM and sensitisation, together with the factors affecting them, are discussed in detail. Mechanical stabilisation, recent research findings on and methods for detecting DIM and sensitisation are also covered.

Keywords: Austenitic stainless steel, metastable austenitic stainless steel, deformation-induced martensite, martensitic transformation, sensitisation, desensitisation, healing, mechanical stabilisation, phase diagram, ε martensite, α' martensite, EPR test, grain boundary engineering.

1. INTRODUCTION

In terms of tonnage produced yearly, austenitic stainless steels (AusSSs) are the largest group in the stainless steel family. The most widely used are those in the AISI300 series, a system of Fe-Cr-Ni alloys. The grade containing about 18wt%Cr and 8wt% Ni (often called the 18-8 steel) may be regarded as the basis for this series. Because nickel is rather pricey ($15000 per tonne in the year 2005 [1]), its replacement using nitrogen and manganese is commonplace nowadays [2]. The 200 series is basically a system of Fe-Cr-Mn alloys [2]. Because of the rising popularity of the high-N, low-Ni (or Ni-free) AusSS, a separate chapter (Chapter **7**) is devoted to this class. Overall, AusSS possesses good corrosion resistance, mechanical properties and fabricability.

In the AISI300 series (see Table **1**), type 301 (17Cr-7Ni) possesses good fabricability and the ability for strengthening *via* cold-work because of the deformation-induced martensitic transformation. Types 302 and 304 contain higher amounts of alloying elements. Type 304 is widely used in a range of applications, especially those involving high temperatures. The nickel content makes type 305 even more stable than type 304. Types 309, 310 and 314 are also adaptable for high-temperature applications. Types 316 and 317, because of their high alloying contents, have good corrosion resistance and high-temperature strengths.

Table 1: The compositions of the steels in the AISI300 series, and some in the 200 series are listed in Table **3.1**. (in wt%).

Steel	Cr	Ni	Mo	Mn	N	C	Si	P	S	Ti
201	16-18	3.5-5.5	---	5.5-7.5	0.25 max	0.15	1.00	0.045	0.015	---
201L	16-17	3.5-5.5	---	6.0-8.0	0.20max	0.030	1.00	0.045	0.015	---
202	17-19	4.0-6.0	---	7.5-10.0	0.25 max	0.15	1.00	0.045	0.015	---
204C	16.0-18.0	2.00	1.00	6.5-8.5	0.30max	0.10	2.00	0.040	0.030	---
301	16-18	6-8	---	2max	0.1max	0.15max	1.0max	0.045max	0.3max	---
302	17-19	8-10	---	2max	0.1max	0.15max	0.75max	0.045max	0.03max	---
304	17.5-19.5	8-10.5	---	2.0max	0.10max	0.07max	0.75max	0.045max	0.03max	---
304L	17.5-19.5	8-12	---	2.0max	0.10max	0.03max	0.75max	0.045max	0.03max	---
309	22-24	12-15	---	2.0max	---	0.20	0.75max	0.045max	0.03max	---
309S	22-24	12-15	---	2.0max	---	0.08	0.75max	0.045max	0.03max	---
316	16-18	10-14	---	2.0max	0.1max	0.08max	0.75max	0.045max	0.03max	---
316L	16-18	10-14	---	2.0max	0.1max	0.03max	0.75max	0.045max	0.03max	---
316LN	16-18	10-14	2-3	2.0max	0.1-0.16	0.03max	0.75max	0.045max	0.03max	---
316Ti	16-18	10-14	2-3	2.0max	0.1max	0.08max	0.75max	0.045max	0.03max	5(C+N)min-0.70max

Joseph Ki Leuk Lai, Kin Ho Lo and Chan Hung Shek

2. MICROSTRUCTURE OF AUSTENITIC STAINLESS STEEL

The phase diagram of the Fe-Cr system is shown in Fig. **1**. Since the Cr contents of AusSSs exceed 16wt%, their equilibrium microstructure at room temperature would be fully ferritic, if no other austenitising alloying elements were added to the material. Elements that are most often used to obtain an austenitic microstructure are Ni, Mn, C and N. It has to be noted the leaner steels in the 300 series (Fe-Cr-Ni alloys) may contain ferrite at elevated temperatures. On the other hand, those in this series that are highly alloyed are usually fully austenitic. However, inhomogeneity in the original casting may result in ferrite being present in some regions. The presence of ferritising elements like Mo, Ti may also lead to ferrite in the final microstructure. Hence, when these ferritisers are present, the Ni content should be increased accordingly.

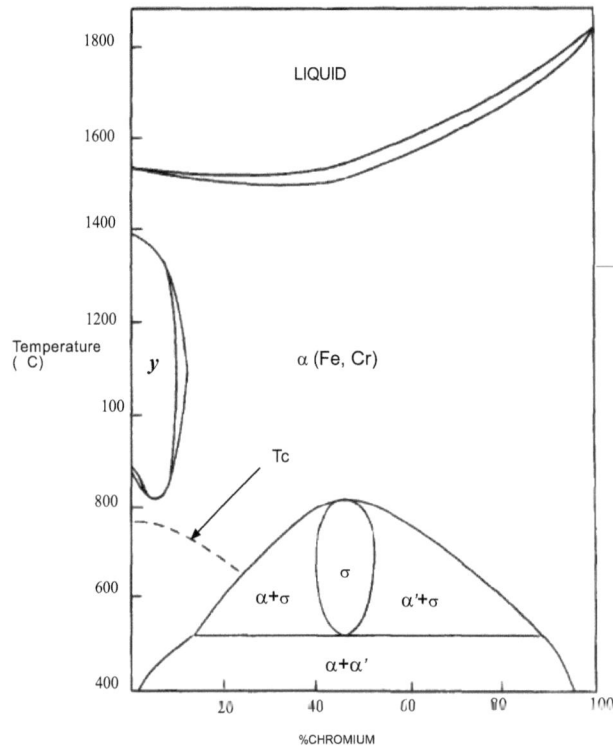

Figure 1: Phase diagram of Fe-Cr alloys [Zwieten ACTMV, Bulloch JH. Some considerations on the toughness properties of ferritic stainless steels – a brief review. Int J Press Vessel Pip 1993; 56(1): 1-31. With permission for reproduction from Elsevier].

For alloys to be welded, the prediction of phase balance is most often done with the Scheffler diagram and the DeLong diagram [3, 4]. Prediction of phase balance is described in detail in Chapter **5**. In essence, the various diagrams and methods proposed by researchers involve converting ferritising elements to an Cr equivalent number, whereas austenitising elements are converted to a Ni equivalent number. The Cr and Ni equivalent numbers suggested by Schaeffler and DeLong (who considered the effect of N also) are as follows, respectively. Subsequent researchers have incorporated the effects of more elements in such expressions.

$$Cr_{eq} = \%Cr + \%Mo + 1.5\%Si + 0.5Cb$$
$$Ni_{eq} = Ni + 30\%C + 30\%N + 0.5Mn$$

3. MARTENSITIC TRANSFORMATION IN METASTABLE AUSTENITIC STAINLESS STEELS

The presence of deformation-induced martensite may cause a multitude of problems, such as delayed cracking of deep-drawn austenitic stainless steel components [5]. Nevertheless, deformation-induced martensite may also enhance the rate of work-hardening and so is good for formability.

The formation of deformation-induced martensite (denoted as α' in this e-Book) is related closely to shear bands, which are planar defects associated with the overlapping of stacking faults on {111}γ. Depending on the nature of the overlapping, twins, the ε-martensite or stacking fault bundles may form. Twins form when stacking faults overlap on successive {111} planes, whereas ε martensite is generated if the overlapping of stacking faults occurs on alternate {111} planes. Stacking fault bundles arise from the irregular overlapping of stacking faults.

The sites of formation for α', the transformation routes of α', modelling of the formation of α', empirical expressions for assessing the stability of austenite against martensitic transformation and methods for detection of α' have all been investigated lately. Even though all of the topics were studied by a lot of early workers, recent studies have led to deeper insights and old results have been critically re-examined.

3.1. Nucleation Sites and Transformation Routes of α'

It is generally accepted that the intersections of the ε martensite may act as the sites of nucleation of α'. For instance, Kruml *et al.* [6] observed that α' had formed at the intersections of ε martensite with the following orientation relationships [6, 7]:

$$\{111\}_\gamma \, / / \{0001\}_\varepsilon \, / / \{110\}_{\alpha'} \, <110>_\gamma \, // <2110>_\varepsilon \, // <111>_{\alpha'}$$

A satisfactory model for α' formation has also been obtained recently by taking the intersections of ε as the embryos of α' [8] (Fig. 2).

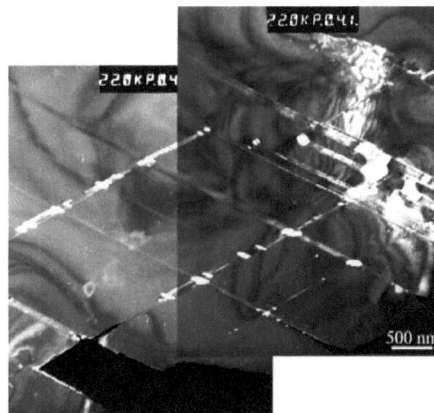

Figure 2: Transmission electron micrograph showing the formation of α' at the intersections of ε martensite [Spencer K, Embury JD, Conlon KT, *et al.* Strengthening *via* the formation of strain-induced martensite in stainless steels. Mater Sci Eng A 387-389; 2004: 873-81. With permission for reproduction from Elsevier].

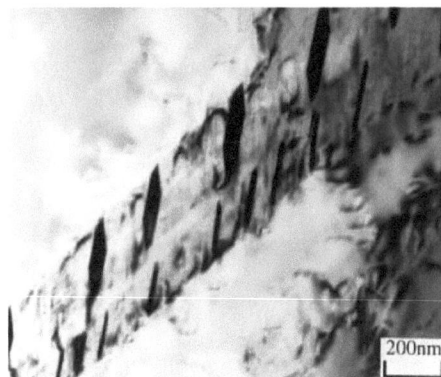

Figure 3: The formation of α' in a single ε martensite in a 304L steel upon impact deformation [Lee WS, Lin CF. The morphologies and characteristics of impact-induced martensite in 304L stainless steel. Scr Mater 43(8); 2000: 777–782. With permission for reproduction from Elsevier].

However, a recent study has concluded that α' is capable of nucleating in a single ε-martensite (instead of at the intersection of two ε-martensite plates) at high-strain rates (Fig. **3**) [406]. A very recent TEM study by Das *et al.* [10] has even found that α' can form in a single shear band in a 304LN steel at low strain rates. This TEM study has also provided evidence that triple points, and the intersection of a shear band and a grain boundary may act as nucleation sites for α'.

As regards transformation routes, both $\gamma \to \varepsilon \to \alpha'$ and $\gamma \to \alpha'$ have been suggested to be feasible. Recent studies by Fukuda *et al.* [11], Huang *et al.* [7] and Das *et al.* [10] have shown that both transformation routes are possible for the same stainless steel.

In the transformation route $\gamma \to \varepsilon \to \alpha'$, the ε martensite acts as the precursor phase of α' [6, 12, 13]. Lee *et al.* [14] have recently verified that this transformation route is applicable to a high-N stainless steel, too. Another recent study by Kireeva and Chumlyakov [15] has shown that the generation of ε using cryogenic pre-deformation at 77K will enable the formation of α' at 300K in a Fe-17%Cr-12%Ni-2%Mn-0.75%Si austenitic stainless steel, which normally does not contain strain-induced martensite at room temperature. This study lends further credence to the notion that ε may act as the precursor phase of α'. However, this notion has been suggested to be unlikely by Hedstrom *et al.* [16]. These authors have suggested that α' embryos would first form at dislocation pile-ups. As their number reaches a certain value, the formation of new embryos would be retarded/suppressed, but those already formed would grow rapidly in an autocatalytic fashion, resulting in a stepwise transformation behaviour. A recent work by Spencer *et al.* [17] shows that the transformation routes does depend on the deformation temperature and the ε-martensite may or may not act as the precursor phase to α'.

Besides the ε-martensite, intersections between mechanical twins and shear bands [18], or the intersections between ε-martensite and slip bands (or twins or grain boundaries) [17] have also been suggested to be nucleation sites for α' [18].

3.2. Factors Affecting the Martensitic Transformation

Factors like the directionality of external loads, external fields, grain size *etc.*, that may influence the martensitic transformation in metastable stainless steels have been looked into in several recent studies.

Kireeva and Chumlyakov [15], after analysing the transformation behaviour of single crystals, have demonstrated that the martensitic transformation is orientation-dependent. In another study, Hedstrom *et al.* [16], using 3DXRD, have probed in-situ the $\gamma \to \varepsilon$ transformation of individual grains in a polycrystalline AISI301 steel. These authors have shown that the $\gamma \to \varepsilon$ transformation takes place in a very localised manner. Of the 47 grains probed, only the one with the highest Schmid factor for the $\{111\} < 1\bar{2}1 >$ slip system underwent the $\gamma \to \varepsilon$ transformation at a tensile strain level of 5%.

The grain size has also been shown to affect the formation of α' [18, 19], with coarser grains leading to higher amounts of α' [19]. The strain rate has been found to affect the austenite-to-martensite transformation through adiabatic heating [20]. And both the onset strain level and the volume fraction of α' are affected by the strain rate [10]. Using artificial neural network, Mirzadeh and Najafizadeh [21] have concluded that increasing strain rate will lower the amount of α'. In the presence of an external magnetic field, the ε to α' transformation has been found to be facilitated [11].

A recent detailed TEM investigation by Spencer *et al.* [17] has revealed that the temperature at which the stainless steel is deformed bears heavily on the transformation route and kinetics, it also affects the spatial distribution of the martensite.

3.3. Effects of Martensite on Mechanical Properties

The influences of the ε martensite have been checked out in several recent studies. The presence of ε martensite, like α', can also lead to strengthening [22]. In a 304 stainless steel, the appearance of ε leads to a rapid decrease in strain hardening, but as soon as α' forms, the rate of strain hardening increases [23]. A uniform distribution of ε produced in high-strain-rate deformation is believed to enhance ductility in the S31254 superaustenitic stainless steel [24]. Damage brought about by cavitation erosion is also mitigated by ε because part of the impact energy will be absorbed by its formation [25].

Not only the presence of deformation-induced martensite affects greatly the properties of metastable austenitic stainless steels, but also changes in it, which occur upon thermal ageing, would exert a great influence [26]. Buono *et al.* [27] have established that strain ageing, whereby interstitial C atoms segregate to dislocations, in ferritic AISI430 steel increases its yield strength. A similar strengthening mechanism has been found to operate in the deformation-induced martensite of austenitic stainless steels, too [26].

3.4. Methods of Detection for α'

A variety of techniques have been developed for the detection of α'. Acoustic emission (AE) activities have been found to correlate well with the amount of α' that forms in differently deformed 304 austenitic stainless steels [28-30], this technique can even detect the presence of pre-existing martensite [31]. Recently, Shaira *et al.* [32] have pointed out that it is possible to classify AE signals by using the k-means algorithm. In their work, these authors showed that the AE signals may be classified into three clusters that are respectively associated with dislocation movement, crack activity and martensitic transformation. Their results make it possible to clearly identify the AE signals that are caused by martensitic transformation.

Changes in magnetic properties due to the formation of α' are also usable for its detection [33]. In fact, a high precision SQUID magnetometer has been shown recently to be a useful means for determining the decomposition temperature of α' [34]. Being able to determine this temperature precisely is of significance in as much as the reversion transformation from α' to γ is used very frequently for grain refinement in meta stable austenitic stainless steels.

Magnetic properties like magnetic coercivity and magnetic permeability associated with magnetic hysteresis loop are frequently used for the detection of α' [35-37]. The magnetic coercive force is affected by the size [35, 37], shape and distribution of α' [35]. In a recent study, Sullivan *et al.* [36] have concluded that the coercivity is more sensitive than the Magnetic Barkhausen Noise (MBN) in characterising work hardening of austenitic stainless steels in the presence of α'. Vertesy *et al.* [38, 39] have demonstrated that the use of a minor magnetic hysteresis loop [38], magnetic adaptive testing [39] and their combination [39] for the detection of α' may be even more effective, because magnetic properties like relative coercivity can be more sensitively detected.

Khan *et al.* [40] have lately demonstrated the usefulness of eddy current in detecting the amount of non-transformed austenite in cold-rolled metastable austenitic stainless steels. It has been found that the eddy current is insensitive to grain size, residual stresses and hardness.

3.5. Austenite Stability

Martensitic transformation of austenitic stainless steels may occur either through plastic deformation or cooling. While it was pointed out in the preceding section that factors like strain rate, stress state, *etc.* may influence this transformation, it may be argued that the most influential factor is composition.

Over the years, the effects of a lot of commonly used alloying elements on austenite stability have been studied quite thoroughly and empirical expressions on assessing the stability of austenitic stainless steels against martensitic transformation abound in the literature. This section will give an overview of these old and new expressions.

3.5.1. The Ms Temperature

The Ms temperature is a very useful parameter for estimating the stability of austenite against martensitic transformation and is affected by a number of factors, such as chemistry of the steel, grain size, structural defects, stress state, existence of pre-existing martensite, *etc.* Amongst these factors, the influence exerted by the steel chemistry has received considerable interest [41-43].

For the AISI300 series, Eichelmanand and Hull [44], and Monkman *et al.* [45] have respectively proposed empirical equations for estimating Ms. The Eichelman equation was later on improved by Hammond [46] such that the effect of Mo was also accounted for. The Hammond equation, which is applicable only within a narrow composition range, is as follows:

$$Ms(K) = 1578 - 41.7Cr - 61.1Ni - 33.3Mn - 27.8Si - 36.1Mo - 1667(C + N)$$

Research aiming at incorporating the effects of a larger number of elements and extending the usable composition range of empirical formulae have been numerous. For example, a recent study by Dai *et al.* [47], using statistical regression analysis for a large number of data in the literature, has proposed empirical formulae for estimating the phase transformation temperatures of ε martensite and α' in austenitic steels (It has to be noted that some of the steels used by these authors in obtaining the expressions did not contain Cr at all. However, the same authors have applied these expressions to austenitic stainless steels [48]). These authors have analysed the martensitic transformation on the basis of the Critical Resolved Shear Stress (CRSS) that is needed to initiate the transformation. The CRSS depends on stacking fault energy and strength of the austenite phase, both of which are affected by steel chemistry. The expressions obtained by these authors are as follows:

$$Ms(K) = A_3 - 199.8(C + 1.4N) - 17.9Ni - 21.7Mn - 6.8Cr - 45.0Si - 55.9Mo -$$

$$1.9(C + 1.4N)(Mo + Cr + Mn) - 14.4[(Ni + Mn)(Cr + Mo + Al + Si)]^{\frac{1}{2}} - 410$$

$$M\varepsilon_s(K) = A\varepsilon - 710.5(C + 1.4N) - 18.5Ni - 12.4Mn - 8.4Cr + 13.4Si - 1.6Mo - 22.7Al +$$

$$11.6(C + 1.4N)(Mo + Cr + Mn) - 3.7[(Ni + Mn)(Cr + Mo + Al + Si)]^{\frac{1}{2}} + 277$$

From these expressions, it may be appreciated that some elements (like Si) may have totally opposite effects on the formation of the ε martensite and α'.

Even more elements has been included in the expression, which was derived by using the thermodynamic approach, proposed by Ishida [49]:

Ms (°C, wt.%) = 545-330C+2Al + 7Co - 14Cr- 13Cu - 23Mn - 5Mo - 4Nb - 13Ni - 7Si + 3Ti + 4V + 0W

The Ishida equation has been entrusted by Xu *et al.* [50], who recently have adopted this equation in their design for nano-precipitation strengthened, ultra-high strength stainless steels.

In addition to the traditional approach of regression/statistical analysis, several recent works have utilised the neural network approach. For example, Capdevila *et al.* [51] have recently adopted this approach, within the Bayesian framework, and derived the following expression for estimating Ms:

$$Ms(K) = 764.2 - 302.6\%C - 30.6\%Mn - 16.6\%Ni - 8.9\%Cr + 2.4\%Mo$$
$$-11.3\%Cu + 8.58\%Co + 7.4\%W - 14.5\%Si$$

The Capdevila expression has been found by the same group of authors to correlate with experimental results reasonably well [52]. Furthermore, Sourmail and Garcia-Mateo [53] have suggested that this expression may perform as well as those obtained through the thermodynamic approach used by Ghosh and Olson [41-43]. Nevertheless, these authors have pointed out that the Capdevila expression can be improved by re-training the neural network model with regard to nitrogen [53]. Using the neural network approach and a large amount of data, Sourmail and Garcia-Mateo [54] have come up with an expression for Ms that is much better than the Capdevila expression.

Besides the formulae listed above, Andrews [55] also proposed an expression for calculating Ms very early:

$$Ms(K) = 273 - 12.1Cr - 17.7Ni - 7.5Mo - 423N$$

The formula proposed by Andrews is simpler than the ones given by Capdevila *et al.* [51] and Hammond [46], yet in a recent study, Wu *et al.* [56] have found that Ms values given by the Andrews formula correlates linearly and well with the amount of retained austenite in martensitic weld metals. The noteworthy point of the study of Wu *et al.* [56] is that their samples contained Mn, Si and C, whose effects on Ms are not considered in the Andrews formula.

Besides steel chemistry, the effect of cooling rate on Ms has also been examined in several recent studies. A higher austenitising temperature and cooling rate have been suggested by Tsai *et al.* [57] to depress Ms, insofar as the higher number of vacancies strengthen the austenite matrix, making it more difficult to transform martensitically. But Leem *et al.* [58] have pointed out that increased quenched-in vacancies may in fact facilitate martensitic transformation. Recently, vacancy clusters have been suggested to enhance martensitic transformation by acting as embryos for transformation [57, 59]. It is to be noted that in Leem's study, the authors did not pointed out whether the quenched-in vacancies form clusters or not. Hence, it is difficult to compare the results of Leem *et al.* [58] with those of Tsai *et al.* [57].

Although this section is devoted specifically to Ms, a comment on the M_d temperature, which is the temperature above which deformation does not set off the martensitic transformation, is in place. It has recently been proposed that the M_d temperature is a better yardstick than is Ms in estimating the stability of austenite against martensitic transformation [60].

3.5.2. Nickel Equivalent Number

Besides Ms (and M_d), some authors have also suggested that the stabilising effects of the various alloying elements on austenite may be quantified by employing the concept of nickel equivalent number. As early as in the 1940s, the use of the nickel-equivalent number (Ni_{eq}) as a yardstick for assessing the stability of austenitic stainless steels against martensitic transformation had been proposed. This early expression is:

$$Ni_{eq} = \%Ni + 0.65\%Cr + 0.98\%Mo + 1.05\%Mn + 0.35\%Si + 12.6\%C$$

Takemoto *et al.* [62], on the other hand, have more recently suggested a different expression of Ni_{eq}:
$$Ni_{eq} = \%Ni + 0.6\%Mn + 0.18\%Cr + 9.69(\%C + \%N) - 0.11\%Si^2$$

A major difference between these two expressions lies in the effect of silicon. It is worth noting that both early [61] and recent [63] studies tend to suggest that Si may decrease the stacking fault energy and thus it favours planar deformation structures like twins, ε-martensite, *etc.*, thereby enhancing the formation of α'.

3.5.3. Mechanical Stabilisation

Using the various expressions introduced above, one will be able to stabilise austenitic stainless steels against martensitic transformation by adjusting the steel composition. Nevertheless, austenite can also be stabilised mechanically by plastic deformation, which itself may initiate martensitic transformation.

Martensitic transformation involves the movement of glissile dislocations at the austenite / martensite interface. When austenitic stainless steels are subjected to very severe plastic deformation, copious amounts of dislocations may hinder the motion of the glissile dislocations, thereby bringing the martensitic transformation to a halt. The halting of deformation-induced martensitic transformation by dislocations is known as mechanical stabilisation [64, 65]. Besides dislocations, solid solution strengthening is also known to impede the motion of the interface [66]. Chatterjee *et al.* [67] have recently developed a simple model for predicting the onset of mechanical stabilisation, which also factors in the effect of solid solution strengthening.

Too many dislocations not only hinder the formation of deformation-induced martensite, but are also detrimental to the formation of quench-induced martensite [68].

4. SENSITISATION AND DESENSITISATION (HEALING)

When exposed between about 450°C and 850°C, austenitic stainless steels are known to suffer from sensitisation. According to the Cr-depletion theory, carbides form along grain and twin boundaries during sensitisation (Fig. **4**), resulting in Cr depletion in their neighbourhoods. The most common carbide is $M_{23}C_6$, here M stands for Cr, Fe, Mo or a combination of them. This carbide is pernicious because it contains 70-80%Cr. Therefore, in a sensitised steel, the austenite near the grain and twin boundaries, and austenite/martensite interfaces is more susceptible to martensitic transformation (sensitisation-induced transformation) and intergranular corrosion attack. While exposure to the sensitising range for extended periods may bring about problems, a large chunk of sensitisation-related failures are found in welded structures [69].

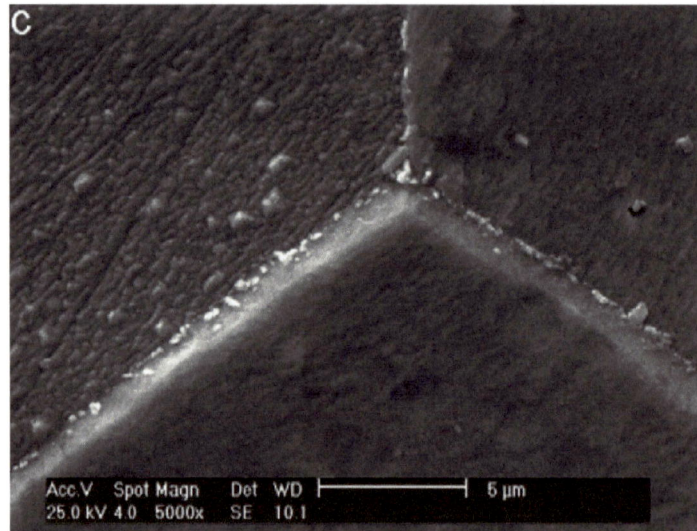

Figure 4: Precipitation of carbides along grain boundaries in an AISI304 steel that had been annealed at 800°C for 6h [Stella J, Cerezo J, Rodriquez E. Characterization of the sensitization degree in the AISI304 stainless steel using spectral analysis and conventional ultrasonic techniques. NDT & E Int 42(4): 2009: 267-74. With permission for reproduction from Elsevier].

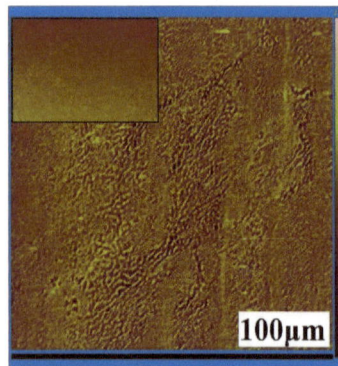

Figure 5: Magnetic force microscopy showing the formation of sensitisation-induced martensite in a 304 steel that had been annealed at 600°C for 2 months (inset: unannealed state) [LO KH, Zeng D, Kwok CT. Effects of sensitisation-induced martensitic transformation on the tensile behaviour of 304 austenitic stainless steel. Mater Sci Eng A 528(3); 2011: 1003-7. With permission for reproduction from Elsevier].

Traditionally, the precipitation of carbides is studied by X-ray diffractometry and electron microscopy. Lately, several studies have deployed Magnetic Force Microscopy (MFM) to this phenomenon. Takaya *et al.* [70] have demonstrated that MFM maybe used as an effective means for monitoring sensitisation. Fig. **5** shows sensitisation-induced martensite in a 304 steel that had been annealed at 600°C for 2 months (note that the inset shows martensite was not easily detectable in the unannealed state). Although the effects of carbide formation on the tensile properties of austenitic stainless steels have been studied [71, 72], the effects of sensitisation-induced martensite are much less investigated. It should be noted that in unsensitised steels, martensite does not just form near grain and twin boundaries.

By isothermally annealing a series of samples at different temperatures for different durations and then performing standard tests for detection of sensitisation (to be discussed in a subsequent subsection), it is possible to construct the Time-Temperature-Sensitisation (TTS) diagram for a given steel (Fig. **6**). It has to be recognised that the TTS diagrams obtained by using different methods (*e.g.*, the Strauss and the Huey methods) can be different. Since the testing conditions are more aggressive in the Strauss test, a steel tested as being sensitised in the Strauss test might be indicated as unsensitised by the Huey test.

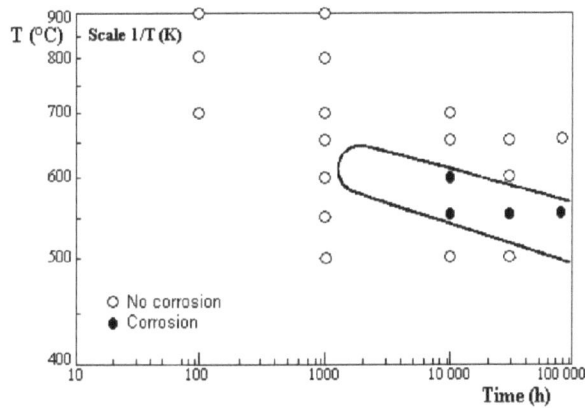

Figure 6: Time-temperature-sensitisation (TTS) diagram for AISI316L [Sahlaoui H, Makhlouf K, Sidhom H, *et al.* Effects of ageing conditions on the precipitates evolution, chromium depletion and intergranular corrosion susceptibility of AISI 316L: experimental and modeling results. Mater Sci Eng A 372(1-2); 2004: 98-108. With permission for reproduction from Elsevier].

Exposure to the sensitising temperature range for extended periods of time may 'heal' sensitised steels (the steels are said to be desensitised). Desensitisation occurs because of Cr diffusion from the grain interiors to the once Cr-depleted regions. Once the originally Cr-depleted grain boundaries are replenished with Cr, intergranular corrosion will be ameliorated. Fig. **7** shows the evolution of the grain-boundary Cr profiles in a 316L steel on annealing at 600°C.

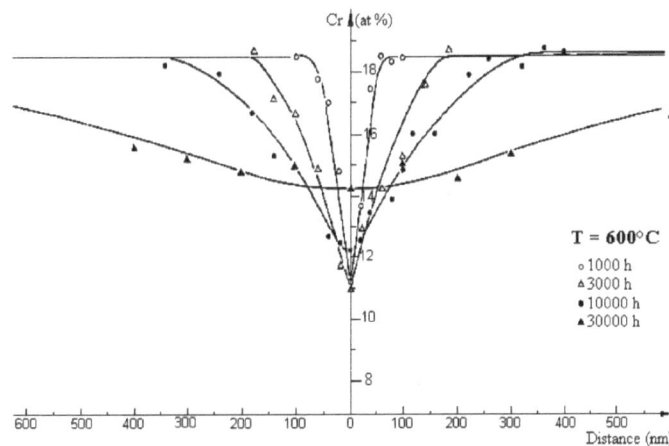

Figure 7: Evolution of the grain-boundary Cr profiles in a 316L steel on annealing at 600°C [Sahlaoui H, Makhlouf K, Sidhom H, *et al.* Effects of ageing conditions on the precipitates evolution, chromium depletion and intergranular corrosion susceptibility of AISI 316L: experimental and modeling results. Mater Sci Eng A 372(1-2); 2004: 98-108. With permission for reproduction from Elsevier].

4.1. Theories on Sensitisation and Its Detection

Sensitisation is one of the problems affecting stainless steels. This phenomenon is also covered in Chapter **9**. The widely accepted Cr-depletion theory as described in the preceding section was put forth by Bain [73]. This theory may be also used to successfully explain desensitisation (healing), the effect of plastic deformation on sensitisation, and stabilisation using Ti and Nb.

Besides this theory, the noble carbide theory and the segregation theory were also proposed to explain sensitisation [74]. In the noble carbide theory, a corrosion cell is thought to be set up between the carbide and its adjacent matrix, the latter being the active potential. Nonetheless, the potentials of matrix and carbides are basically the same in most oxidising media and so such corrosion cells are unviable. Furthermore, this theory cannot explain desensitisation. That is, when the once Cr-depleted matrix becomes replenished with Cr (desensitised or healed), no intergranular corrosion exists even though carbides are still present. As regards the segregation theory, it is suggested that intergranular corrosion is due to the presence of a grain-boundary second phase or soluble impurities, which are attacked by highly oxidising media. However, this theory cannot explain why intergranular corrosion still takes place in less oxidising media.

4.2. Common Methods of Detection of Sensitisation

The most commonly used methods for sensitisation detection are the ASTM Standard A-262 Practices A – F. Lately, electrochemical methods like the single-loop and double-loop electrochemical potentiokinetic reactivation (EPR) tests are also adopted. The double-loop EPR (DLEPR) has been quite popular as it is convenient and quantitative [69, 75]. The various tests have been documented in detail by Dayal *et al.* [76].

Except for the A-262 Practice A test, these tests are destructive. Furthermore, all of these tests are not quantitative. Sometimes, sensitised steels that may perform satisfactorily in service conditions may fail these tests. The main features of the ASTM Standard A-262 Practices A – F are summarised in Table **2**.

Table. 2: Main features of the ASTM Standard A-262 Practices.

Standard	Medium	Test method	Evaluation criterion
A 262-A	10% oxalic acid (room temp)	Electrolytic etching ($1(A/cm^2)$, 90s)	Microstructure
A 262-B	$50\%H_2SO_4+25(g/L)Fe_2(SO_4)_3$ (boiling)	Exposure for 120h	Weight loss
A 262-C	HNO_3 (boiling)	Exposure for 48h for 5 times	Average weight loss
A 262-E	$16\%H_2SO_4+100(g/L)CuSO_4+Cu$ (boiling)	Exposure for 15 or 24h	U-bend test, tensile properties
A 262-F	$50\%H_2SO_4+100(g/L)CuSO_4$	Exposure for 120h	Weight loss or U-bend test

The A-262 Practice A test is just qualitative, but it is useful for screening purposes. In this test, chromium carbides will be dissolved, leaving behind an etched microstructure that may be used for evaluation of the severity of sensitisation. If the etched microstructure is 'step' (Fig. **8**) or 'dual' (meaning a mix of 'steps' and 'ditches'), then the steel is thought not to be prone to intergranular corrosion. On the other hand, a 'ditch' microstructure (Fig. **9**) may or may not result in intergranular corrosion. So, it is advisable to perform the ASTM A262 Practices B-F tests in order to be certain. It is noticeable that even unsensitised steel may exhibit a 'ditch' microstructure after the A-262 Practice A test because of the aggressiveness of the test solution.

In the A-262 Practice B test, the intermetallic sigma phase in types 321 and 347 steels are attacked, but the sigma phase in the Mo-containing type 316 is not. In the A-262 Practice A test, the sigma phase of Mo-containing steels is also not attacked. The A-262 Practice C test (the Huey test) is used only when the nitric acid is expected to be present. In as much as electrical resistivity and tensile properties are changed by sensitisation, they may be used for sensitisation detection. This is the underlying principle of the A-262 Practice E test (Strauss test), which is not quantitative. The A-262 Practice F test is adapt for use for Mo-containing steels and steels containing very low carbon levels.

Figure. 8: A typical 'step' microstructure after the ASTM A262 Practice A test [Parvathavarthini N, Dayal RK. Time-temperature-sensitization diagrams and critical cooling rates of different nitrogen containing austenitic stainless steels. J Nucl Mater 399(1): 2010: 62-7. With permission for reproduction from Elsevier].

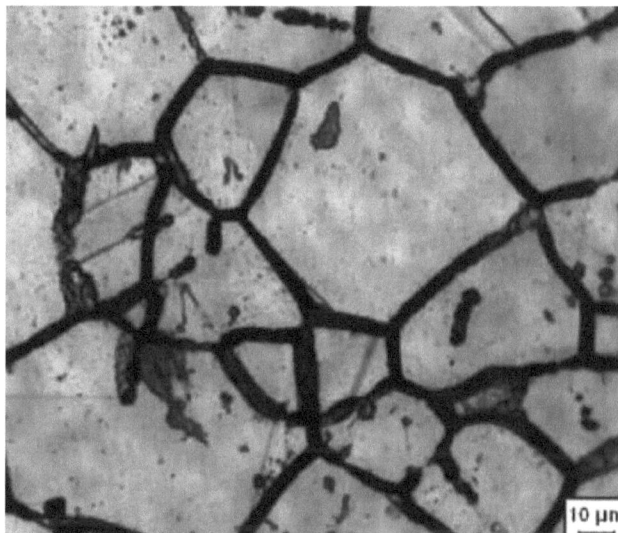

Figure. 9: A 'ditch' microstructure after the ASTM A262 Practice A test (the steel is the same as the one used in Figure **3.8**, except that it had been annealed at 923K for 50h) [Parvathavarthini N, Dayal RK. Time-temperature-sensitization diagrams and critical cooling rates of different nitrogen containing austenitic stainless steels. J Nucl Mater 399(1): 2010: 62-7. With permission for reproduction from Elsevier].

The EPR tests are rapid, quantitative and non-destructive. The single-loop EPR test is conducted using deaerated 0.5M H_2SO_4 + 0.01M KSCN at room temperature. The sample is first made anodic (at +200mV) and held in this range for 2 minutes. The potential is then reversed (reactivation) at a constant rate (6V/h) up to the corrosion potential (-400mV). A protective chromium oxide film will form on unsensitised grain boundaries, but the sensitised grain boundaries will be susceptible to corrosion attack during reactivation, which is reflected as an anodic peak in the potential-vs-current plot. Because the area under the anodic peak is directly proportional to the electric charge released during reactivation, it may be used to quantify the degree of sensitisation (DOS) (Fig. **10**). It has been suggested that the reactivation electric charge should be normalised by dividing it by the total grain boundary area [77]. However, since corrosion attack may not just occur along grain boundaries, and even if it does, it may not be uniformly and continuously distributed along grain boundaries. Furthermore, the width of the attack zones should be considered. Bruemmer *et al.* [78] have demonstrated that a volume depletion parameter correlates well with DOS.

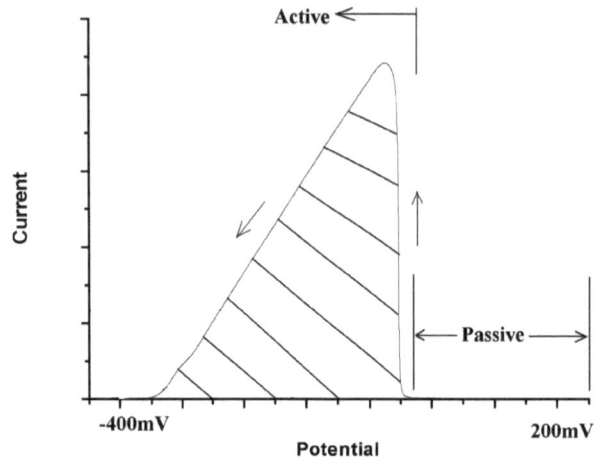

Figure 10: Schematic representation of the single-loop EPR test.

In the double-loop EPR test, an anodic scan from the open-circuit potential is done before the reactivation scan from the passivation potential. The ratio of the reactivation peak current (I_R) to the peak current in the anodic scan (I_A) is indicative of DOS (Fig. **11**). The double-loop EPR test is not sensitive to surface finish.

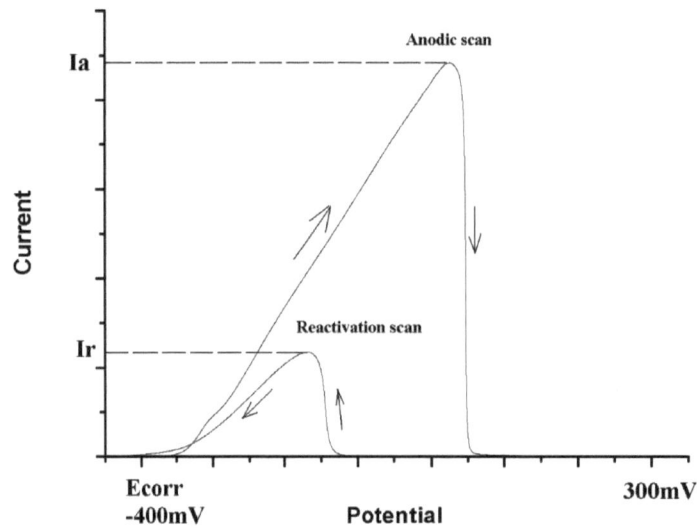

Figure 11: Schematic representation of the double-loop EPR test.

There are some points to note for the EPR tests. Firstly, temperature exerts a big influence on EPR results. Secondly, since carbides may also precipitate along twin boundaries, it is advisable to perform optical microscopy to ascertain that the reactivation peak or the reactivation electric charge are really attributable to grain boundary carbides. Thirdly, KSCN may be too aggressive for unsensitised steels, it has been proposed that thioacetamide may be more suitable for low-level sensitisation [79].

Other detection methods include local electrochemical impedance spectroscopy (LEIS) [80], magnetic methods like the detection of change in the Curie or Neel temperature [81], atomic force microscopy (AFM) [82], ultrasound [83], and eddy current [84].

4.3. Factors Affecting Sensitisation

Steel chemistry plays an important role in sensitisation. A useful method is to quantify the propensity to sensitisation by using the effective chromium content [85]. Grain size and plastic deformation also affect sensitisation behaviour. A small grain size means more numerous grain boundaries for Cr diffusion and shorter diffusion paths. As the grain size decreases, the time required for desensitisation becomes shorter [86]. Plastic deformation generates a profusion of dislocations and these are diffusion paths for Cr. The DOS and amount of plastic deformation is not directly proportional, because desensitisation will be enhanced by dislocations [87]. Deformation-induced martensite also has to be taken into account. All of these factors are detailed in Chapter **9**.

4.4. Effects of Sensitisation on Mechanical Behaviour

The effects of sensitisation on tensile behaviour may be seen in Fig. **12**. Annealing at 600°C for 2 months produced a very significant change in the tensile behaviour of the AISI304 steel. The yield strength is reduced, but the ultimate tensile strength is nearly unaltered. The strain-hardening behaviour is rather different. The tensile curve of the sensitised sample is sigmoidal, whereas that of the as-received sample is parabolic. Nonetheless, the sigmoidal shape is not dramatic, implying that the increase in strain-hardening rate is not significant. The small change into a sigmoidal shape of the tensile curve is likely to be due to the dispersion of fine carbides, but not to martensite as discussed in the preceding section.

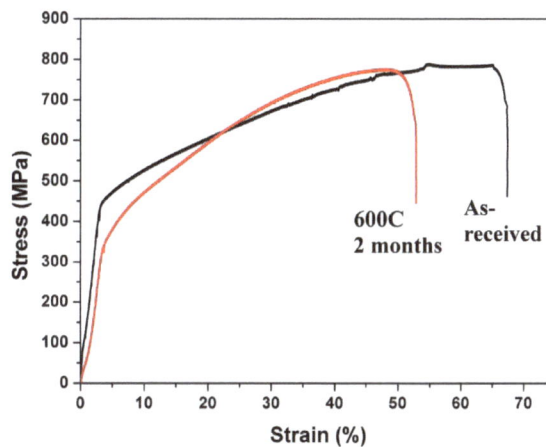

Figure. 12: Comparison of the tensile curves of unsensitised and sensitised AISI304 [LO KH, Zeng D, Kwok CT. Effects of sensitisation-induced martensitic transformation on the tensile behaviour of 304 austenitic stainless steel. Mater Sci Eng A 528(3); 2011: 1003-7. With permission for reproduction from Elsevier].

Besides the strain-hardening behaviour, there are two obvious features in Fig. **12.** They are the reduction in yield strength and the near invariance of the ultimate tensile strength. These two features have also been observed by Ghosh *et al.* [71] in a 304LN steel and by Shankar *et al.* [88] in a 316LN steel. The reduction in yield strength is mainly associated with the decrease in solid-solution strengthening elements in the matrix as a result of the precipitation of carbides [71].

The analysis of the near invariance in ultimate tensile strength is more subtle, as deformation-induced martensite will form during plastic deformation and pre-existing sensitisation-induced martensite will be broken. In the 304LN steel having no sensitisation-induced martensite, both dislocation multiplication and deformation-induced martensitic transformation during plastic deformation are thought to compensate for the loss in solid-solution strengthening elements, such that the ultimate tensile strength remains almost unchanged even after sensitisation [71]. In the sensitised steel, there is one more factor that has to be factored in, which is sensitisation-induced martensite that forms near grain and twin boundaries, and austenite/martensite interfaces. However, Lo *et al.* [89] have shown that sensitisation-induced martensite does not affect the tensile behaviour too much in the 304 steel [89].

In addition to tensile properties, sensitisation affects other mechanical properties. Fracture and ductility have been found to deteriorate significantly with the degree of sensitisation [90]. Overall, a steel will be embrittled by sensitisation as microvoids are more easily nucleated in grain boundaries due to the presence of carbides [91]. Sensitised steels are more vulnerable to stress-corrosion cracking [92].

4.5. Methods for Mitigating Sensitisation: Grain Boundary Engineering

Methods aimed at preventing sensitisation include reduction of the carbon content or the addition of strong carbide formers like Ti and Nb. Another commonly employed method is the manipulation of grain boundary structures. Numerous studies have shown that low-Σ coincidence site lattice (CSL) boundaries, usually taken as $\Sigma 3^n$ with n = 1, 2 and 3, possess good resistance to carbide precipitation [93, 94]. Carbides first precipitate out on high energy interfaces like random grain boundaries, then on non-coherent twins which are of lower energy. Coherent twin boundaries, on the other hand, are resistant to carbide formation [95]. The possibility of obtaining very favourable properties by the generation of high fractions of low-Σ CSLs in alloys is known as Grain Boundary Engineering (GBE) [96].

Methods for generating low-Σ CSL boundaries have been explored by a few groups and these include thermomechanical treatments such as strain annealing [97] and strain recrystallisation [96]. Iterative strain annealing and recrystallisation is also used to increase the fraction of low-Σ CSL boundaries without excessive grain growth [98]. A simple one-step thermomechanical process involving a pre-strain of 3% and a subsequent annealing at 1240K for 72h has been found to generate a high fraction of CSLs in a 316 austenitic stainless steel [99]. For a 304 steel, Fang *et al.* [100] have demonstrated that a pre-strain of 6~10% combined with a subsequent annealing at 1173K for 24-96h can produce a high population of $\Sigma 3^n$ (n=1~3) with a significant interruption of high angle grain boundaries. A novel method, using laser melting and annealing, for obtaining a high fraction of CSL boundaries on the surface of 304 has been developed by Yang *et al.* [101]. Therefore, it seems that the best processing route for achieving a high fraction of CSLs varies from grade to grade. Laser surface melting [102] and shot-peening [103] followed by proper annealing have also been found to improve the surface properties of austenitic stainless steels by the introduction of low-Σ CSL grain boundaries.

While the methods for obtaining low-Σ CSL boundaries almost invariably involve plastic deformation, Kumar *et al.* [104] have asserted that too much plastic deformation in a 304L steel may retard the formation of $\Sigma 3$ boundaries. A similar effect of plastic deformation was later confirmed in a 304 stainless steel by Shimada *et al.* [105]. They explained that small pre-strains would tend to encourage twin formation but without generating random grain boundaries, whereas too much plastic deformations would promote recrystallisation, with attendant formation of random boundaries [105].

While the resistance to carbide precipitation of CSL boundaries has been pointed out in a large number of studies, Singh *et al.* [106] have indicated that an increase in the frequency of CSL boundaries does not necessarily translate into a better resistance to sensitisation. Although positive reports on the low-Σ CSL boundaries are numerous, a recent study by Santos *et al.* [107] has reported unfavourably on the susceptibility of the $\Sigma 3$ (twins) boundary of a gas-nitrided AISI304L to corrosion-erosion damage.

Besides the amounts of CSLs, their distribution has been found to be important [105, 108]. When the CSLs are uniformly distributed in 304 stainless steels, a more significant enhancement of the resistance to intergranular corrosion can be attained because the network of susceptible random boundaries will be broken up. Twin-related variants and the fractions of triple junctions consisting of two or three special boundaries are also important in reducing the connectivity of random grain boundaries [109].

REFERENCES

[1]   News and Updates In: J Mater (JOM) 2006; 58: p.6.
[2]   High Nitrogen Steels and Stainless Steels - Manufacturing, Properties and Applications (U.K.Mudali and B.Raj, eds). Alpha Sci. Narosa Publishing House. 2004; p. 232.

[3]   Schaeffler AL. Constitution diagram for stainless steel weld metal. Metal Prog 1949; 56(11): 680-680B.

[4]   DeLong WT, Ostrom GA, Szumachowski ER. Measurement and calculation of ferrite in stainless steel weld metal. Weld J 1956; 35(11): 521s-28s.

[5]   Berrahmoune MR, Berveiller S, Inal K, *et al.* Delayed cracking in 301LN austenitic steel after deep drawing: Martensitic transformation and residual stress analysis. Mater Sci Eng A 2006; 438-440 (Sp.Iss.SI): 262-6.

[6]   Kruml T, Polak J, Degallaix S. Microstructure in 316LN stainless steel fatigued at low temperature. Mater Sci Eng A 2000; 293(1-2): 275-80.

[7]   Huang CX, Yang G, Gao YL, *et al.* Investigation on the nucleation mechanism of deformation-induced martensite in an austenitic stainless steel under severe plastic deformation. J Mater Res 2007; 22(3): 724-9.

[8]   Spencer K, Embury JD, Conlon KT, *et al.* Strengthening *via* the formation of strain-induced martensite in stainless steels. Mater Sci Eng A 2004; 387-9: 873-81.

[9]   Lee WS, Lin CF. The morphologies and characteristics of impact-induced martensite in 304L stainless steel. Scr Mater 2000; 43(8): 777-82.

[10]   Das A, Sivaprasad S, Ghosh M, *et al.* Morphologies and characteristics of deformation induced martensite during tensile deformation of 304 LN stainless steel. Mater Sci Eng A 2008; 486(1-2): 283-6.

[11]   Fukuda T, Kakeshita T, Kindo K. Effect of high magnetic field and uniaxial stress at cryogenic temperatures on phase stability of some austenitic stainless steels. Mater Sci Eng A 2006; 438-440(Sp.Iss.SI): 212-7.

[12]   Gey N, Petit B, Humbert M. Electron backscattered diffraction study of epsilon/alpha martensitic variants induced by plastic deformation in 304 stainless steel. Metall Mater Trans A 2005; 36(12): 3291-9.

[13]   Humbert M, Petit B, Bolle B, *et al.* Analysis of the γ-ϵ-α variant selection induced by 10% plastic deformation in 304 stainless steel at -60°C. Mater Sci Eng A 2007; 454-55: 508-17.

[14]   Lee TH, Oh CS, Kim SJ. Effects of nitrogen on deformation-induced martensitic transformation in metastable austenitic Fe-18Cr-10Mn-N steels. Scr Mater 2008; 58(2): 110-3.

[15]   Kireeva IV, Chumlyakov YI. The orientation dependence of γ-α' martensitic transformation in austenitic stainless steel single crystals with low stacking fault energy. Mater Sci Eng A 2008; 481-82 (Sp.Iss.SI): 737-41.

[16]   Hedstrom P, Lienert U, Almer J, *et al.* Stepwise transformation behavior of the strain-induced martensitic transformation in a metastable stainless steel. Scripta Mater 2007; 56(3): 213-6.

[17]   Spencer K, Veron M, Zhang KY, *et al.* The strain induced martensite transformation in austenitic stainless steels Part 1-Influence of temperature and strain history. Mater Sci Technol 2009; 25(1): 7-17.

[18]   Choi JY, Jin W. Strain induced martensite formation and its effect on strain hardening behavior in the cold drawn 304 austenitic stainless steels. Scr Mater 1997; 36(1): 99-104.

[19]   Raman SGS, Padmanabhan KA. Tensile deformation-induced martensitic transformation in AISI304LN austenitic stainless steel. J Mater Sci Lett 1994; 13(5): 389-92.

[20]   Talonen J, Nenonen P, Pape G, *et al.* Effect of strain rate on the strain-induced $\gamma \rightarrow \alpha'$ martensite transformation and mechanical properties of austenitic stainless steels. Metall Mater Trans 2005; 36(2): 421-32.

[21]   Mirzadeh H, Najafizadeh A. Correlation between processing parameters and strain-induced martensitic transformation in cold worked AISI 301 stainless steel. Mater Charact 2008; 59(11): 1650-4.

[22]   Holmberg H, Nilsson JO, Liu L. Development of low cost nonmagnetic stainless spring steels. ISIJ Intl 1990; 30(8): 594-9.

[23]   De AK, Speer JG, Matlock DK, *et al.* Deformation-induced phase transformation and strain hardening in type 304 austenitic stainless steel. Metall Mater Trans A 2006; 37(6): 1875-86.

[24]   Wu CC, Wang SH, Chen CY, *et al.* Inverse effect of strain rate on mechanical behavior and phase transformation of superaustenitic stainless steel. Scr Mater 2007; 56(8): 717-20.

[25]   Xiaojun Z, Procopiak LAJ, Souza NC, *et al.* Phase transformation during cavitation erosion of a Co stainless steel. Mater Sci Eng A 2003; 358(1-2): 199-204.

[26]   Rathbun RW, Matlock DK, Speer JG. Strain aging behavior of austenitic stainless steels containing strain induced martensite. Scr Mater 2000; 42(9): 887-91.

[27]   Buono VTL, Gonzalez BM, Andrade MS. Strain aging of AISI 430 ferritic stainless steel. Scr Mater 1997; 38(2): 185-90.

[28]   Mukhopadhyay CK, Kasiviswanathan KV, Jayakumar T, *et al.* Acoustic emission during tensile deformation of annealed and cold-worked AISI type 304 austenitic stainless steel. J Mater Sci1993; 28(1): 145-54.

[29]   Mukhopadhyay CK, Kasiviswanathan KV, Jayakumar T, *et al.* Acoustic emission from aging-induced α'-martensite in cold-worked AISI type 304 stainless steel. Scr Metall Mater 1994; 30(3): 303-7.

[30]   Mukhopadhyay CK, Kasiviswanathan KV, Jayakumar T, *et al.* Study of aging-induced α'-martensite formation in cold-worked AISI type 304 stainless steel. J Mater Sci 1995; 30(18): 4556-60.

[31]   Mukherjee P, Barat P. Acoustic emission studies on welded and thermally treated AISI 304 stainless steel during tensile deformation. Scr Mater 1997; 37(8): 1193-8.

[32]   Shaira M, Godin N, Guy P, *et al.* Evaluation of the strain-induced martensitic transformation by acoustic emission monitoring in 304L austenitic stainless steel: Identification of the AE signature of the martensitic transformation and power-law statistics. Mater Sci Eng A 2008; 492(1-2): 392-9.

[33]   Tavares SSM, Da Silva MR, Neto JM, *et al.* Ferromagnetic properties of cold rolled AISI 304L steel. J Magn Magn Mater 2002; 242-45: 1391-4.

[34]   Mumtaz K, Takahashi S, Echigoya J, *et al.* Use of saturation magnetism measurements to detect martensite formation in austenitic stainless steel after compressive deformation at high temperatures. J Mater Sci Lett 2002; 21(15): 1199-201.

[35]   Mumtaz K, Takahashi S, Echigoya J, *et al.* Detection of martensite transformation in high temperature compressively deformed austenitic stainless steel by magnetic NDE technique. J Mater Sci 2003; 38(14): 3037-50.

[36]   O'Sullivan D, Cotterell M, Meszaros I. The characterisation of work-hardened austenitic stainless steel by NDT micro-magnetic techniques. NDT & E Int 2004; 37(4): 265-9.

[37]   Zhang L, Takahashi S, Kamada Y, *et al.* Magnetic properties of SUS 304 austenitic stainless steel after tensile deformation at elevated temperatures. J Mater Sci 2005; 40(9-10): 2709-11.

[38]   Vertesy G, Meszaros I, Tomas I. Nondestructive indication of plastic deformation of cold-rolled stainless steel by magnetic minor hysteresis loops measurement. J Magn Mater 2005; 285(3): 335-42.

[39]   Vertesy G, Tomas I, Meszaros I. Non-destructive indication of plastic deformation of cold-rolled stainless steel by magnetic adaptive testing. J Magn Magn Mater 2007; 310(1): 76-82.

[40]   Khan SH, Ali F, Khan AN, *et al.* Eddy current detection of changes in stainless steel after cold reduction. Comp Mater Sci 2008; 43(4): 623-8.

[41]   Ghosh G, Olson GB. Kinetics of fcc-> bcc heterogeneous martensitic nucleation.1. the critical driving force for athermal nucleation. Acta Mater 1994; 42(10): 3361-70.

[42]   Ghosh G, Olson GB. Kinetics of fcc → bcc heterogeneous martensitic nucleation. 2. thermal activation. Acta Mater 1994; 42(10): 3371-9.

[43]   Ghosh G, Olson GB. The isotropic shear modulus of multicomponent Fe-base solid solutions. Acta Mater 2002; 50(10): 2655-75.

[44]   Eichelman GH, Hull FC. The Effects of Composition on the Temperature of Spontaneous Transformation of Austenite to Martensite in 18-8 Type Stainless Steels. Trans Am Soc Metal 1953; 45: 77-104.

[45]   Monkman FC, Cuff FG, Grant NJ. An empirical relationship between rupture life and minimum creep rate in creep-rupture tests. Metal Prog 1957; 71: 94-102.

[46]   Hammond CM. The development of maraging stain- less steels containing *cobalt.* Cobalt 1964; 25: 195-202.

[47]   Dai QX, Cheng XN, Zhao NT, *et al.* Design of martensite transformation temperature by calculation for austenitic steels. Mater Charact 2004; 52(4-5): 349-54.

[48]   Yuan ZZ, Dai QX, Zhang Q, *et al.* Microstructural thermostability of high nitrogen austenitic stainless steel. Mater Charact 2007; 58(1): 87-91.

[49]   Ishida K. Calculation of the effect of alloying elements on the Ms temperature in steels. J Alloy Compd 1995; 220(1-2): 126-31.

[50]   Xu W, Rivera-Diaz-Castillo PEJ, Van der Zwaag S. Designing nanoprecipitation strengthened UHS stainless steels combining genetic algorithms and thermodynamics. Comput Mater Sci 2008; 44(2): 678-89.

[51]   Capdevila C, Caballero FG, De Andres CG. Determination of Ms temperature in steels: A Bayesian neural network model. ISIJ Int 2002; 42(8): 894-902.

[52]   Caballero FG, Alvarez LF, Capdevila C, *et al.* The origin of splitting phenomena in the martensitic transformation of stainless steels. Scr Mater 2003; 49(4): 315-20.

[53]   Sourmail T, Garcia-Mateo C. Critical assessment of models for predicting the M-s temperature of steels. Comput Mater Sci 2005; 34(4): 323-34.

[54]   Sourmail T, Garcia-Mateo C. A model for predicting the M-s temperatures of steels. Comput Mater Sci 2005; 34(2): 213-218.

[55]   Andrews KW. Empirical formulae for the calculation of some transformation temperatures. J Iron Steel Inst 1965; 203: 721-7.

[56] Wu W, Hwu LY, Lin DY, *et al.* The relationship between alloying elements and retained austenite in martensitic stainless steel welds. Scr Mater 2000; 42(11): 1071-6.

[57] Tsai MC, Chiou CS, Du JS, *et al.* Phase transformation in AISI 410 stainless steel. Mater Sci Eng A 2002; 332(1-2): 1-10.

[58] Leem DS, Lee YD, Jun JH, *et al.* Amount of retained austenite at room temperature after reverse transformation of martensite to austenite in an Fe-13%Cr-7%Ni-3%Si martensitic stainless steel. Scr Mater 2001; 45(7): 767-72.

[59] Holmquist M, Nilsson JO, Stigenberg AH. Isothermal formation of martensite in a 12Cr-9Ni-4Mo maraging stainless steel. Scr Metall Mater 1995; 33(9): 1367-73.

[60] Mumtaz K, Takahashi S, Echigoya J, *et al.* Temperature dependence of martensitic transformation in austenitic stainless steel. J Mater Sci Lett 2003; 22(6): 423-7.

[61] Rhodes CG, Thompson AW. Composition dependence of stacking fault energy in austenitic stainless steels. Metall Trans A 1977; 8(12): 1901-6.

[62] Takemoto T, Murata Y, Tanaka T. Effects of alloying elements and thermomechanical treatments on mechanical and magnetic properties of Cr-Ni austenitic stainless steel. ISIJ Int 1990; 30(8): 608-14.

[63] Li JC, Zhao M, Jiang Q. Alloy design of FeMnSiCrNi shape-memory alloys related to stacking-fault energy. Metall Mater Trans A 2000; 31(3); 581-4.

[64] Tsai MC, Chiou CS, Du JS, *et al.* Phase transformation in AISI 410 stainless steel. Mater Sci Eng A 2002; 332(1-2): 1-10.

[65] Wang HS, Yang JR, Bhadeshia HKDH. Characterisation of severely deformed austenitic stainless steel wire. Mater Sci Technol 2005; 21(11): 1323-8.

[66] Ghosh G, Olson GB. Computational thermodynamics and the kinetics of martensitic transformation. J Phase Equilib 2001; 22(3): 199-207.

[67] Chatterjee S, Wang HS, Yang JR, *et al.* Mechanical stabilisation of austenite. Mater Sci Technol 2006; 22(6): 641-4.

[68] Zhang L, Takahashi S, Kamada Y. Quench-induced martensitic transformation in austenitic stainless steel after tensile deformation at elevated temperature. Scr Mater 2007; 57(8): 711-4.

[69] Moura V, Kina YA, Tavares SSM, *et al.* Investigation of cracks and sensitization in an AISI 304L stainless steel exposed to 500-600 °C. Eng Fail Anal 2009; 16(1): 545-51.

[70] Takaya S, Suzuki T, Matsumoto Y, *et al.* Estimation of stress corrosion cracking sensitivity of type 304 stainless steel by magnetic force microscope. J Nucl Mater 2004; 327(1): 19-26.

[71] Ghosh S, Kain V, Ray A, *et al.* Deterioration in Fracture Toughness of 304LN Austenitic Stainless Steel Due to Sensitization. Metall Mater Trans A 2009; 40(12): 2938-49.

[72] Kain V, Chandra K, Adhe KN, *et al.* Effect of cold work on low-temperature sensitization behaviour of austenitic stainless steels. J Nucl Mater 2004; 334(2-3): 115-32.

[73] Bain FC, Aborn RH, Rutherford JJ. The nature and prevention of intergranular corrosion in austenitic stainless steel. Trans Am Soc Steel Treat 1933; 21: 481-509.

[74] Cowan RL, Tedmon CS. In: Advances in Corrosion Science and Technology (Fontana MG and Staehle RW. Editors). New York. Plenum Press. Vol.3. 1972. pp.293-400.

[75] Lima AS, Nascimento AM, Abreu HFF, *et al.* Sensitization evaluation of the austenitic stainless steel AISI 304L, 316L, 321 and 347. J Mater Sci 2005; 40(1): 138-44.

[76] Dayal RK, Parvathavarthini N, Baldev R. Influence of metallurgical variables on sensitisation kinetics in austenitic stainless steels. Int Mater Rev 2005; 50(3): 129-55.

[77] Clark WAT, Mozhi AT, MacDonald DD. Report DOE/ER/10972-T2. April 1983.

[78] Bruemmer SM, Charlot LA, Atteridge DG. Sensitization development in austenitic stainless steel - measurement and prediction of thermomechanical history effects. Corrosion 1988; 44(7): 427-34.

[79] Fang Z, Wu YS, Zhang L, *et al.* Application of the modified electrochemical potentiodynamic reactivation method to evaluate intergranular corrosion susceptibility of stainless steels. Corrosion 1998; 54(5): 339-46.

[80] Neto PDL, Farias JP, Herculano LFG, *et al.* Determination of the sensitized zone extension in welded AISI 304 stainless steel using non-destructive electrochemical techniques. Corrosion Sci 2008; 50(4): 1149-55.

[81] Zhang L, Kamada Y, Kikuchi H, *et al.* Magnetic transition temperatures of some model alloys for simulating radiation induced segregation in austenitic stainless steel. J Magn Mag Mater 2004; 271(2-3): 402-8.

[82] Huang YL, Kinsella B, Becker T. Sensitisation identification of stainless steel to intergranular stress corrosion cracking by atomic force microscopy. Mater Lett 2008; 62(12-13): 1863-6.

[83] Stella J, Cerezo, Rodriquez E. Characterization of the sensitization degree in the AISI304 stainless steel using spectral analysis and conventional ultrasonic techniques. NDT & E Int 2009; 42(4): 267-74.

[84]   Shaikh H, Sivaibharasi N, Sasi B, *et al.* Use of eddy current testing method in detection and evaluation of sensitisation and intergranular corrosion in austenitic stainless steels. Corros Sci 2006; 48(6): 1462-82.

[85]   Parvathavarthini N, Dayal RK. Influence of chemical composition, prior deformation and prolonged thermal aging on the sensitization characteristics of austenitic stainless steels. J Nucl Mater 2002; 305(2-3): 209-19.

[86]   Beltran R, Maldonada JG, Murr LE, *et al.* Effects of strain and grain size on carbide precipitation and corrosion sensitization behavior in 304 stainless steel. Acta Mater 1997; 45(10): 4351-60.

[87]   Singh R, Ravikumar B, Kumar A, *et al.* The effects of cold working on sensitization and intergranular corrosion behavior of AlSl 304 stainless steel. Metall Mater Trans A 2003; 34(11): 2441-7.

[88]   Shankar P, Shaikh H, Sivakumar S, Venugopal S, Sundararaman D, Khata H.S. J Nucl Mater 1999; 264(1-2): 29-34.

[89]   Lo KH, Zeng D, Kwok CT. Effects of Sensitisation-induced Martensitic Transformation on the Tensile Behaviour of 304 Austenitic Stainless Steel. Mater Sci Eng A 2011; 528(3): 1003-7.

[90]   Ghosh S, Kain V, Ray A, *et al.* Deterioration in Fracture Toughness of 304LN Austenitic Stainless Steel Due to Sensitization. Metall Mater Trans A 2009; 40(12): 2938-49.

[91]   Tavares SSM, Fonseca MPC, Maia A, *et al.* Influence of the starting condition on the kinetics of sensitization and loss of toughness in an AISI 304 steel. J Mater Sci 2003; 38(17): 3527-33.

[92]   Abou-Elazm A, Abdel-Karim R, Elmahallawi I, *et al.* Correlation between the degree of sensitization and stress corrosion cracking susceptibility of type 304H stainless steel. Corros Sci 2009; 51(2): 203-8.

[93]   Crawford DG, Was GS. The role of grain boundary misorientation in intergranular cracking of Ni-16Cr-9Fe in 360°C argon and high-purity water. Metall Trans A 1992; 23(4): 1195-206.

[94]   Palumbo G, Aust KT. Structure-dependence of intergranular corrosion in high purity nickel. Acta Metall Mater 1990; 38(11): 2343-52.

[95]   Trillo EA, Murr LE. A TEM investigation of $M_{23}C_6$ carbide precipitation behaviour on varying grain boundary misorientations in 304 stainless steels. J Mater Sci 1998; 33(5): 1263-71.

[96]   Palumbo G, Lehockey EM, Lin P. Applications for grain boundary engineered materials. Jom 1998; 50(2): 40-43.

[97]   King WE, Schwartz AJ. Toward optimization of the grain boundary character distribution in OFE copper. Scr Mater 1998; 38(3): 449-55.

[98]   Thaveeprungsriporn V, Sinsrok P, Thong-Aram D. Effect of iterative strain annealing on grain boundary network of 304 stainless steel. Scri Mater 2001; 44(1): 67-71.

[99]   Michiuchi M, Kokawa H, Wang ZJ, *et al.* Twin-induced grain boundary engineering for 316 austenitic stainless steel. Acta Mater 2006; 54(19): 5175-84.

[100] Fang X, Zhang K, Guo H, *et al.* Twin-induced grain boundary engineering in 304 stainless steel. Mater Sci Eng A 2008; 487(1-2): 7-13.

[101] Yang S, Wang ZJ, Kokawa H, *et al.* Grain boundary engineering of 304 austenitic stainless steel by laser surface melting and annealing. J Mater Sci 2007; 42(3): 847-53.

[102] Yang S, Wang ZJ, Kokawa H, *et al.* Reassessment of the effects of laser surface melting on IGC of SUS 304. Mater Sci Eng A 2008; 474(1-2): 112-9.

[103] Alyousif OM, Engelberg DL, Marrow TJ. Surface grain boundary engineering of shot-peened type 304 stainless steel. J Mater Sci 2008; 43(4): 1270-7.

[104] Kumar BR, Das SK, Mahato B, *et al.* Effect of large strains on grain boundary character distribution in AISI 304L austenitic stainless steel. Mater Sci Eng A 2007; 454-455: 239-44.

[105] Shimada M, Kokawa H, Wang ZJ, *et al.* Optimization of grain boundary character distribution for intergranular corrosion resistant 304 stainless steel by twin-induced grain boundary engineering. Acta Mater 2002; 50(9): 2331-41.

[106] Singh R, Chowdhury SG, Kumar BR, *et al.* The importance of grain size relative to grain boundary character on the sensitization of metastable austenitic stainless steel. Scr Mater 2007; 57(3): 185-8.

[107] Dos Santos JF, Garzon CM, Tschiptschin AP. Improvement of the cavitation erosion resistance of an AISI 304L austenitic stainless steel by high temperature gas nitriding. Mater Sci Eng A 2004; 382(1-2): 378-86.

[108] Bi HY, Kokawa H, Wang ZJ, *et al.* Suppression of chromium depletion by grain boundary structural change during twin-induced grain boundary engineering of 304 stainless steel. Scr Mater 2003; 49(3): 219-23.

[109] Kumar M, King WE, Schwartz AJ. Modifications to the microstructural topology in f.c.c. materials through thermomechanical processing. Acta Mater 2000; 48(9): 2081-91.

CHAPTER 4

Martensitic Stainless Steels

Abstract This chapter is on martensitic stainless steels. A brief introduction to the compositions and uses of the various grades of steels in the ferritic class is given first. The purposes of and the changes in properties associated with typical heat treatments (austenisation and tempering), together with their underlying mechanisms, are then introduced. Recent results on the effects of heat-treatment parameters (temperature, cooling rate, *etc.*) are presented. A recently-developed tempering map for martensitic stainless steel and a new constitution diagram for ferritic-martensitic weld metals are also presented.

Keywords: Martensitic stainless steel, austenitisation, austenitising temperature, tempering, tempering map, martensitic transformation, retained austenite, constitution diagram, strengthening, carbide.

1. INTRODUCTION

Martensitic Stainless Steels (MSSs) typically contain 12~17%Cr, 0~4%Ni and 0.1~1.0%C (C<0.015% for the supermartensitic grades). Alloying elements like Mo, V, Nb, Al and Cu are added for enhancement of specific properties. Mo improves pitting corrosion resistance and Cu machinability [1], for example. The high-nitrogen grades are getting more popular as they possess higher strengths, toughness [2, 3] and pitting corrosion resistance [4]. Nitrogen may also bring about a finer prior austenite grain size [5]). The hardness of the martensitic class mainly stems from carbon, whereas hardenability is imparted by other alloying elements.

The martensitic class is used in a variety of fields. These include turbine blades, surgical instruments, cutlery, bearings, compressors, razors, *etc.* New steels possessing specific properties to suit particular applications have been made. For instance, steels possessing suitable workability, drivability *etc.*, for use as self-drilling and tapping screws have been developed [5, 6]. The steel, designated as XD15NW, has been developed with the aim of replacing AISI440C for cryogenic aerospace bearings because of its improved tribological and fatigue properties [7].

Like the ferritic class, the martensitic class is also susceptible to 475°C-embrittlement. It suffers from hydrogen embrittlement, too [8]. A steel may acquire hydrogen during melting, casting and heat treatment. For example, during casting, reaction of the steel with water vapour may generate hydrogen that is trapped in the metal during solidification. Alternatively, hydrogen may be produced as a corrosion product in some environments [9, 10]. This kind of embrittlement typically reveals itself during slow deformation and the fracture is usually intergranular [8, 11]. Annealing at low temperatures (between 200°C and 370°C, for instance) drives out hydrogen from the steel. The lowering in quantities of trapping sites like dislocations and twin boundaries on annealing makes hydrogen entry to the steel more difficult, thereby ameliorating embrittlement [9, 11].

The compositions of some of the commonly used MSSs are listed in Table **1**. AISI403, AISI410, AISI416 and its sibling with Se addition are the low-C varieties. AISI410 is the most widely used among them. AISI430, compared with AISI410, possesses better forgeability. AISI416 and AISI416Se are more readily machinable. Except for these, their mechanical properties are largely identical. AISI414, AISI422 and AISI431 contain nickel, which improves toughness, counterbalances the ferritising effects of molybdenum, vanadium, *etc.* and betters corrosion resistance in neutral chlorides and weakly oxidising acids. The three varieties of AISI440 all contain appreciable levels of carbon and are thus brittle. They have found applications in which abrasion resistance is a prime concern. Retained austenite may form in these steels on grounds of their high chromium contents. If the retained austenite transforms martensitically on quenching, then cracking may occur.

AISI403, AISI410 and AISI416 are considered the low-C grades. 416 is similar to 410 except that it is more readily machinable. 403 has a bit lower Cr content in order to avoid ferrite formation. These low-C grades have minimum hardness if they are annealed between 845°C and 870°C for about 2 hours per inch of thickness and then air-cooled. If it is desired to have partial softening and better machinability, then annealing is carried out between 730°C and 790°C for several hours and air-cooled. Annealing between 925°C and 1010°C for about several hours per inch of thickness, followed by air cooling, yields a high strength. AISI414, AISI422 and AISI431 are also of low C contents, but they contain nickel compared with AISI403, AISI410 and AISI416. In 414 and 431, Ni improves toughness. In 422 and 431, Ni ensures that the high-temperature microstructure remains austenitic (422 also contains Mo, V and W, all of them are ferritising and they improve high-temperature properties). AISI420, 420F and AISI440 are of high C contents. These grades, when annealed at 900°C and then slowly air-cooled to 590°C, will have minimum hardness for forming operations. If it is desired to harden these grades, then they are first heated to 790°C. Following an isothermal hold at this temperature, they are then heated to between 995°C and 1050°C and subsequently cooled in air or in oil. Because of the high C contents, retained austenite may form. This austenite can then transform isothermally to martensite at room temperature, resulting in a size change or cracking. Therefore, a subzero quench or a double tempering treatment, with cooling to at least room temperature between treatments, might be required. In the quenched and stress-relieved state, the high-C grades may have a high yield strength (up to 1895MPa) and hardness (up to 600 in the Brinell scale for 44°C), but they are brittle. Tempering may improve ductility, but corrosion resistance may be compromised.

Table 1: Compositions of some commonly used martensitic stainless steels (in wt%).

Designation	Cr	C	Mn	Si	P	S	Mo	Ni	Se	V	W
403	11.5-13.0	0.15	1.0	0.5	0.04	0.03	---	---	---	---	---
410	11.5-13.5	0.15	1.0	1.0	0.04	0.03	---	---	---	---	---
414	11.5-13.5	0.15	1.0	1.0	0.04	0.03	---	1.25-2.50	---	---	---
416	12.0-14.0	0.15	1.25	1.0	0.06	0.15 min	0.6 (optional)	---	---	---	---
416Se	12.0-14.0	0.15	1.25	1.0	0.06	0.06	---	---	0.15 min	---	---
420	12.0-14.0	0.15 min	1.0	1.0	0.04	0.03	---	---	---	---	---
420F	12.0-14.0	0.15 min	1.25	1.0	0.06	0.15 min	---	---	---	---	---
422	11.0-13.0	0.20-0.25	1.0	0.75			0.75-1.25	0.5-1.0	---	0.15-0.30	0.75-1.25
431	15.0-17.0	0.20	1.0	1.0	0.04	0.03	---	1.25-2.50	---	---	---
440A	16.0-18.0	0.60-0.75	1.0	1.0	0.04	0.03	0.75	---	---	---	---
440B	16.0-18.0	0.75-0.95	1.0	1.0	0.04	0.03	0.75	---	---	---	---
440C	16.0-18.0	0.95-1.20	1.0	1.0	0.04	0.03	0.75	---	---	---	---

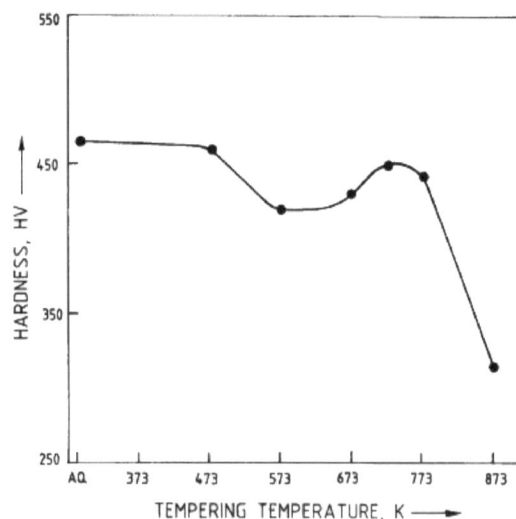

Figure 1: Variation in hardness with tempering temperature of a 16Cr-2Ni martensitic stainless steel [Balan KP, Reddy AV, Sarma DS. Austenite precipitation during tempering in 16Cr-2Ni martensitic stainless steels. Scr Mater 1998; 39(7): 901-5. With permission for reproduction from Elsevier].

MSSs are usually fully austenitic between 950°C and 1000°C. Their high alloying contents impart to them sufficient hardenability such that they may acquire a martensitic microstructure on air-cooling, even for larger sizes. Desired properties are then obtained by subsequent tempering treatments.

As mentioned, the subsequent tempering treatment affects properties of the martensitic class greatly (Fig. 1). Some typical values of properties of the various martensitic steels before and after tempering treatments are shown in Table **2**. The temperature above which re-austenitisation and hardening begin again sets the upper limit for tempering.

Table 2: Mechanical properties of representative martensitic stainless steels before and after tempering treatments.

Designation	Condition	Yield strength (MPa)	Tensile strength	%Elongation	Hardness (Brinell)	Izod impact toughness at room temperature
403	Annealed	275	515	30	155	135
	Quenched and tempered	415-1035	620-1310	30-15	180-390	135-27
410	Annealed	275	515	30	155	135
	Quenched and tempered	415-1035	620-1310	30-15	180-390	135-27
414	Annealed	690	825	20	245	81
	Quenched and tempered	725-1035	825-1380	20-15	260-415	108-54
416	Annealed	275	515	30	155	108
	Quenched and tempered	415-1035	620-1310	25-10	180-390	81-27
420	Annealed	345	655	25	200	81
	Quenched and tempered	550-1380	760-1655	25-5	250-550	Low
431	Annealed	690	860	20	260	95
	Quenched and tempered	690-1240	965-1515	20-10	270-440	95-34
440A	Annealed	415	725	20	215	Low
	Quenched and tempered	550-1655	825-1790	10-2	To 570	Low
440B	Annealed	450	760	15	225	Low
	Quenched and tempered	620-1725	895-1860	10-2	To 600	Low
440C	Annealed	480	795	10	230	Low
	Quenched and tempered	620-1860	895-1930	10-2	To 600	Low

Usually, the best corrosion properties in typical oxdising media are obtained in the stress-relieved condition, but impact toughness may be inadequate. Stress-corrosion resistance in the stress-relieved condition is also not good (although general corrosion resistance is good) because of the high hardness. In the tempered condition, stress-corrosion resistance is improved, but general corrosion resistance is degraded.

2. CONSTITUTION OF MARTENSITIC STAINLESS STEELS

As mentioned in Chapter **2**, a fully austenitic microstructure may be obtained at high temperatures in Fe-Cr alloys when Cr is less than about 11wt%. As stainless steels, the martensitic class must contain at least

11wt%Cr and so it is obvious that austenitising alloying elements (C, N and Ni, *e.g.*) should be added in order to expand the high-temperature austenitic field. Besides the austenitic region, the duplex ferrite plus austenite region is also expanded by alloying and so retained ferrite may be present in some of the martensitic grades.

The requisite element that imparts stainlessness, *i.e.*, Cr, is a ferritising element and so the high-temperature austenite region shrinks as the Cr content goes up. To impart temper resistance to the steels, molybdenum and vanadium are added. To counter the ferritising effects of Mo and V, austenitising elements like Ni, Co, Cu and Mn must be added, too. Equations have been developed for assessment of the ferrite-forming abilities of the commonly used elements. Some of these equations are listed below for reference. Nevertheless, use of these equations should be with caution for specific cases.

Thielemann: $Cr_{eq} = Cr + 2.1W + 2.8Ta + 4.2Mo + 4.5Cb + 5.2Si + 7.2Ti + 11V + 12Al - 40(C + N) - 3Ni - 2Mn$

Aggen: $Cr_{eq} = Cr + 1.5Si + 7.2Ti + 2.5Al + 3Cb + 10Zr + 2V - 40(C + N) - 3Ni - 2(Mn + Cu)$

The Aggen equation is more applicable to the martensitic class as it includes a 10 to 12%Cr base.

Alloying must also be chosen carefully in order not to depress the start-temperature of the martensitic transformation Ms to below room temperature, or else retained austenite may form, thereby lowering strength. Carbide precipitation during tempering may raise the Ms temperature of the retained austenite, which may then transform to martensite (untempered martensite), resulting in cracking. The effects of different elements in depressing the Ms temperature for a 12%Cr steel are shown in Table **3**. A good alloying element in this regard should make the ratio of decrease in ferrite per wt% to decrease in Ms per wt% as large as possible. Economy aside, cobalt is considered a good replacement for Ni as it is not effective in depressing the martensitic temperature Ms [12]. Nitrogen, which is frequently used nowadays, is also an effective depressor of Ms [13].

Table 3: Abilities of different elements in depressing the Ms temperature for a 12%Cr steel.

Element	C	Mn	Ni	Cr	Mo	W	Si
Depression of Ms (°C/wt%)	-474	-33	-17	-17	-21	-11	-11

The as-quenched microstructure after austenitisation may be obtained by using the appropriate Cr-equivalent and Ni-equivalent numbers, together with the pertinent phase diagram. For instance, the equivalent numbers proposed by Schneider [14], together with the Schaeffler diagram, may be used [15].

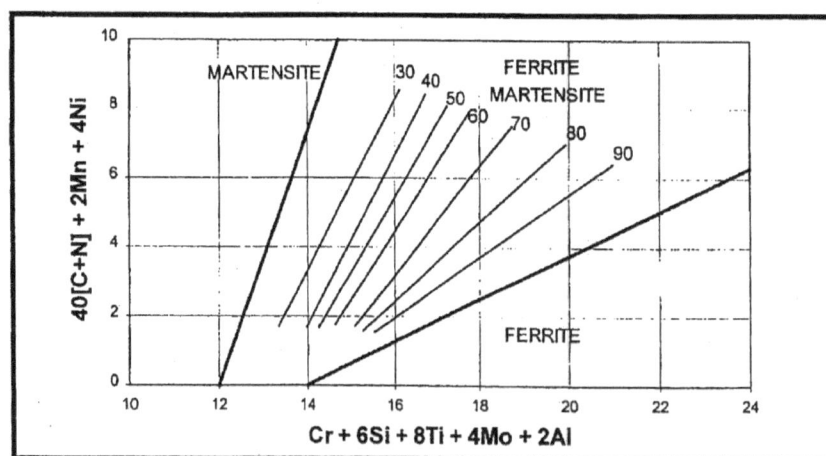

Figure 2: Constitution diagram for ferritic-martensitic stainless steel weld metals [Balmforth MC, Lippold JC. A preliminary ferritic-martensitic stainless steel constitution diagram. Weld J 1998; 77(1): 1s-7s. With permission for reproduction from American Welding Society].

For a prediction of microstructures in martensitic stainless steel welds, the usual choices are still the oft-used WRC-1992 diagram, the Schaeffler diagram and the like [16]. However, these diagrams have been found to be not that accurate. For instance, a fully martensitic microstructure predicted by the Schaeffler diagram might actually contains some ferrite, which is easily sensitised, thereby lowering strength. A recent, very accurate constitution diagram specifically for use on ferritic-martensitic stainless steel weld metals has been developed by Balmforth and Lippold [17, 18] (Fig. **2**). The diagram is well suited for conventional arc welding processes, but not for high-energy sources like laser and electron beam [18].

For martensitic stainless steel weld metals, the Ms temperature may be estimated by the equation proposed by Gooch, which is often used in practice [16, 19]:

$$Ms(^{\circ}C) = 540 - (497\%C + 6.3\%Mn + 36.3\%Ni + 10.8\%Cr + 46.6\%Mo)$$

3. HEAT TREATMENT OF MARTENSITIC STAINLESS STEELS

3.1. Effects of Austenitisation Treatment

MSSs are usually used in the quenched-and-tempered state. Tempering is necessary in order for the steels to attain reasonable ductility. MSSs are first heated to the fully austenitic field (>980°C) for dissolution of carbides (as mentioned, the high-temperature microstructure might be duplex, depending on steel chemistry). Quenching in different media (water, air, oil, *e.g.*) will then be performed. Usually, it is desired to have a completely martensitic microstructure in the as-quenched state. However, because MSSs contain appreciable levels of carbon, carbides usually form upon quenching. Furthermore, austenite and ferrite may be retained in the as-quenched microstructure. Therefore, the final microstructure and properties depend on a lot of factors like cooling rate, austenitising temperature, *etc.* For instance, while retained austenite may be beneficial to fatigue resistance, it may promote precipitation of carbides during tempering and hence will decrease pitting corrosion resistance [20]. Also, if the retained austenite transforms to martensite upon quenching after tempering, cracking may result. It has to be noted that the surface of the steel may easily contain retained ferrite because of decarburisation.

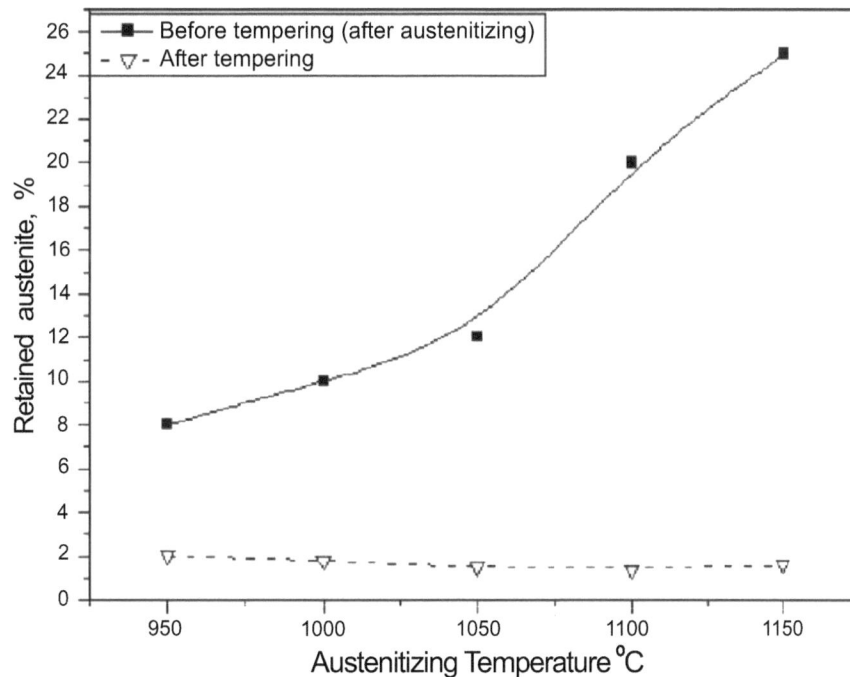

Figure 3: Effect of austenitising temperature on the amount of retained austenite in the AISI431 steel weld [Rajasekhar A, Reddy GM, Mohandas T, *et al.* Influence of austenitising temperature on microstructure and mechanical properties of AISI431 martensitic stainless steel electron beam welds. Mater Des 2009; 30(5): 1612-24. With permission for reproduction from Elsevier].

As the austenitising temperature goes up, more carbides will be dissolved and hence the austenite matrix will be more enriched with Cr and C, directly affecting the martensitic temperature (Ms) of the matrix [21-23]. Below a certain austenitising temperature (around 1025°C for the AISI420 steel [21]), corrosion resistance will be enhanced as the temperature goes up because of the dissolution of carbides and the resulting enrichment of the matrix with Cr [24]. However, the beneficial effect of carbide dissolution is negated if the austenitising temperature goes up to such an extent that the matrix is so enriched with C and other alloying elements that its Ms point is depressed to a very low value. In this case, some of the austenite will be retained on quenching [21] (Fig. **3**). Also, too much carbon saturation in the matrix may bring about high internal lattice stresses [21] and pinning of clustered vacancies to partial dislocations [22], both of them may depress the Ms temperature (also refer to Fig. **4**). Kim and Kim [25] have found in a 12%Cr steel that the MC carbides are harder to dissolve than the $M_{23}C_6$ carbides, the latter is dissolved above 1000°C. Since most Cr is tied up by $M_{23}C_6$, there is no need to use an excessively high austenitising temperature. Longer austenitising times may reduce carbide precipitation [26].

Recent research has pointed out that double austenitisation may bring about dissolution of carbides without significant grain growth [27]. In a 16Cr-2Ni steel, double austenitisation ($1070°C \rightarrow$ oil quenching $\rightarrow 1000°C \rightarrow$ oil quenching $1070°C$) has been shown to produce very good mechanical properties. The higher temperature dissolves carbides, whereas the lower temperature results in a finer grain size. For grades having very high carbon contents, double austenitisation plus double tempering has also been found to be beneficial. In a MSS containing 0.69%C, a fine grain size, together with good strength and toughness, may be obtained by adopting the following heat treatment [28]:

$$1523K \rightarrow \text{quickly cool to and tempered at } 923K \rightarrow \text{quick cooling} \rightarrow 1173K \rightarrow$$
$$\text{quickly cool to and temepred at } 923K \rightarrow \text{quick cooling}$$

The steel produced in this way possesses higher ductility and tensile strength than the 403, 410, 420J2 and 420J1 steels [28].

The first tempering at 923K results in decomposition of austenite into ferrite and finely-dispersed $M_{23}C_6$. The subsequent 1123K treatment cannot dissolve the carbides and so they restrain the growth of austenite, thereby producing fine austenite grains.

The austenitising temperature at which complete carbide dissolution is attained [29] and the start and finish temperatures of austenitisation [30, 31] depend on the heating rate. The reverse transformation from martensite to austenite may occur either by diffusion or by a diffusionless shear process [30]. As mentioned, complete carbide dissolution may in fact be detrimental to corrosion resistance [23] and grain growth may be significant at such high austenitising temperatures [32]. Since the dissolution of carbides during austenitisation exerts such a big influence, being able to monitor this dissolution will be valuable in property optimisation. Recently, Caballero *et al.* [33] have demonstrated that thermoelectric power may be employed for such a purpose.

3.2. Effects of Cooling Rate

In addition to austenitising temperature, the cooling rate also has a dramatic effect on the amount of retained austenite [22], in as much as high cooling rates suppress carbide formation and so the C-enriched retained austenite cannot transform martensitically [34]. The influence of cooling rate on Ms is particularly pronounced for grades whose carbon contents exceed 0.20 (Fig. **4**).

After austenitisation, if the austenite is deformed before cooling to form martensite, then retained austenite may form because mechanical stabilisation prevents some austenite from transforming martensitically [35].

3.3. Effects of Tempering Treatment

Without a tempering treatment, MSSs lack ductility and toughness. The high-Cr grades are usually just tempered at low temperatures for stress relief, because they are designed to have good corrosion resistance and so precipitation of chromium carbides is not desired. For the low-Cr grades, the tempering temperature depends on their desired final properties. For instance, the 0.3C-12Cr steel is tempered to a hardness of about 550HV for use as cutlery.

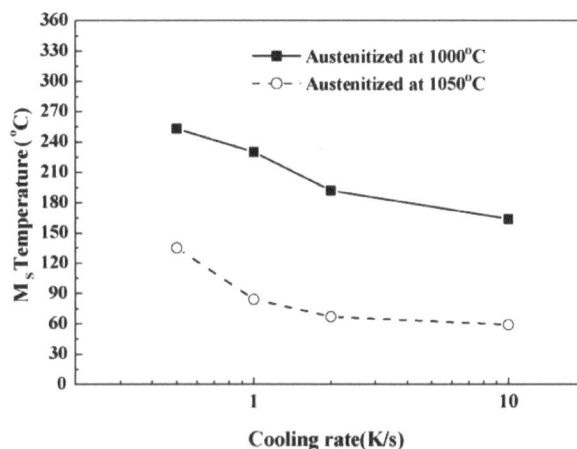

Figure 4: Effect of austenitising temperature and cooling rate on the Ms point of a 14Cr-3Mo martensitic stainless steel containing 0.3%C [Park JY, Park YS. The effects of heat-treatment parameters on corrosion resistance and phase transformations of 14-Cr-3Mo martensitic stainless steel. Mater Sci Eng A 2007; 449-51: 1131-4. With permission for reproduction from Elsevier].

Tempering is usually conducted between 200°C and 700°C. However, the range of 475°C to 550°C must be used with caution, because severe embrittlement of the material may occur due to 475°C embrittlement (please also refer to Chapter **2**). In general, the precipitation of the sigma phase in the Fe-Cr system is sluggish, except for steels that are heavily deformed [36]. The martensitic class is also prone to sensitisation [37], especially in the heat affected zones of weld metals [38]. Even the very low-C supermartensitic grades (typically <0.015%) are not immune to this problem [39].

Like the Ms temperature, the A_{c1} temperature (the temperature over which austenitisation occurs) depends on steel chemistry and this temperature sets the upper limit at which tempering can be carried out. The A_{c1} temperature may be estimated by the equation proposed by Lovejoy [40]:

$$A_{C1}(°C) = 310 + 35Cr + 3.5(Cr - 17)^2 + 60Mo + 73Si + 170Cb + 290V + 620Ti + 750Al$$
$$+ 1400B - 250C - 280N - 115Ni - 66Mn - 18Cu$$

This equation does not take into account nitrogen. However, since nitrogen is an austenite-forming element, it is expected to decrease the A_{c1} temperature. Cobalt also depresses A_{c1}, but not as much as nickel does [41].

The following phenomena may occur during tempering [25, 42, 43] relief of stresses, precipitation of new carbides, coarsening of undissolved carbides, annihilation and rearrangement of dislocations, formation and growth of subgrains (these subgrains form from martensite laths). All of these changes accompanying tempering are affected by the prior austenitising temperature. For instance, finer $M_{23}C_6$ carbides and a higher dislocation density (retained within tempered martensite) result from high austenitising temperatures. The austenitising temperature also affects the morphology of the as-quenched martensite and the prior austenite grain size. Insofar as carbides preferentially form on prior austenite grain boundaries and sub-grain boundaries [42], which form from the prior martensite boundaries, the precipitation behaviour of carbides is influenced directly by the austenitisation treatment. These in turn will affect properties (*e.g.*, strength of tempered martensite [44]). As the austenitising temperature increases, the yield and ultimate tensile strengths tend to increase because of fine carbides and a high dislocation density. Nonetheless, the toughness decreases appreciably on the grounds of the large prior austenite grain size [44].

Lim *et al.* [45] have proposed a tempering map for the AISI 403 steel (Fig. **5**). In the annealing zone, hardness decreases but ductility increases. In the secondary hardening zone, strength is maintained or improved (because of heterogeneous precipitation of fine carbides in the martensite matrix). As its name suggests, the steel is sensitised if tempered in the sensitisation zone. In the healing zone, the steel will be desensitised, but it will suffer from a considerable loss of strength.

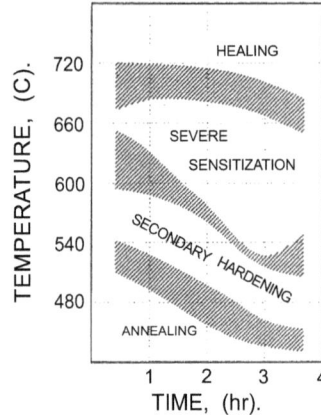

Figure 5: Tempering map for the AISI403 martensitic stainless steel [Lim LC, Lai MO, Ma J, *et al.* Tempering of AISI 403 stainless steel. Mater Sci Eng A 1993; 171(1-2): 13-9. With permission for reproduction from Elsevier].

Lula suggested the use of a subzero quench, followed by double tempering treatments with cooling to room temperature as a proper tempering process for MSSs [46]. Recently, Yang *et al.* [47] have shown that a subzero quench to -196°C, plus a 2 hour tempering treatment at 600°C could lead to a more complete transformation of retained austenite than with a single 4-hour tempering at 600°C for the AISI440C steel.

Besides retained austenite, the tempering treatment should be controlled in such a way that coarse carbides do not form because they tend to detract from fatigue performance. Fig. **6.** shows the microstructure of a substandard blade of a cutting device in a meat processing plant. The figure shows that the substandard blade contained large carbides, which acted as initiation sites for brittle fracture (Fig. **7**). Further, carbides containing Cr may decrease the corrosion resistance. In a 12.5Cr MSS, Falleiros *et al.* [37] found that carbides formed below 450°C were mostly iron carbides (of the M_3C type) and so Cr was not depleted around them. Lim *et al.* [45] also found in an AISI403 steel containing 0.099%C that it was not susceptible to sensitisation below 480°C. In a 12%Cr steel, it was found that M_3C would gradually dissolve and M_7C_3 would form above 400°C [41]. In a 0.63C-12.7Cr steel, Lin and Lin [48] found that M_7C_3 and $M_{23}C_6$ both form on tempering at 300°C, with the latter getting more prevalent as the temperature went up to 500°C and 600°C. The precipitation in N-containing MSSs is even more complicated and may involve MN, M_2N, ζ -nitride, ε -nitride, ε -carbide, cementite, M_7C_3 [49, 50].

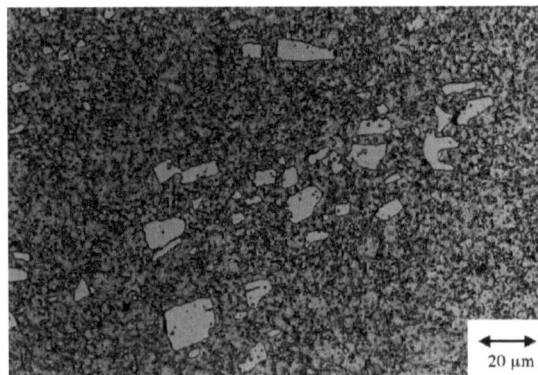

Figure 6: Coarse carbides in a substandard cutting blade made of a 11%Cr martensitic stainless steel [Neri MA, Colas R. Analysis of a martensitic stainless steel that failed due to the presence of coarse carbides. Mater Charact 2001; 47(3-4): 283-9. With permission for reproduction from Elsevier].

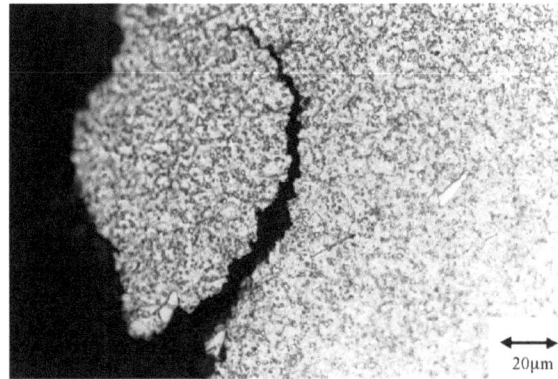

Figure 7: A fatigue crack that initiated in a coarse carbide [Neri MA, Colas R. Analysis of a martensitic stainless steel that failed due to the presence of coarse carbides. Mater Charact 2001; 47(3-4): 283-9. With permission for reproduction from Elsevier].

Measurement of the changes in magnetic parameters like coercive force, saturation magnetisation *etc.* has been shown to be a good means for monitoring the tempering of MSSs [51].The precipitation of fine and coarse carbides and the attendant change in matrix composition lead to the changes in magnetic properties.

4. STRENGTHENING OF MARTENSITIC STAINLESS STEELS

For MSSs, the most important strengthening mechanism is martensitic transformation. To obtain maximum strength, it is important to properly control the steel composition such that a 100% martensitic microstructure is obtained after quenching from the austenitisation temperature. The strength of the martensite is directly proportional to the carbon content up to about 0.6%. Over this level, the steel will be so saturated with C that carbides will precipitate from the melt as primary carbides. Substitutional alloying elements (Cr and Ni, *e.g.*) can solid-solution strengthen the steels. Reduction of the prior austenite grain size is certainly conducive to strengthening. Cold working of annealed martensitic steels may bring about some additional strengthening. Ausworking, *i.e.*, deformation of the metastable austenite before it transforms to martensite, may significantly improve strength and toughness of many steels. A further increase in strength may be obtained by continuing the deformation after martensite has formed.

On tempering, the precipitation of carbides may also strengthen the steels. The low carbon 0.1C-12Cr steel is used for illustration purposes. This steel, when tempered below about 500°C, will form small Cr_7C_3 and some fine Cr_2C. Molybdenum and vanadium encourage the formation of Cr_2C. Nitrogen favours the formation of $Cr_2(CN)$ at the expense of Cr_7C_3. The precipitation of the M_2X carbide/carbonitrides leads to intense hardening. When the steel is overaged, $M_{23}C_6$ forms and then coarsens, it also gradually replaces the strengthening M_2X carbide. Since molybdenum and vanadium may stabilise M_2X, the temper resistance is enhanced when they are present. Niobium also stabilises M_2X and may refine the prior austenite grain size. Nickel, on the other hand, lowers the temper resistance.

REFERENCES

[1]   Geng H, Wu X, Wang H, *et al.* Effects of copper content on the machinability and corrosion resistance of martensitic stainless steel. J Mater Sci 2008; 43(1): 83-7.

[2]   Speidel MO. HNS 88-High Nitrogen Steels. Institute of Metals. London. P.92.

[3]   Horovitz MB, Neto FB, Garbogini A, *et al.* Nitrogen bearing martensitic stainless steels: microstructure and properties. ISIJ Int 1996; 36(7): 840-5.

[4]   Ono AA, Alonso N, Tschiptschin AP. The corrosion resistance of nitrogen bearing martensitic stainless steels. ISIJ Intl 1996; 36(7): 813-7.

[5]   Leda H. Nitrogen in martensitic stainless steel. J Mater Process Technol 1995; 53(1-2): 263-72.

[6]   Calliari I, Zanesco M, Dabala M, *et al.* Investigation of microstructure and properties of a Ni-Mo martensitic stainless steel. Mater Des 2008; 29(1): 246-50.

[7]   Girodin D, Manes L, Moraux JY, *et al*. Characterization of the XD15N high nitrogen martensitic stainless steel for aerospace bearing. In: Proceedings of the 4th International Conference on Launcher Technology "Space Launcher Liquid Propulsion". Dec 2002, Liefe, Belgium.

[8]   Lancha AM, Serrano M, Lapena J, *et al*. Failure analysis of a river circulating pump shaft from a NPP. Eng Fail Anal 2001; 8(3): 271-91.

[9]   Rodriguez JGG, Martinez GB, Bravo VMS. Effect of heat treatment on the stress corrosion cracking behaviour of 403 stainless steel in NaCl at 95°C. Mater Lett 2000; 43(4): 208-14.

[10]   Rodriguez JGG, Bravo VMS, Villafane AM. Hydrogen embrittlement of type 410 stainless steel in sodium chloride, sodium sulfate, and sodium hydroxide environments at 90°C. Corrosion 1997; 53(6): 499-504.

[11]   Rodriguez JGG, Bravo VMS, Villafane AM. Stress corrosion cracking of type 403 stainless steel in sodium chloride at 95° C under different heat treatment conditions. Corrosion 1999; 55(10): 991-6.

[12]   Balan KP, Rao KVR, Reddy AV, *et al*. Determination of potency factor of cobalt for estimation of nickel equivalent in 16Cr-2Ni martensitic stainless steel. Mater Sci Technol 1999; 15(7); 798-802.

[13]   Lo KH, Lai JKL. On the cryogenic magnetic transition and martensitic transformation of the austenite phase of 7MoPLUS duplex stainless steels. J Magn Magn Mater 2010; 322(16): 2335-9.

[14]   Schneider H. Investment casting of high-hot strength 12% chrome steel. Foundry Trade J 1960; 108: 562-63.

[15]   Balan KP, Reddy AV, Sarma DS. Austenite precipitation during tempering in 16Cr-2Ni martensitic stainless steels. Scr Mater 1998; 39(7): 901-5.

[16]   Nebhnani MC, Bhakta UC, Gowrisankar I, *et al*. Failure of a martensitic stainless steel pipe weld in a fossil fuel power plant. Eng Fail Anal 2002; 9(3): 277-86.

[17]   Balmforth MC, Lippold JC. A new ferritic-martensitic stainless steel constitution diagram. Weld J 2000; 79(12): 339s-45s.

[18]   Balmforth MC, Lippold JC. A preliminary ferritic-martensitic stainless steel constitution diagram. Weld J 1998; 77(1): 1s-7s.

[19]   Dupont JN, Kusko CS. Martensite Formation in Austenitic/Ferritic Dissimilar Alloy Welds. Weld J 2007; 86(2): 51s-4s.

[20]   Bayati H, Elliott R. Austempering process in high manganese alloyed ductile cast iron. Mater Sci Technol 1995; 11(2): 118-29.

[21]   Candelaria AF, Pinedo CE. Influence of the heat treatment on the corrosion resistance of the martensitic stainless steel type AISI420. J Mater Sci Lett 2003: 22(16): 1151-3.

[22]   Park JY, Park YS. The effects of heat-treatment parameters on corrosion resistance and phase transformations of 14Cr-3Mo martensitic stainless steel. Mater Sci Eng A 2007: 449-51: 1131-4.

[23]   Park JY, Park YS. Effects of austenitizing treatment on the corrosion resistance 14Cr3Mo martensitic stainless steel. Corros 2006; 62(6): 541-7.

[24]   Choi YS, Kim JG, Park YS, *et al*. Austenitizing treatment influence on the electrochemical corrosion behavior of 0.3C-14Cr-3Mo martensitic stainless steel. Mater Lett 2007; 60(1): 244-7.

[25]   Kim HD, Kim IS. Effect of austenitising temperature on microstructure and mechanical properties of 12%Cr steel. ISIJ Int 1994; 34(2): 198-204.

[26]   Vitek JM, Klueh RL. Precipitation reactions during the heat treatment of ferritic steels. Metall Trans A 1983; 14(6): 1047-55.

[27]   Balan KP, Reddy AV, Sarma DS. Effect of single and double austenitisation treatments on the microtructure and mechanical properties of 16Cr-2Ni steel. J Mater Eng Perform 1999; 8(3): 385-93.

[28]   Tsuchiyama T, Ono Y, Takaki S. Microstructure control for toughening a high carbon martensitic stainless steel. ISIJ Int 2000; 40(Suppl): S184-8.

[29]   Andres CGD, Caruana G, Alvarez LF. Control of M23C6 carbides in 0.45C-13Cr martensitic stainless steel by means of three representative heat treatment parameters. Mater Sci Eng A 1998; 241(1-2): 211-5.

[30]   Leem DS, Lee YD, JJun JH, *et al*. Amount of retained austenite at room temperature after reverse transformation of martensite to austenite in an Fe-13%Cr-7%Ni-3%Si martensitic stainless steel. Scr Mater 2001; 45(7): 767-72.

[31]   Kaluba WJ, Kaluba T, Tillard R. The autenitizing behaviour of high-nitrogen martensitic stainless steels. Scr Mater 1999; 41(12): 1289-93.

[32]   Andres CGD, Alvarez LF, Lopez V, *et al*. Effects of carbide-forming elements on the response to thermal treatment of the X45Cr13 martensitic stainless steel. J Mater Sci 1998; 33(16): 4095-100.

[33]   Caballero FG, Capdevila C, Alvarez LF, *et al*. Thermoelectric power studies on a martensitic stainless steel. Scr Mater 2004; 50(7): 1061-6.

[34]   Alvarez LF, Garcia C, Lopez V. Continuous cooling transformation in martensitic stainless steels. ISIJ Int 1994; 34(6): 516-21.

[35]   Tsai MC, Chiou CS, Du JS, *et al.* Phase transformation in AISI 410 stainless steel. Mater Sci Eng A 2002; 332(1-2): 1-10.

[36]   Link HS, Marshall PW. The formation of sigma phase in 13%-16% chromium steel. Trans Am Soc Metal 1952; 44: 549-59.

[37]   Falleiros NA, Magri M, Falleiros IGS. Intergranular corrosion in a martensitic stainless steel detected by electrochemical tests. Corrosion 1999; 55(8): 769-78.

[38]   Nakamichi H, Sato K, Miyata Y, *et al.* Quantitative analysis of Cr-depleted zone morphology in low carbon martensitic stainless steel using FE-(S)TEM. Corrosion Sci 2008; 50(2): 309-15.

[39]   Smith L. Sensitization of martensitic stainless steels. Mater Perform 2002; 41(12): 54-5.

[40]   Lovejoy PT. Structure and constitution of wrought martensitic stainless steels. In: Handbook of Stainless Steels. Peckner D, Bernstein (Eds), McGraw-Hill, USA.

[41]   Irvine KJ, Crowe DJ, Pickering FB. The physical metallurgy of 12% chromium steels. J Iron Steel Inst 1960; 195: 386-405.

[42]   Eggeler G. The effect of long-term creep on particle coarsening in tempered martensite ferritc steels. Acta Metall 1989; 37(12): 3225-34.

[43]   Abe F, Araki H, Noda T. The effect of tungsten on dislocation recovery and precipitation behavior of low-activation martensitic 9Cr steels. Metall Trans A 1991; 22(10): 2225-35.

[44]   Iwabuchi Y. Temper embrittlement of type 13-Cr-4Ni cast steel. Trans Iron Steel Inst Jpn 1987; 27(3): 211-7.

[45]   Lim LC, Lai MO, Ma J, *et al.*. Tempering of AISI 403 stainless steel. Mater Sci Eng A 1993; 171(1-2): 13-9.

[46]   Lula RA. Stainless Steel. America Society for Metals. Metals Park. Ohio. 1988.

[47]   Yang JR, Yu TH, Wang CH. Martensitic transformation in AISI440C stainless steel. Mater Sci Eng A 2006; 438-40(Suppl): 276-80.

[48]   Lin CH, Lin Y. Microstructure and mechanical properties of 0.63C-12.7Cr martensitic stainless steel. Chin Hua J Sci Eng 2009; 7(2): 41-6.

[49]   Berns H, Gavriljuk V. Tempering of martensitic stainless steel with 0.6 w/o nitrogen and/or carbon. J Phys IV 1997; 7(C5): 263-8.

[50]   Toro A, Misiolek WZ, Tschiptschin AP. Correlations between microstructure and surface properties in a high nitrogen martensitic stainless steel. Acta Mater 2003; 51(12): 3363-74.

[51]   Tavares SSM, Fruchart D, Miraglia S, *et al.* Magnetic properties of an AISI 420 martensitic stainless steel. J Alloy Compd 2000; 312(1-2): 307-14.

CHAPTER 5

Duplex Stainless Steels

Abstract: This chapter is devoted to duplex stainless steels (DSSs). This chapter begins with an introduction to the different types of DSSs (ordinary, lean, superduplex and hyperduplex), their applications and compositions of the commonly used grades. Their typical microstructures and methods for phase prediction via different constitution diagrams are discussed. The precipitate phases that may form in DSSs and their implications on mechanical/corrosion properties are presented (a more in-depth discussion on these precipitates is presented in Chapter **8**). The superplasticity of DSSs is discussed in detail. Both the conventional method and a new one for producing the microduplex structure are introduced. Recent developments in explaining the underlying mechanisms leading to the superplastic behaviour in DSSs are discussed.

Keywords: Duplex stainless steel, wrought duplex stainless steel, cast duplex stainless steel, superduplex stainless steel, hyperduplex stainless steel, lean duplex stainless steel, microduplex microstructure, superplasticity, phase prediction, constitution diagram, ferrite potential, ferrite factor.

1. INTRODUCTION

A Duplex Stainless Steel (DSS) is a duplex alloy whose two constituent phases are both stainless steels. That is, both constituent phases contain at least about 11wt% Cr. The two phases are usually present in substantial volume fractions and as two large distinct volumes. The most commonly used duplex stainless steels are those comprising ferrite (α) and austenite (γ), which form the basis of this chapter. In addition to duplex stainless steels (ferritic-austenitic, martensitic-ferritic, and ferritic-martensitic), triplex stainless steels (ferritic-austenitic-martensitic) are also available. The existence of Fe-Cr-Ni DSSs was reported by Bain and Griffiths in 1927 [1]. These authors also published isothermal sections of the Fe-Cr-Ni system for 900°C, 1050°C, 1200°C and 1300°C.

Typically, DSSs contain 17 to 30wt% Cr and 3 to 13wt%Ni. To guard against oxidation, Mn and Si in the range of 0.5 to 2.0wt% may be added to the steel. Newer varieties, like 7MoPLUS and 2205, contain raised levels of N and Mo for enhancement of strengths, general corrosion resistance and pitting corrosion resistance. Their pitting resistance equivalent numbers (PREN=%Cr + 3.3%Mo + 16%N) are usually between 33 and 36. Carbon is commonly kept at a low level, although some carbon-strengthened DSSs may contain up to 0.3 wt%C. Superduplex stainless steels, with even higher alloying levels and a minimum of 0.25wt%N, have a PREN that is at least 40. Table **1.** shows some of the commercially available DSSs, together with their compositions. The typical ferrite fractions and the state of the steel (wrought or cast) are also shown in the last column, but it has to be noted that the volume fraction of ferrite (and that of austenite) depends heavily on the exact heat treatment and composition [2]. Ordinary DSSs find applications in a variety of fields. The millennium bridge in York, England was built with 2205. This grade is also a standard material for making chemical tankers. The pulp-and-paper industry is also a heavy user of DSSs.

Besides ordinary duplex and superduplex DSSs, recent interests are on lean DSSs [3] and hyperduplex DSSs [4]. LDX2101 is a new lean DSS, which was designed with the aim to compete with traditional austenitic stainless steels 304/304L and 316 [3, 4]. In the oil and gas industry, the use of lean DSSs as pipe racks, cable trays, insulation claddings has been found to be promising as replacements for carbon steels and AISI316. Fig. **1** shows how DSSs, lean and conventional, compare with traditional austenitic stainless steels in terms of strength and corrosion resistance. Lean DSSs are less prone to sigma phase formation than their conventional duplex counterparts [3], as they contain less Ni and Mo, but higher levels of N, Mn and Cr. It has been suggested that lean DSSs are those almost Mo-free and contain less than 3wt% Ni. However, LDX2101 contains about 0.3wt% Mo. Crevice corrosion may be more severe in lean DSSs because of their low Ni levels. Also, their high Mn contents may detract from pitting corrosion resistance.

Joseph Ki Leuk Lai, Kin Ho Lo and Chan Hung Shek

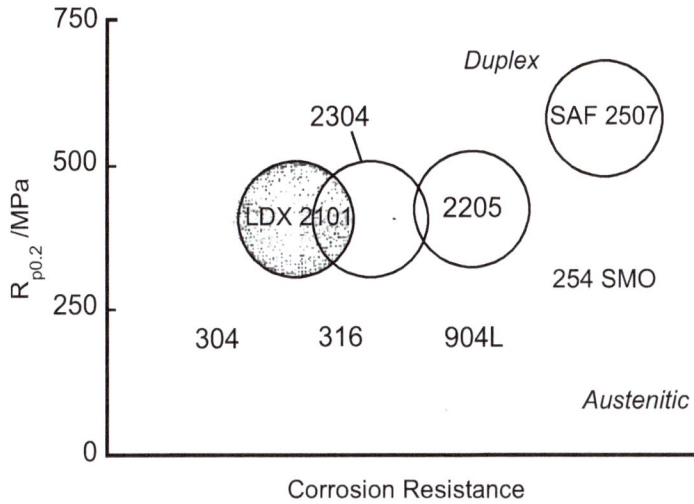

Figure 1: Comparison of duplex stainless steels and austenitic stainless steels [Liljas M, Johansson P, Liu H, *et al.* Development of a lean duplex stainless steel. Steel Res Int 2008; 79(6): 466-73. Copyright Wiley-VCH Verlag GmbH & Co. Reproduced with permission].

For hyper DSSs (S32707), they contain significant amounts of nitrogen and are highly alloyed, such that their PREN may reach 49 [5]. Fig. **2**. indicates the compositional differences among the various types of DSSs (note particularly their N contents). The hyper DSS UNSS32707 may rival titanium and Ni-based alloys. One of the typical uses of hyperduplex DSSs is in heat exchangers that are used in very hostile environments (involving seawater, *e.g.*). For severe applications that the superDSSs might just be adequate, the hyperDSSs may be a wiser choice. For instance, the hyper DSSs have been found to be more suitable than the superDSSs for use as subsea umbiblical cords, because of the very high water pressure and severe corrosive environment [6].

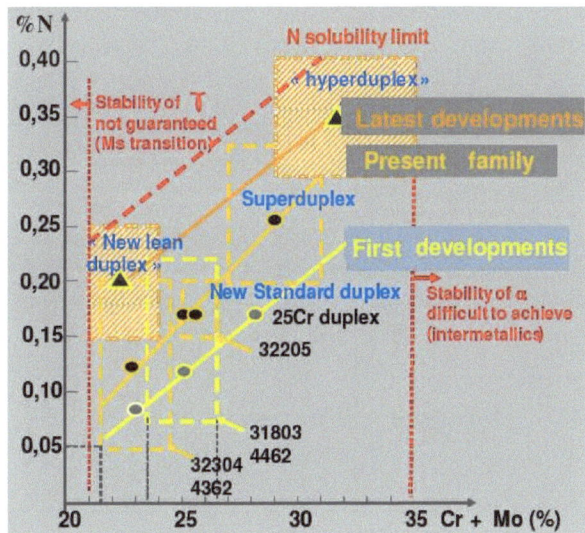

Figure 2: A map showing the compositional differences among the various types of DSSs [Charles J. Duplex stainless steels – a review after DSS'07 held in Grado. Steel Res Int 2008; 79(6): 455-65. Copyright Wiley-VCH Verlag GmbH & Co. Reproduced with permission].

The Fe contents of most commonly used DSSs centre at about 70wt% and the compositions of DSSs are designed in such a way that the alloys fall into the ($\alpha + \gamma$) field. However, whether or not a Fe-Cr-Ni alloy containing 70wt% Fe is duplex depends also on the temperature. A duplex microstructure may either be obtained during solidification (cast DSSs) or by hot working between about 1000°C and 1200°C (wrought DSSs).

Table 1: Compositions (in wt%)and ferrite contents of some commonly used duplex stainless steels.

Designation	Fe	Cr	Ni	Mo	Mn	C	Si	S	P	Others	%Ferrite
Paralloy 3FL	Bal.	21-24	4.5-6	2.5-3.5	≦1.0	≦0.04	≦1.0			N=0.1-0.2	50 (cast)
Ferralium 225	Bal	24-27	4.5-6.5	2-4	≦2.0	≦0.08	≦2.0	≦0.04	≦0.04	N≧1.0 Cu=1.3-4.0	50 (wrought)
7Mo	Bal	23-28	2.5-5	1-2	≦1.0	≦0.08	≦0.75	≦0.03	≦0.04		85 (wrought)
U52	Bal	24.5-25.5	6.5-7.5	2.5-3.5	≦2.0	≦0.03	≦1.0	≦0.03	≦0.04	N≦2.0 Cu=0.5	50 (wrought)
Zeron25	Bal	21-23	4.5-6.5	2.5-3.5	<2.0	<0.03	<1.0				55-65 (wrought)
22Cr-5Ni-2Mo-N	Bal	25.0	5.0	2.0	2.0	0.5	0.025	0.5		Cu~1 N=0.15	
SAF2205	Bal	22	5.5	3.0	≦2.0	0.03	≦0.8	≦0.02	≦0.03	N~0.14	45
X2CrNiMo225	Bal	21-23	4.5-6.5	2.5-3.5	2.0	≦0.03	1.0			N=0.08-0.20	45
DP1	Bal	18.0-19.0	4.25-5.25	2.5-3.0	1.2-2.0	≦0.03	1.4-2.0	≦0.03	≦0.03		50
LDX	Bal	21.5	1.5	0.3	5	0.03				N=0.22 Cu=0.3	
S32707		32	7	3.5		0.03max				N=0.4	

2. CAST DUPLEX STAINLESS STEELS

2.1. Solidification Modes of Cast DSSs

If the alloy in the molten state is cooled extremely slowly (approaching equilibrium solidification, which, in theory, never occurs in practice), then a duplex microstructure is obtainable only when the composition is designed in such a way that the alloy is within the ($\alpha + \gamma$) field at the melting point. In almost all cases, either α or γ may form first because most cast alloys and duplex weld metals have a composition that falls outside this pseudo-eutectic range. The first solid that forms may actually be well outside the pseudo-eutectic range.

For instance, the nominal Cr and Ni contents of the CF series of cast DSSs are 19wt% and 9wt%, respectively (CF$_3$, *e.g.*). Under equilibrium solidification, the whole alloy should be composed solely of α. But the microstructures of the CF series are actually duplex, with γ being the predominant phase (Fig. **3**). The root cause of the duplex microstructure has to do with the non-equilibrium solidification nature that is encountered in practice.

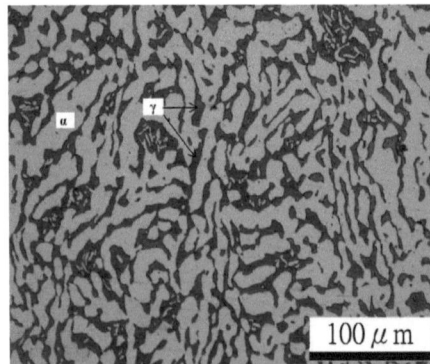

Figure 3: Typical microstructure of cast duplex stainless steels.

α forms first if the composition is in the α side of the eutectic range. This α will be enriched with Cr and lacking in Ni compared with the bulk alloy composition (and even more so relative to the eutectic composition). As Ni is progressively ejected to and Cr absorbed from the remaining liquid, the liquid composition will be driven towards the eutectic as solidification proceeds. The situation is similar if the first solid that forms is γ. In both scenarios, the last material to solidify will be duplex, although the bulk composition of the alloy may be outside the (α+γ) field. It is this non-equilibrium solidification nature that causes most cast alloys (like the CF series mentioned above) and weld metals to be duplex.

2.2. Phase Prediction of Cast DSSs

Besides Cr and Ni, other alloying elements such as Mn, Mo, C and N, may also influence phase proportions, even though they may be present in just small quantities. For weld metals, the Schaeffler diagram and its modified version are frequently utilised for phase prediction. The axes of these diagrams are expressed in Cr equivalent and Ni equivalent numbers. Cr is a ferritiser and is assigned a coefficient 1 in the Cr equivalent number. Mo is also assigned a coefficient 1, and this means Mo is as effective as Cr in stabilising ferrite. The coefficients of other elements are interpreted in a similar fashion.

The modified Schaeffler diagram also contains the Welding Research Council (WRC) numbers, which indicates ferrite fraction in terms of a magnetically determined ferrite content. The modified Schaeffler diagram tends to yield a higher ferrite content than that obtained from the original Schaeffler diagram. The Schaeffler and the modified Schaeffler diagrams are, strictly speaking, inapplicable to castings. Nonetheless, they do give an indication of what phases are most likely to be present in cast alloys. For cast alloys, it has been suggested that the Schoefer diagram should be used.

2.3. Ferrite Potential and Ferrite Factor

Besides the afore-mentioned diagram, phase balance in cast DSSs may be predicted by using the concept of the ferrite potential (FP), too. The FP was originally proposed by Wolf to describe the amounts of ferrite and austenite just after solidification [7].

$$FP = 5.26 \cdot (0.74 - \frac{Ni_{eq}}{Cr_{eq}})$$

$$Ni_{eq} = Ni + 0.31Mn + 22C + 14.2N + Cu$$

$$Cr_{eq} = Cr + 1.5Si + 1.4Mo + 3Ti + 2Nb$$

In addition to phase prediction, the FP has also been used for assessing the tendency of hot cracking in ferritic stainless steel. The presence of some ferrite (at least 5%) may alleviate hot cracking, which results from the rupture of liquid films separating already solidified material, or the rupture of isolated solid bridges. Hot cracking may occur in both castings and weld metals. Ray *et al.* [8] adopted the FP to predict the sensitivity/tendency to cracking for continuously cast ferritic stainless steels (clinking) [8]. These authors stated that a lower FP (<3.5) would result in more austenite to prevent ferrite grain growth and so should lead to a finer grain structure and higher strength, which are desirable for minimising clinking.

The ferrite factor (FF), also known as the Kaltenhauser Ferrite Factor (KFF), indicates the amount/stability of austenite at high temperatures [9]. Consequently, the FF may also be used to assess the susceptibility of ferritic stainless steels to high-temperature cracking [10, 11].

$$FF = Cr + 6Si + 8Ti - 40(C + N) - 2Mn - 4Ni$$

$$KFF = Cr + 6Si + 8Ti + 4Mo + 2Al + 4Nb - 40(C + N) - 2Mn - 4Ni$$

Austenite may exist at higher temperatures for steels having higher values of the FF. In general, it is desired to have FF<12. Barbe *et al.* [12] have stated that both the FP and the FF are utilisable as indicators for the presence of cracks in ingots. The FF may also be used to predict the amount of retained ferrite in the high-temperature heat affected zones of welded ferritic stainless steels [12].

3. WROUGHT DUPLEX STAINLESS STEELS

In general, the compositions of wrought DSSs are on the Cr-rich side of the pseudo-binary eutectic. For this reason, they mostly solidify as ferrite first and then austenite will form upon cooling. Therefore, As the temperature goes up, the ferrite content increases. Fig. **4**. shows that the DSS 7MoPLUS is almost entirely ferritic at 1300°C, whereas ferrite accounts for about 70% at 1100°C (Fig. **5**). Between 1000°C and 1200°C where most wrought DSSs are hot worked, they have a duplex ($\alpha + \gamma$) microstructure. In Fig. **4**, there are numerous small spot in the grain interiors. These spots are mostly nitrides (carbides are of secondary importance because of the low C contents in most DSSs), which may detract from corrosion and toughness properties [4]. Nitrides are especially troublesome in the heat-affected zones of DSS weld metals. The fast formation of nitrides may be ascribed to the high-N contents and lack of time for austenite formation during rapid cooling.

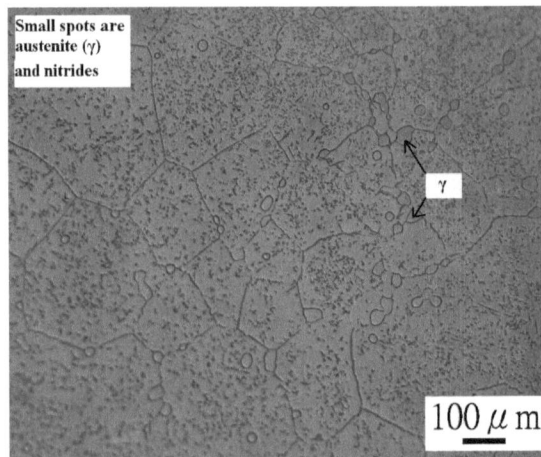

Figure 4: The microstructure of the DSS 7MoPLUS is mainly ferritic after solution-treatment at 1300°C for 1 hour.

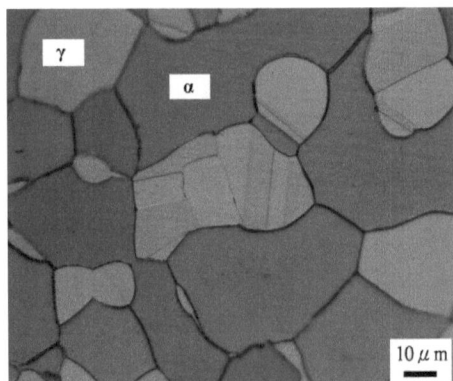

Figure 5: The microstructure of the DSS 7MoPLUS is almost entirely ferritic after solution-treatment at 1100°C for 1 hour.

As regards phase prediction of wrought DSSs, the Schaeffler diagram, the modified Schaeffler diagram and the Schoefer diagram are not useable for wrought alloys (although the Schaeffler might give a rough indication on what phases are likely to be in the microstructure [4]). For wrought DSSs, a Schaeffler type diagram has been developed by Pryce and Andrews at 1150°C [13]. But this diagram is strictly only applicable at 1150°C. Nevertheless, this diagram may give a general idea of phase proportions at other temperatures.

The presence of ferrite, while being able to mitigate hot cracking, may also bring about other beneficial effects to mechanical properties. For example, a duplex microstructure may enable the alloy to deform superplastically, *i.e.*, the alloys may exhibit very large neck-free, uniform elongation. Superplasticity of DSSs, because of its importance, will be discussed in a separate section.

4. PRECIPITATE PHASES IN DUPLEX STAINLESS STEELS

A jumble of phases may form in DSSs upon exposure to elevated temperatures. Fig. **6.** is the Temperature-Time-Precipitation (TTP) diagram for the DSS U50. Chapter **8.** gives a detailed account of the precipitates in Fig. **6.** and the problems caused by them.

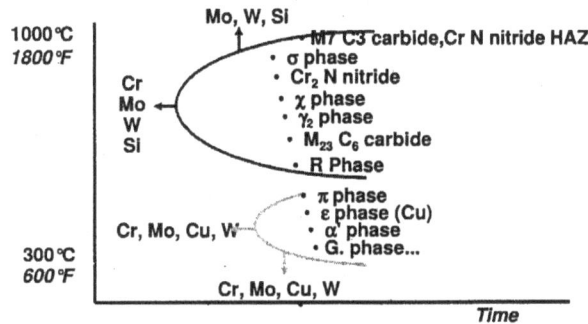

Figure 6: The temperature-time-precipitation diagram for the DSS U50 [Charles J. Duplex stainless steels – a review after DSS'07 held in Grado. Steel Res Int 2008; 79(6): 455-65. Copyright Wiley-VCH Verlag GmbH & Co. Reproduced with permission]

In most cases, these phases are considered deleterious because properties are degraded. Consequently, it is advisable to limit the service temperature of DSSs to about 250°C. For example, severe embrittlement and loss of corrosion resistance is associated with the precipitation of the sigma phase and spinodal decomposition (*i.e.*, 475-°C embrittlement). Nevertheless, both of these 'deleterious' phenomena have been found to be utilisable for temperature monitoring. The technique that utilises spinodal decomposition is called Feroplugs, while the technique that takes advantage of the sigma phase is designated as Sigmaplugs. Both of these 'plugs' are described in more detail in Chapter **10.**

Compared with ferritic and austenitic stainless steels, DSSs are particularly prone to sigma phase formation. Fig. **7.** shows that even exposure to elevated temperature for just a few minutes is sufficient for the sigma phase to form in conventional DSSs. Lean DSSs, such as LDX2101, are more resistant to sigma formation because they contain less Cr and Mo.

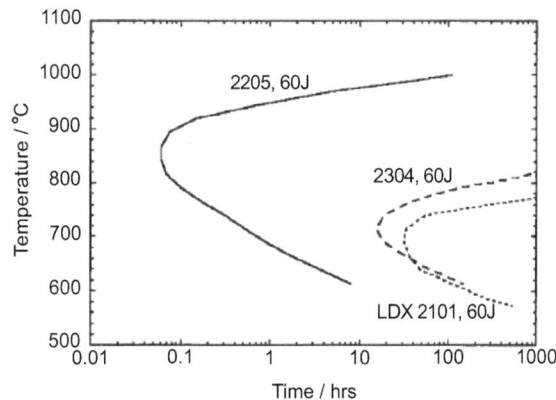

Figure 7: Iso-impacy energy curves for three DSSs [Liljas M, Johansson P, Liu H, *et al.* Development of a lean duplex stainless steel. Steel Res Int 2008; 79(6): 466-73. Copyright Wiley-VCH Verlag GmbH & Co. Reproduced with permission].

5. MECHANICAL PROPERTIES OF DUPLEX STIANLESS STEELS

The yield strengths of DSSs are usually higher than those of austenitic stainless steels. For example, the yield strength of the 304 steel is about 200~250MPa, but the yield strength of 7Mo is about 565MPa and that of SAF2205 is in the range of 410 to 450MPa. The room-temperature ultimate tensile strengths of DSSs are usually between 600MPa and 800MPa. Compared with annealed austenitic stainless steels, the ductilities of DSSs are lower (about 20 to 30% in terms of elongation, compared with the 50 to 60% for annealed 304), because the ferrite phase is prone to cleavage fracture.

The mechanical properties of DSSs are a reflection of the mechanical properties of ferrite and austenite, with a pronounced effect exerted by the former. Ferritic stainless steels in general have higher yield strengths than do austenitic stainless steels (at the same interstitial level and in the annealed state). And so ferrite may increase the yield strength when it is added to austenite. Hence, as the volume fraction of ferrite increases, the yield strength of a DSS increases. However, the ultimate tensile strength decreases with increasing ferrite fraction. This is because ferrite does not work-harden as well as austenite does. Also, the presence of ferrite may constrain the austenite from transforming to martensite, which is a strengthening phase.

Upon exposure to high temperatures, the unstable phase, *i.e.*, the phase in which precipitates form, is ferrite. As a result, the high-temperature mechanical properties of DSSs are also affected to a large extent by the ferrite phase. For example, when spinodal decomposition occurs, it is the ferrite phase that spinodally decomposes, whereas the austenite phase remains largely unaffected. Fig. **8**. shows that the hardness of the austenite phase does not change much during spinodal decomposition, but the ferrite phase is progressively embrittled. The overall embrittlement is thus almost solely associated with the embrittlement of the ferrite phase [14]. The same is also true when it comes to embrittlement due to sigma phase precipitation [15].

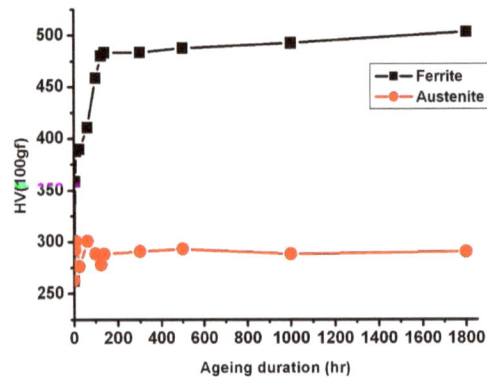

Figure 8: Hardness changes in the ferrite and austenite phases during spinodal decomposition.

As regards the sigma phase, its presence in DSSs might be beneficial under some circumstances [16]. In cold-rolled DSSs whose ferrite grains are elongated in the rolling direction, the morphology of the resulting sigma phase after prolonged thermal ageing is also elongated in the same direction, *i.e.*, the sigma phase forms bands. The band-like sigma plus secondary austenite ($\gamma_2 + \sigma$) microstructure may improve creep strength because movement of dislocations is hindered. Of course, the morphology of the band-like ($\gamma_2 + \sigma$) microstructure should be properly controlled through a thermomechanical treatment in order to bring about this enhancement in creep strength.

6. SUPERPLASTICITY OF DUPLEX STAINLESS STEELS

6.1. Treatments for Obtaining Superplastic Microstructures

Microduplex stainless steels, whose microstructure is a homogeneous dispersion of fine, equiaxed, ferrite and austenite (Fig. **9**), may exhibit superplasticity. Superplasticity is the occurrence of a very large neck-free elongation (may be in excess of 1000%) at a relatively high homologous temperature ($\geqq 0.5$) [17]. Besides superplasticity, microduplex stainless steels also possess very good toughness, strength and fatigue properties because of their fine microstructure. A variety of applications using DSSs involve superplastic operations. For instance, the sink in the toilet of Boeing 737 aircraft is fabricated by superplastic forming of the DSS NAS SuperDux64 [18].

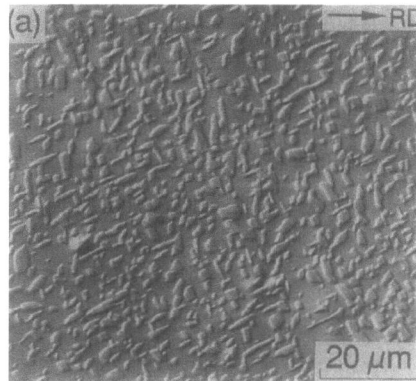

Figure 9: A typical microduplex structure [Tsuzaki K, Huang X, Maki T. Mechanism of continuous recrystallization during superplastic deformation in a microduplex stainless steel. Acta Mater 1996; 44(11): 4491-9. With permission for reproduction from Elsevier].

In order to obtain a microduplex microstructure, it is customary to subject DSSs to a high-temperature solution-treatment (*e.g.*, at 1573K), water-quench them to obtain a supersaturated ferritic structure, cold-roll them heavily and then re-anneal them in the ($\alpha + \gamma$) field (at 1273K, *e.g.*). The amount of cold deformation is important, and so is the annealing temperature in the two-phase field. Annealing of the deformed, supersaturated ferritic microstructure leads to very rapid subgrain formation (recovery) before the precipitation of austenite. Maki *et al.* [19] gave a very thorough account on the microstructural changes when different annealing temperatures are used. The details are as follows: when the annealing temperature is right, then a lot of fine austenite will form uniformly along subgrain boundaries and at subgrain junctions (Fig. **10**). The subgrains cannot grow effectively because of the pinning effects exerted by the austenite. That is, the fine austenite will eventually show up in almost all of the subgrain junctions and the subgrains remain at nearly the same size as shown in Fig. **10**. So a microduplex structure results.

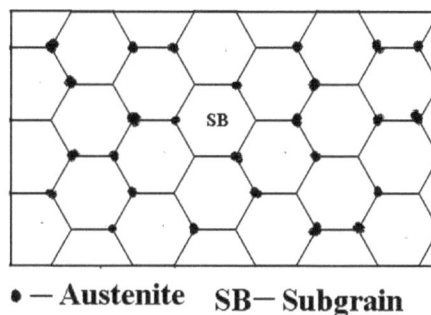

● — **Austenite**   SB— **Subgrain**

Figure 10: Precipitation of fine austenite at subgrain junctions in the ferrite matrix [Figure drawn with reference to Maki T, Furuhara T, Tsuzaki K. Microstructure development by thermomechanical processing in duplex stainless steel. ISIJ Int 2001; 41(6): 571-9].

At higher annealing temperatures, the size of the subgrains is bigger and the volume fraction of austenite is reduced (Fig. **11**). Hence the ferrite subgrain can recrystallise and grow easily, leading to a coarse duplex microstructure (Fig. **11**). At even higher annealing temperatures, the ferrite subgrains can recrystallise and grow well before austenite can form. Subsequent to ferrite grain growth, austenite then forms along the grain boundaries. Therefore, the microstructure is composed of very coarse ferrite grains and film-like austenite (Fig. **12**).

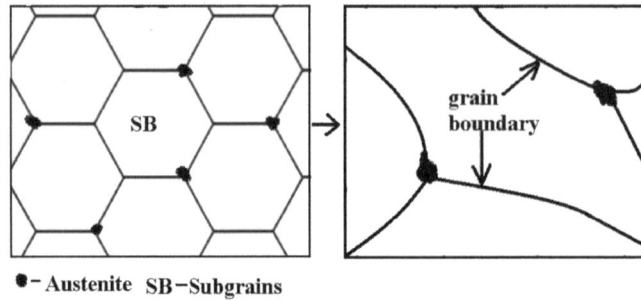

●– Austenite SB–Subgrains

Figure 11: Fewer austenite will form at higher temperatures and subgrains can grow [Figure drawn with reference to Maki T, Furuhara T, Tsuzaki K. Microstructure development by thermomechanical processing in duplex stainless steel. ISIJ Int 2001; 41(6): 571-9].

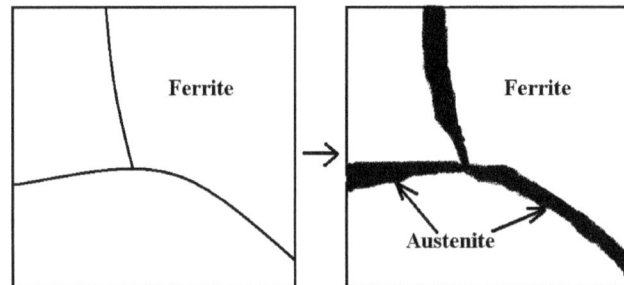

Figure 12: Growth of ferrite precedes film-like austenite formation along grain boundary [Figure drawn with reference to Maki T, Furuhara T, Tsuzaki K. Microstructure development by thermomechanical processing in duplex stainless steel. ISIJ Int 2001; 41(6): 571-9].

While the afore-mentioned thermomechanical treatment is often employed to produce a microduplex structure in DSSs, Maki *et al.* [19] have recently shown that it is possible to obtain a microduplex microstructure in as-hot-forged, coarse-grained duplex stainless steels without the high-temperature solution-treatment. Process **(A)** in Fig. **13**. is the conventional method for producing a microduplex structure as elaborated in the preceding paragraphs, whereas process **(B)** is the method proposed by Maki *et al.* [19]. After going through process **(B)**, the originally elongated austenite grains become fragmented and equiaxed. The resulting microstructure resembles a microduplex structure.

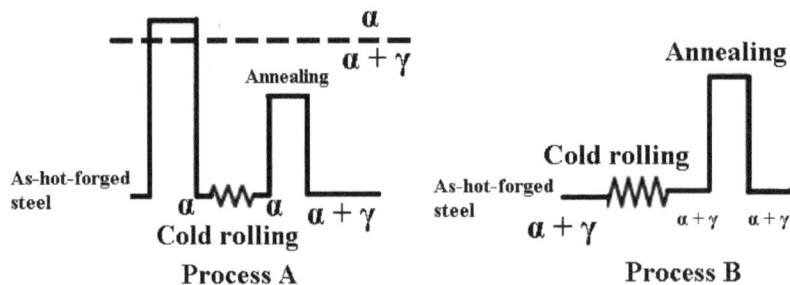

Figure 13: Process A is the conventional method for obtaining a microduplex structure and Process B is the scheme proposed by Maki *et al.* for achieving a microduplex structure without solution-treatment [Figure drawn with reference to Maki T, Furuhara T, Tsuzaki K. Microstructure development by thermomechanical processing in duplex stainless steel. ISIJ Int 2001; 41(6): 571-9].

6.2. Mechanisms Leading to Superplasticity

The superplasticity of duplex stainless steels was discovered in the 1960s [20, 21]. Since then, different mechanisms for the superplastic behaviour of duplex stainless steels have been put forth. In general, superplasticity in duplex stainless steels and duplex steels is attributed to grain boundary sliding [20-24]. Later studies by Maehara and Ohmori [25], and Maehara [26, 27] proposed that it was the balance between dynamic continuous recrystallisation (sometimes referred to simply as dynamic recrystallisation in some publications) and strain hardening in the neighbourhood of the softer phase (α in ($\alpha + \gamma$) or γ in ($\gamma + \sigma$)) that led to superplasticity, not grain sliding. Tsuzaki *et al.* [28], on the other hand, argued that the role of dynamic continuous recrystallisation during superplastic deformation was to maintain a fine-grained microstructure by retarding grain growth of the softer phase, such that grain sliding could take place during superplastic deformation.

Some of the recent work on the superplasticity of duplex stainless steels has been devoted to clarifying further the mechanism(s) involved. In a study by Han and Hong [29] on a duplex stainless steel having a ($\gamma + \sigma$) microstructure, it has been concluded that superplasticity is due to dynamic-recrystallisation-assisted grain sliding. However, dynamic continuous recrystallisation, in addition to generating a fine, equiaxed microstructure at the early stage, also transforms the low-angle γ / γ grain boundaries into high-angle ones, which expedites grain sliding.

Miyamoto *et al.* [30], while emphasising the importance of the different rates of sliding of different types of grain boundaries [30, 31], have gone further to introduce the concept of cooperative interphase (α / γ) grain boundary sliding. In this mechanism, a large amount of sliding involving a few tens of grains takes place cooperatively on preferred planes of α / γ grain boundaries [30]. The interphase grain sliding stops when it is obstructed by other grains. Grain sliding resumes when the obstructing grains are no longer able to hold the sliding in place or when other planes suitable for sliding kick into action [30]. The removal of the obstruction to sliding by other grains involves the sliding of γ / γ and γ / γ homophase grain boundaries. The homophase boundaries slide much slower than do the α / γ heterophase grain boundaries [30, 31].

The importance of the heterophase grain boundaries to superplasticity has also been recognised by Nieh *et al.* [31], who suggested that the superplastic strain rate can be increased if the total area of the heterophase boundaries is increased [31]. Nonetheless, if the inhomogeneous deformation due to the different sliding rates of the various type of boundaries cannot be properly accommodated, then cavities will form, especially in the α / γ heterophase grain boundaries [31] and grain triple junctions [32].

As just mentioned, dynamic continuous recrystallisation of the softer phase in the duplex structure plays a very important role in superplastic deformation. The increase in grain boundary misorientation, which is crucial to superplasticity, during dynamic recrystallisation has been closely examined by Tsuzaki *et al.* [33]. These authors have found that at the early stage of superplastic deformation, deformation occurs mainly through the usual intragranular slips rather than grain boundary sliding, since the misorientations amongst the softer phase are not high. Therefore, the strain incompatibility between the soft phase and the hard phase of the duplex structure is accommodated by intragranular slips. The absorption of accommodation dislocations into (sub)grain boundaries will gradually increase the misorientations among (sub)grain boundaries. Eventually the originally low-angle boundaries become high-angle ones that are capable of supporting grain boundary sliding [33, 34].

So, the increase in misorientation may in fact take place without grain boundary sliding. Tsuzaki *et al.* [33] have thus suggested that the term 'continuous recrystallisation' is a misnomer, in as much as the process is in fact 'recovery'. The importance of intragranular slips is also stressed by Song *et al.* [35, 36].

REFERENCES

[1]   Bain EC, Griffiths WE. Introduction to the Iron-Chromium-Nickel Alloys. Trans Amer Inst Mech Eng 1927; 75: 166-72.

[2]   Lo KH, Shek CH, Lai JKL, *et al.* Effects of pre-treatment on the ac magnetic susceptibility and ageing behaviour of duplex stainless steels. Mater Sci Eng A 2007; 452: 78-86.

[3]   Liljas M, Johansson P, Liu H, *et al.* Development of a lean duplex stainless steel. Steel Res Int 2008; 79(6): 466-73.

[4]   Charles J. Duplex stainless steels – a review after DSS'07 held in Grado. Steel Res Int 2008; 79(6): 455-65.

[5]   Holmquist M. Hyperduplex stainless steel as alternative to Ni-based alloys. Euro Chem Eng 16 April 2010.

[6]   Chai G, Kivisakk U, Tokaruk J, *et al.* Hyper duplex stainless steel for deep subsea applications. Stainless Steel World http://www.stainless-steel-world.net/pdf/SSW_0903_SANDVIK.pdf (assessed on 20-Apr, 2010).

[7]   Wolf M. in: Proceedings of the 1st European Conference on Continuous Casting, Florence, Italy (1991) p.2489.

[8]   Ray SK, Mukhopadhyay B, Bhattacharyya SK. Prediction of crack-sensitivity of continuously cast slabs of AISI 430 ferritic stainless steel. ISIJ Int 1996; 36(5): 611-2.

[9]   Greeff ML, Toit M. Looking at the sensitization of 11-12% chromium EN 1.4003 stainless steels during welding. Weld J 2006; 85(11): 243s-51s.

[10] Wai SW, Cortie MB, Robinson FPA. A study of high-temperature cracking in ferritic stainless steels. Mater Sci Eng A 1992; 158(1): 21-30.

[11] Toit M, Rooyen GTV, Smith D. Heat-affected zone sensitization and stress corrosion cracking in 12% chromium type 1.4003 ferritic stainless steel. Corrosion 2007; 63(5): 395-404.

[12] Barbe L, Bultinck I, Duprez L, *et al.* Influence of composition on crack sensitivity of ferritic stainless steel. Mater Sci Technol 2002; 18(6): 664-72.

[13] Pryce L, Andrews KW. Practical estimation of composition balance and ferrite in stainless steels. J Iron Steel Inst 1960; 195: 415-17.

[14] Lo KH, Lai JKL. On the cryogenic magnetic transition and martensitic transformation of the austenite phase of 7MoPLUS duplex stainless steels. J Magn Magn Mater 2010; 322(16): 2335-9.

[15] Lopez N, Cid M, Puiggali M. Influence of σ-phase on mechanical properties and corrosion resistance of duplex stainless steels. Corros Sci 1999; 41(8): 1615-31.

[16] Shek CH, Li DJ, Wong KW, *et al.* Creep properties of aged duplex stainless steels containing σ phase. Mater Sci Eng A 1999; 266(1-2): 30-6.

[17] Gibson R. Structure and constitution of wrought microduplex stainless steels. In: Handbook of Stainless Steels. Peckner D, Bernstein (Eds). McGraw-Hill, USA.

[18] Osada K. Commercial applications of superplastic forming. J Mater Process Technol 1997; 68(3) 241-5.

[19] Maki T, Furuhara T, Tsuzaki K. Microstructure development by thermomechanical processing in duplex stainless steel. ISIJ Int 2001; 41(6): 571-9.

[20] Humphries CW, Ridley N. Cavitation in alloy steels during superplastic deformation. J Mater Sci 1974; 9(9): 1429-35.

[21] Hayden HW, Brophy JH. The interrelation of grain size and superplastic deformation in Fe-Cr-Ni alloys. Trans Am Soc Metal 1967; 61: 542-9.

[22] Jimenez JA, Frommeyer G, Carsi M, *et al.* Superplastic properties of a delta/gamma stainless steel. Mater Sci Eng A 2001; 307(1-2): 134-42.

[23] Gibson RC, Hayden HW, Brophy JH. Properties of stainless steels with a microduplex structure. Trans Amer SoC Metal 1968; 61: 85-93.

[24] Smith CI, Norgate B, Ridley N. Superplastic deformation and cavitation in a microduplex stainless steel. Metal Sci 1976; 10: 182-8.

[25] Maehara Y, Ohmori Y. Microstructural change during superplastic deformation of ferritic-austenite duplex stainless steel. Metal Trans A 1987; 18(4): 663-72.

[26] Maehara Y. Superplastic deformation mechanism of delta/gamma duplex stainless steels. Trans Iron Steel Inst Jpn 1987; 27(9): 705-12.

[27] Maehara Y. high-strain rate superplasticity of a 25wt% Cr7wt% Ni3wt% Mo0.14wt% N duplex stainless steel. Metal Trans A 1991; 22(5): 1083-91.

[28] Tsuzaki K, Matsuyama H, Nagao M, *et al.* High strain rate superplasticity and role of dynamic recrystallization in a superplastic duplex stainless steel. Mater Trans JIM 1990; 31(11): 983-94.

[29] HanYS, Hong SH. Microstructural changes during superplastic deformation of Fe-24Cr-7Ni-3Mo-0.14N duplex stainless steel. Mater Sci and Engi A1999; 266(1-2): 276-84.

[30] Miyamoto H, Mimaki T, Hashimoto S. Superplastic deformation of micro-specimens of duplex stainless steel. Mater Sci Eng A 2001; 319-21(Suppl): 779-83.

[31] Nieh TG, Lesuer DR, Syn CK. Characterization of a commercial superplastic stainless steel, Superdux64. Mater Sci Eng A 1995; 202(1-2): 43-51.

[32] Hayden HW, Floreen S, Goodell PG. The deformation mechanisms of superplasticity. Metall Trans 1973; 3(4): 833-42.

[33]   Tsuzaki K, Xiaoxu H, Maki T. Mechanism of dynamic continuous recrystallization during superplastic deformation in a microduplex stainless steel. Acta Mater 1996; 44(11): 4491-9.

[34]   Jimenez JA, Carreno F, Ruano OA, *et al*. High temperature mechanical behaviour of delta-gamma stainless steel. Mater Sci Technol 1999; 15(2): 127-31.

[35]   Song JL, Blackwell PL. Superplastic behaviour of commercial duplex stainless steel SAF 2304. Mater Sci Technol 1999; 15(11): 1285-92.

[36]   Song JL, Bate PS. Plastic anisotropy in a superplastic duplex stainless steel. Acta Mater 1997; 45(7): 2747-57.

CHAPTER 6

Precipitation-Hardening Stainless Steels

Abstract: This chapter focuses on precipitation-hardening stainless steels (PHSSs). The three main types of PHSSs, *i.e.*, the semiaustenitic class, the martensitic class and the austenitic class, are described in detail. The duplex class is also briefly discussed. The different heat treatments, together with their effects on properties, that are commonly used for attaining precipitation strengthening in PHSSs are presented. Recent developments in the understanding of the precipitation sequences and the precipitates in PHSSs are introduced.

Keywords: Semiautenitic class, duplex class, martensitic class, precipitation-hardening, heat treatment, austenitisation, austenitic conditioning, conditioning treatment, precipitation sequence, Cu precipitate, overageing.

1. INTRODUCTION

Precipitation-Hardening Stainless Steels (PHSSs) are commonly classified as austenitic, martensitic and semiaustenitic (recently, a duplex grade has been developed and is discussed below). It is possible to attain high strengths, but at the same time without seriously compromising corrosion resistance and ductility as in other classes of stainless steels, in the precipitation-hardening class. The strengthening mechanisms are chiefly associated with the formation of martensite and precipitation hardening, the latter being the final hardening mechanism. The following are believed to be the main contributors to strengthening associated with precipitation hardening [1]: **1**. the coherency strain between the strengthening precipitates, **2**. the state of ordering in the strengthening precipitates, **3**. the difference between the shear moduli of the precipitates and the matrix, and **4**. the fine particles of reverted austenite

Most PHSSs contain titanium, aluminium, copper or molybdenum for achieving strengthening and hardening. For example, the A286 steel contains about 2% of titanium, the 17-7 PH steel contains about 1% of aluminium. The precipitation-hardening processes are dependent on both temperature and time. A higher temperature may hasten the attainment of maximum strength level in a shorter time. The precipitation processes are quite complex in the precipitation-hardening class. In fact, new precipitation phenomena and sequences have been proposed very recently, with the advent of more sophisticated experimental techniques.

The precipitation reactions that occur during a hardening treatment may be different for the different types of precipitation-hardening stainless steels. Therefore, instead of enumerating in detail the precipitation reactions in all of the commonly used PHSSs, the precipitation reactions that occur in the epitomes of each of the sub-classes (semiaustenitic, martensitic, and austenitic) will be discussed. For instance, the 17-7PH steel be used singled out for discussion on the semiaustenitic class, while the 13-8PH steel will be selected for the martensitic class.

2. SEMIAUSTENITIC PRECIPITATION-HARDENING STAINLESS STEELS

The compositions of four typical steels in this class are listed in Table **1**. In the solution-treated or annealed state (often referred to as condition A), the semiaustenitic class is predominantly austenitic, with some ferrite present (ranging from 5 to 20%). Through proper thermomechanical treatments, martensite may form in the microstructure. The semiaustenitic class is ductile in the as-solution-treated state and so can be formed without difficulty. Subsequent to forming operations, the steel may be strengthened to the desired level by a martensitic transformation and precipitation-hardening.

The austenite-forming and ferrite-forming elements must be carefully balanced or else the desired properties are not obtained. This is because if the austenite is too stable, then it will not transform martensitically. On the other hand, if the austenite is too unstable, then martensite may form even at room temperature. In this case, the steel is martensitic precipitation-hardening, instead of semiaustenitic precipitation-hardening. The quenching following solution-treatment should be fast enough to avoid a coarse grain size. The final properties of the steel will be better if a fine grain size is achieved at this stage. To obtain good properties, the solution treatment should be conducted properly. Too low a solution treatment temperature may result in incomplete homogenisation of the austenite, leading to an excess of retained austenite in the quenched state and affecting adversely the final properties [2].

Table 1: Compositions of commonly used semiaustenitic PHSSs.

Steel	C	Mn	Si	Cr	Ni	Mo	Al	N
17-7PH	0.07	0.50	0.30	17.0	7.1	---	1.2	0.04
PH15-7Mo	0.07	0.50	0.30	15.2	7.1	2.2	1.2	0.04
PH14-8Mo	0.04	0.02	0.02	15.1	8.2	2.2	---	0.10
AL17-7	0.07	0.50	0.25	17.0	7.0	---	1.25	---

2.1. Heat treatments of the Semiaustenitic Class

The heat treatment for the semiaustenitic class is mainly composed of: **1.** austenitic conditioning, **2.** martensitic transformation and **3.** precipitation-hardening.

As mentioned, in the annealed or solution-treated state, the microstructure of the semiaustenitic class is a mixture of austenite and ferrite. The steel should be heated up (conditioned) such that martensite may form upon subsequent treatment.

During conditioning, carbides ($Cr_{23}C_6$) may form and so carbon is taken out of solution, which raises the start temperature of martensitic transformation (Ms). The precipitation of carbides begins preferentially at the interface between ferrite and austenite. If the carbon content is high, then grain boundary carbides also form.

If the conditioning temperature is high, then less carbides will form and the Ms temperature may be depressed to such an extent that cooling below room temperature is necessary in order to kick-start and complete the transformation to martensite. Consequently, the temperature at which the austenite is conditioned affects Ms, which then affects the precipitation behaviour during precipitation-hardening and hence the final properties.

Of equal importance is the finish temperature of martensitic transformation, M_f, which is approximately about 150°F lower than Ms. The transformation is more complete if the quenching temperature is closer to M_f. Alternatively, holding the steel at a certain quenching temperature for a longer time allows for a more complete transformation.

For the semiaustenitic class, refrigerating to subzero temperatures is frequently employed. The refrigerated-and-aged condition is identified by the designation RH (TH if refrigeration is not used), while the cold-worked-and-aged condition is commonly designated by CH.

The time duration for which a steel is conditioned also affects the final properties. This is understandable because longer times may allow more complete dissolution of carbides and any pre-existing phases like martensite and ferrite.

Following the conditioning treatment, the steel is subjected to precipitation-hardening treatment for attainment of the desired properties. Precipitation-hardening takes place between about 250°F and 1250°F for the semiaustenitic class. In most cases, precipitation-hardening is conducted between about 900°F and 1200°F. During precipitation-hardening, the martensite is tempered and stress-relieved for improved toughness, ductility and corrosion resistance. Some martensite may transform back to austenite. Also, additional hardening is obtained because of the precipitation of intermetallic compounds. The final properties depend on the contributions of these phenomena. Coarsening of the strengthening precipitates (overageing) will occur when the hardening temperature is very high. Although strengths decrease due to overageing, toughness in general increases.

Besides quenching and deep cooling, martensitic transformation may also be brought about by cold working [3]. The final properties hinge on the amount of cold work. If the steel in the annealed or solution-treated condition is plastically deformed (for example, during forming operations), then large amounts of martensite may form in parts of the steel that are heavily deformed. If the steel is subsequently conditioned at a low temperature (say, at 1400°F), then the martensite in the heavily deformed parts might not be taken totally back into solution. Consequently, the properties of these severely deformed parts may be unsatisfactory and the desired precipitation-hardening may not be achievable. Usually, cold-working after precipitation-hardening is not recommended.

Upon prolonged exposure between 700°F and 800°F, severe embrittlement may occur because of continued precipitation of intermetallics and 885°F-embrittlement (475°C-embrittlement). Problems associated with these effects are detailed in Chapters **2** and **5**.

3. MARTENSITIC PRECIPITATION-HARDENING STAINLESS STEELS

The compositions of some of the commonly used steels in this class are shown in Table **2**. As regards corrosion resistance, the martensitic class is comparable to austenitic stainless steels in most media. The martensitic PHSSs outperform ordinary martensitic stainless steels in almost all aspects. While work-hardenable austenitic stainless steels and microduplex stainless steels may develop high strengths, they are often produced in plate form because of the necessity for cold-working. In this respect, martensitic PHSSs offer higher flexibility in the shapes of the final products, with high strengths, ductility, corrosion resistance and toughness being attainable.

Table 2: Compositions of commonly used martensitic PHSSs.

Steel	C	Mn	Si	Cr	Ni	Mo	Al	Cu	Ti	Cb	Nb
17-4PH	0.04	0.30	0.60	16.0	4.2	---	---	3.4	---	0.25	---
15-5PH	0.04	0.30	0.40	15.0	4.5	---	---	3.4	---	0.25	---
13-8PH	0.04	0.03	0.03	12.7	8.2	2.2	1.1	---	---	---	---
FV520B	0.042	0.578	0.355	13.80	5.540	1.470	---	1.560	---	---	0.391
Corrax	0.03	0.3	0.3	12.0	9.2	1.4	1.6	---	---	---	---

At room temperature, the as-solution-treated microstructure of this class is almost always martensitic. Nevertheless, some retained austenite may be present, depending on the prior solution treatment condition. Improper solution-treatment (the temperature too low or soaking time too short) may lead to incomplete homogenisation of the high-temperature austenite. So, the enrichment of austenite-forming elements in some regions of the austenite may not be eliminated. These chemically inhomogeneous regions may form in the as-solution-treated steel, thereby affecting properties [4-6]. For instance, toughness has been found to improve, whereas yield strength, ultimate tensile strength and fatigue limit are reduced [4]. Ferrite may also be present if solution-treatment is not done properly [6]. Reversion of martensite to austenite is also possible during the precipitation-hardening process if the temperature is high enough such that the diffusion of nickel in the martensitic matrix is facilitated [4].

To ensure a martensitic starting microstructure, the composition must be carefully chosen. One indicator for checking whether a martensitic microstructure can be obtained after solution-treatment is the Ms point. In the literature, a litany of empirical formulae for calculating Ms have been proposed. The one proposed by

Ishida [7] has recently been adopted in the design of an ultra high-strength martensitic PHSS by Xu [8]. This Ishida formula is as follows:

$$M_S(K) = 818 - 33000\%C + 200\%Al + 700\%Co - 1400\%Cr - 1300\%Cu - 2300\%Mn - 500\%Mo$$
$$- 400\%Nb - 1300\%Ni - 700\%Si + 300\%Ti + 400\%V$$

It has to be noted that even if a steel is quenched below the Ms point, some austenite may be retained because of chemical and mechanical stabilisations [9]. Between the Ms and the M_f points, solutes like carbon tend to segregate to untransformed austenite, thereby lowering its M_f point even more. Eventually, untransformed austenite may remain between martensite laths and this is known as chemical stabilisation. As regards mechanical stabilisation, dislocations and strains generated by austenite that has already transformed to martensite may prevent the motion of glissile interfaces, thereby halting the remaining austenite from transforming [10].

Compared with the ordinary 410 martensitic stainless steel, the hardnesses of the as-quenched martensite in martensitic PHSSs are lower on grounds of: **1**. the lower C contents and **2**. the tying-up of C by alloying elements like Ti and Cb. Consequently, the martensitic PHSSs are readily machinable and workable before the precipitation-hardening treatment.

The 17-4PH steel is primarily hardened by the addition of copper. Insofar as the solubility of copper is much higher in austenite than in ferrite, a supersaturation of copper is attained in the martensite matrix after austenitisation. Upon annealing between 800°F and 1100°F, a strengthening copper-rich phase will form in the steel. Fig. **1**. shows how the hardness of the 17-4 PH steel changes with annealing time at different annealing temperatures. The decrease in hardness is associated with overageing.

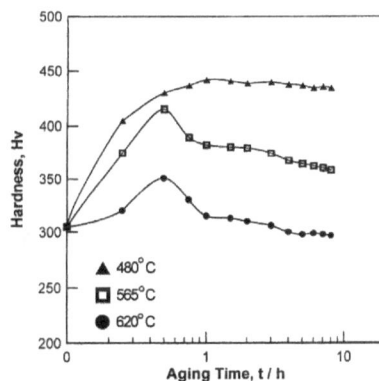

Figure 1: Effects of annealing time and temperature on the hardness of the 17-4 PH steel [Hsiao CN, Chiou CS, Yang JR. Aging reactions in a 17-4 PH stainless steel. Mater Chem Phys 2002; 74(2): 134-42. With permission for reproduction from Elsevier].

3.1. The Precipitation Sequence in Martensitic PHSSs

Due to limitations on the resolution of instruments, the strengthening precipitates and their formation sequences as described in past studies might not be detailed enough. In these early studies, overaged samples were used in order to form precipitates that were big enough for analyses. For example, samples used in early studies were usually annealed for hours [11], whereas the latest studies have been able to deal with precipitates that form after annealing for just a few minutes [12]. The problem with using overaged samples is that precipitates that form in overaged materials might be different from those that form in the early stage of ageing. Recent researchers, using more sophisticated equipment that was unavailable to their early colleagues, have been able to reveal better the formation sequence of the strengthening precipitates. Different precipitation sequences have been proposed by different workers. In addition, the various strengthening phases in different types of precipitation-hardening stainless steels have been better characterised.

Both the 17-4 and 15-5 PH stainless steels are strengthened by copper(-enriched) precipitates [13]. The 15-5 stainless steel contains about 3wt% of copper and the sequence of formation of practically pure Cu precipitates [14] is usually modelled after those in the Fe-Cu and Fe-Cu-X systems. Early studies concluded that coherent bcc Cu precipitates would form first upon ageing, they would then attain a fcc structure upon overageing. The overaged Cu precipitates were found to adopt the K-S orientation relationships with the martensitic matrix [15, 16]. More recent studies have revealed that the bcc to fcc transformation actually goes through an intermediary martensitic precipitate having a twinned-9R structure [17, 18], followed by an untwining process to a precipitate having a 3R structure [17, 19], as follows:

$$bcc \rightarrow twinned \quad 9R \xrightarrow{\text{untwining}} 3R \rightarrow fcc$$

The orientation relationships between the smaller 3R precipitates and the matrix are approximately K-S and the 3R precipitates have a distorted fcc lattice. But for larger 3R precipitates, their lattice is fcc and they are aligned to the matrix according to the K-S relationship [17]. Nevertheless, Bajguirani and Jenkins [20], using HRTEM, have found that the Cu precipitates of 15-5 PH stainless steel that forms upon annealing at 480°C for 2h do not go through the untwining process, *i.e.*,

$$bcc \rightarrow twinned \quad 9R \rightarrow 3R \rightarrow fcc$$

Further studies by Bajguirani *et al.* [21, 22] on a 15-5 PH steel have revealed that, in addition to the hardening around 450°C, a second hardening occurs at about 650~700°C. This hardening is thought to be due to a spherical, fcc Cu precipitate, which is coherent with the matrix at the beginning of ageing and then becomes semi-coherent upon further ageing.

The precipitation sequence of the strengthening precipitates, which is Cu(-enriched) [23], in 17-4 PH stainless steel is also modelled after the Fe-Cu and Fe-Cu-X systems. A study by Hsiao *et al.* [24] has shown that the precipitation sequence of the Cu-(enriched) strengthening phase is:

$$bcc \rightarrow twinned \quad 9R \rightarrow fcc$$

It is worth noticing that no intermediary precipitate of the 3R structure was mentioned in the study by Hsiao *et al.* [24]. Additionally, Murayama *et al.* [25] have found another type of Cu precipitate in 17-4PH upon ageing at 400℃. In this precipitate, the content of Cu is just about 20 at%.

While the afore-mentioned studies involve ageing above 400℃, recent studies by Wang *et al.* [26, 27] have focused on the various precipitates that may form in the 17-4 PH steel during low-temperature ageing. A Cu precipitate (referred to as ε -Cu in [26, 27]), reverted austenite, carbides and the G phase have been found to form in the low temperature range (350°C) and the martensitic matrix has been shown to undergo spinodal decomposition [25-27]. The ε -Cu, which adopts the K-S orientation relationships with the martensitic matrix, acts as nucleation sites for the G phase and shares with it cube-cube orientation relationships [25, 26].

Unlike 15-5 and 17-4, the 13-8 PH steel is strengthened by the ordered phase NiAl which is believed to have a B2 (CsCl) structure [1, 28]. As mentioned, early studies were constrained by the limited resolution of experimental equipment (see [1], for example). Such a limitation was noted by Guo *et al.* [12] and these authors pointed out that the NiAl precipitates in overaged samples might differ from those that form in the early stages. By using APFIM, these authors have shown that at the early stage of ageing at 510°C, 13-8 will contain NiAl-enriched zones whose composition is far from the stoichiometric NiAl phase [12]. This agrees with the results obtain from Thermo-Calc® [29]. While stoichiometric NiAl is of the ordered B2 structure and is stable at room temperature for Al contents ranging from 41.5 to 55at% [30], Guo *et al.* [12] have not been able to conclude that the NiAl-enriched zones are of the ordered B2 structure, in as much as Ni and Al make up less than half of the composition (the other atoms are Cr and Fe) in these zones.

In a 13Cr–8Ni–2.5Mo–2A PH steel, the good resistance to coarsening of the strengthening β-NiAl phase (of the B2 structure) has been attributed to Mo and Cr segregation to the precipitate–matrix interface [31], no such explanation was given in early studies [32].

A recent research by Frandsen *et al.* [33] has shown that by properly choosing the nitriding potential and temperature, the surface of martensitic PHSSs may be transformed to expanded martensite for even better wear resistance, while the bulk of the steels is precipitation-hardened concurrently with the surface treatment. What is striking about this study is that an improvement in both surface and bulk properties may be obtained simultaneously in a one-step treatment.

4. AUSTENITIC PRECIPITATION-HARDENING STAINLESS STEELS

Unlike the semiaustenitic and martensitic classes, the austenitic PHSSs are fully austenitic after solution-treatment or the cold-working process that may follow. Hence, the only strengthening mechanism is the precipitation-hardening process. The compositions of two of the steels in the austenitic class are shown in the Table **3**.

Table 3: Compositions of commonly used martensitic PHSSs.

Steel	C	Mn	P	S	Si	Cr	Ni	Mo	Al	V	Ti	B
A-286	0.05	1.45	0.03	002	0.50	14.75	25.25	1.30	0.15	0.30	2.15	0.005
17-10PH	0.10	0.60	0.30	0.04max	0.50	17.0	11.0	---	---	---	---	---

In most media, the corrosion resistance of the austenitic class is comparable to the 300 series austenitic stainless steels. The weldability of A-286 is better than 17-10PH because of the former's lower P content. In terms of mechanical properties, both steels are comparable. This austenitic class, because of their stable austenitic microstructures, has found use in cryogenic applications. They are also used in jet engines, turbine wheels and as fasteners.

The solution-treatment temperature is usually high for this class of steels. The solution-treatment temperature affects the grain size, which then affects the final mechanical properties. A finer grain size improves room-temperature properties, whereas a coarser grain size leads to better creep strength. The fully austenitic PHSSs may be used at temperatures that are not allowed for the semiaustenitic and martensitic steels.

The coherent fcc, $Ni_3(Al, Ti)$ intermetallic compound (γ') is mainly responsible for the strengthening in A-286. The presence of γ' is also believed to enhance resistance to stress corrosion cracking [34]. However, on prolonged annealing at this temperature, other phases like the η phase will form in favour of the fcc $Ni_3(Al, Ti)$. The η phase is undesirable as it detracts from mechanical properties like creep strength [35]. Coarsening of γ' deteriorates strengths, too.

The low carbon content of A-286 means it is corrosion-resistant, tough and ductile. However, the low carbon content also means the tribological properties of A-286 are not good. However, this problem may be remedied by surface treatment like plasma nitriding that imparts the S phase to the surface region [36].

Besides annealing, strengthening precipitates may also form through irradiation. Recently, it has been found that by irradiating the austenitic steel Kh16N15M3 having about 1%Ti and then subjecting it to fast neutron irradiation between 480℃ and 500℃, a PHSS with Ni_3Ti as the strengthening phase will form [37], as in a Fe–20Ni–23Co–0.07Al–0.17Ti PH stainless steel [38].

5. DUPLEX PRECIPITATION-HARDENING STAINLESS STEELS

While the commonly used PHSSs are martensitic, proprietary martensitic-ferritic duplex PHSS wires having very good tensile strength have been developed [39]. These steels contain titanium and the strengthening phase is the orthorhombic $Ni_3(Mo,Ti)$. The transformation route is as follows:

$$Martensite \, / \, Ferrite \rightarrow Hexagonal \quad Omega \rightarrow Orthorombic \quad Ni_3(Mo,Ti)$$

The hexagonal omega phase is a metastable phase having a diffraction pattern that is analogous to that of the hexagonal omega phase in several titanium alloys [39].

REFERENCES

[1]   Seetharaman V, Sundararaman M, Krishnan R. Precipitation hardening in a PH 13-8 Mo stainless steel. Mater Sci Eng 1981; 47(1): 1-11.

[2]   Xu XL, Yu ZW. Metallurgical analysis on a bending failed pump-shaft made of 17-7PH precipitation-hardening stainless steel. J Mater Process Technol 2009; 198(1-3): 254-9.

[3]   Craig RG, Slesnick HJ, Peyton FA. Application of 17-7 precipitation-hardenable stainless steel in dentistry. J Dent Res 1965; 44: 587-95.

[4]   Nakagawa T, Miyazaki T, Yokota H. Effects of aging temperature on the microstructure and mechanical properties of 1.8Cu-7.3Ni-15.9Cr-1.2Mo-low C, N martensitic precipitation hardening stainless steel. J Mater Sci 2000; 35(9): 2245-53.

[5]   Nakagawa T, Miyazaki T. Effect of retained austenite on the microstructure and mechanical properties of martensitic precipitation hardening stainless steel. J Mater Sci 1999; 34(16): 3901-08.

[6]   Abdelshehid M, Mahmodieh K, Mori K, *et al.* On the correlation between fracture toughness and precipitation hardening heat treatments in 15-5PH stainless Steel Eng Fail Anal 2007; 14(4): 626-31.

[7]   Ishida K. Calculation of the effect of alloying elements on the Ms temperature in steels. J Alloy Compd 1995; 220(1-2): 126-31.

[8]   Xu W, Castillo PEJRDD, Zwaag SVD. A combined optimization of alloy composition and aging temperature in designing new UHS precipitation hardenable stainless steels. Comput Mater Sci 2009; 45(2): 467-73.

[9]   Thomas G. Retained austenite and tempered martensite embrittlement. Metall Trans A 1978; 9(3): 439-50

[10] Chatterjee S, Wang HS, Yang JR, *et al.* Mechanical stabilisation of austenite. Mater Sci Technol 2006; 22(6): 641-44.

[11] Antony KC. Aging reactions in precipitation hardenable stainless steel. J Metal Dec 1963, pp.922-927.

[12] Guo Z, Sha W, Vaumousse D. Microstructural evolution in a PH13-8 stainless steel after ageing. Acta Mater 2003; 51(1): 101-16.

[13] Ozbaysal K, Inal OT. Thermodynamics and structure of solidification in the fusion zone of CO_2 laser welds of 15-5 PH stainless steel. Mater Sci Eng A 1990; 130(2): 205-17.

[14] Bajguirani HRH, Servant C, Lyon O. ASAXS study of the copper enriched precipitation in the 15-5PH alloy. Nanostruct Mater 1994; 4(7): 833-50.

[15] Hornbogen E, Glenn RC. Metallographic study of precipitation of copper from alpha iron. Trans Metall Soc AIME 1960; 218: 1064-70.

[16] Pizzini S, Robeerts KJ, Phythian WJ, *et al.* A fluorescence EXAFS study of the structure of copper-rich precipitates in Fe-Cu and Fe-Cu-Ni alloys. Philos Mag Lett 1990; 61(4): 223-9.

[17] Othen PJ, Jenkins ML, Smith GDW. A high-resolution electron-microscopy studies of the structure of Cu precipitates in alpha-Fe. Philos Mag A 1994; 70(1): 1-24.

[18] Othen PJ, Jenkins ML, Smith GDW, *et al.* Transmission electron-microscope investigations of the structure of Copper precipitates in thermally aged Fe-Cu and Fe-Cu-Ni. Philosl Mag Lett 1991; 64(6): 383-91.

[19] Duparc HAH, Doole RC, Jenkins ML, *et al.* A high-resolution electron-microscopy study of copper precipitate in Fe-1.5wt%Cu under electron irradiation. Philos Mag A 1995; 71(6): 325-33.

[20] Bajguirani HRH, Jenkins ML. High-resolution electron microscopy analysis of the structure of copper precipitates in a martensitic stainless steel of type PH 15-5. Philos Mag Lett 1996; 73(4): 155-62.

[21] Bajguirani HRH. The effect of ageing upon the microstructure and mechanical properties of type 15-5 PH stainless steel. Mater Sci Eng A 2002; 338(1-2): 142-59.

[22] Bajguirani HRH, Servant C, Cizeron G. TEM investigation of precipitation phenomena occurring in PH-15-5 alloy. Acta Metall Mater 1993; 41(5): 1613-23.

[23] Viswanathan UK, Banerjee S, Krishnan R. Effects of aging on the microstructure of 17-4 PH stainless steel. Mater Sci Eng A 1988; 104: 181-9.

[24] Hsiao CN, Chiou CS, Yang JR. Aging reactions in a 17-4 PH stainless steel. Mater Chem Phys 2002; 74(2): 134-42.

[25] Murayama M, Katayama Y, Hono K. Microstructural evolution in a 17-4 PH stainless steel after aging at 400C. Metall Mater Trans A 1999; 30(2): 345-53.

[26] Wang J, Zou H, Li C, *et al.* The microstructure evolution of type 17-4PH stainless steel during long-term aging at 350C. Nucl Eng Des 2006; 326(24): 2531-6.

[27] Wang J, Zou H, Li C, *et al.* The effect of microstructural evolution on hardening behavior of type 17-4PH stainless steel in long-term aging at 350C. Mater Charact 2006; 57(4-5): 274-80.

[28] Irvine KJ. The development of high-strength steels. J Iron Steel Inst 1962; 200: 820-6.

[29] Guo Z, Sha W. Thermodynamic calculation for precipitation hardening steels and titanium aluminides. Intermet 2002; 10(10): 945-50.

[30] Liu ZG, Frommeyer G, Kreuss M. Atom probe FIM investigations on the intermetallic NiAl phase with B2 superlattice structure. Surf Sci 1991; 246(1-3): 272-7.

[31] Ping DH, Ohnuma M, Hirakawa Y, *et al.* Microstructural evolution in 13Cr-M-2.5Mo-2Al martensitic precipitation-hardened stainless steel. Mater Sci Eng A 2005; 394(1-2): 285-95.

[32] Hochanadel PW, Robino CV, Edwards GR, *et al.* Heat treatment of investment cast PH 13-8 Mo stainless steel 1. mechanical properties and microstructure. Metall Mater Trans A 1994; 25(4): 789-98.

[33] Frandsen RB, Christiansen T, Somers MAJ. Simultaneous surface engineering and bulk hardening of precipitation hardening stainless steel. Surf Coat Technol 2006; 200(16-17): 5160-9.

[34] Fournier L, Savoie M, Delafosse D. Influence of localized deformation on A-286 austenitic stainless steel stress corrosion cracking in PWR primary water. J Nucl Mater 2007; 366(1-2): 187-97.

[35] DeCicco H, Luppo MI, Raffaeli H, *et al.* Creep behavior of an A286 stainless steel. Mater Charact 2005; 55(2): 97-105.

[36] Esfandiari M, Dong H. Improving the surface properties of A286 precipitation-hardening stainless steel by low-temperature plasma nitriding. Surf Coat Technol 2007; 201(14): 6189-96.

[37] Sagaradze VV, Goshchitskii BN, Arbuzov VL, *et al.* Precipitation-hardening austenitic steel for fast neutron reactors. Metal Sci Heat Treat 2003; 45(7-8): 293-9.

[38] Cheng IL, Thomas T. Structure and properties of Fe–Ni–Co–Al–Ti precipitation hardening stainless steel. Trans Am Soc Metal 1968; 61: 14-21.

[39] Ayer R, Bendel LP, Zackey VF. Metastable precipitate in a duplex martensite plus ferrite precipitation-hardening stainless steel. Metall Trans A 1992; 23(9): 2447-53.

CHAPTER 7

High-Nitrogen Stainless Steels

Abstract: This chapter is devoted to stainless steels containing high-levels of nitrogen. Ni-free stainless steels and those containing raised levels of manganese are also covered. The problems associated with nickel and the reasons for developing low-Ni/Ni-free stainless steels using nitrogen and manganese are introduced. New constitution diagrams for phase prediction of the new high-N (Ni-free) stainless steels are presented. Recently developed methods for introducing very high levels of nitrogen to stainless steels are discussed. The effects of nitrogen on mechanical and corrosion properties, together with the underlying mechanisms, are introduced. Although high-N stainless steels typically contain low/no Ni, this chapter also discusses briefly high-N, high-Ni grades, whose production has been suggested rather recently.

Keywords: High-nitrogen steel, nitride, Ni-free stainless steel, Ni-containing stainless steels, high-Ni stainless steels, phase prediction, fatigue, corrosion, ductile-to-brittle transition, Schaeffler diagram, modified Schaeffler diagram.

1. INTRODUCTION

The beneficial effects of adding nitrogen to steels, stainless or not, are manifold. The benefits include: **(1)** high yield and tensile strengths **(2)** good ductility **(3)** enhanced resistance to martensitic transformation **(4)** good strain-hardening behaviour **(5)** high resistance to pitting and crevice corrosion **(6)** low/no magnetism

By properly adjusting the composition and judiciously choosing the right thermomechanical treatments, very high strength levels may be achieved because of solid solution-strengthening (N is a more potent solid-strengthening element than substitutional elements), grain refinement and the high work-hardening that are characteristic of high-nitrogen stainless steels. A remarkable feature of high-N stainless steels is that their increase in strength at room temperature does not come by at the expense of toughness.

The absence of a martensitic transformation means heavily cold-worked high-N austenitic stainless steels may remain non-magnetic, even at cryogenic temperatures. Cold-working, while increasing strengths, does decrease ductility. Nevertheless, the combination of strengths and ductility in high-N stainless steels is far better than that achievable with low-alloy steels.

According to Speidel [1], high-N martensitic/ferritic stainless steels are those whose N contents exceed 0.08wt%. And for austenitic stainless steels, their N contents must be higher than 0.4wt% [1]. To date, it seems that there still does not exist a generally agreed N content above which a stainless steel can be viewed as 'high-N' [2].

With the advent of technologies like the Argon Oxygen Decarburisation (AOD) and high-pressure metallurgy, stainless steels having very low-Ni contents (or no Ni whatsoever) are common. The Cromanite from Columbus Joint Venture (South Africa) having a nominal composition of 19Cr-10Mn-1Ni-0.5N is an example.

The bulk of research on high-N stainless steels is on the austenitic class. A lot of recent research has also focused on producing high-N stainless steels of the duplex class (as discussed subsequently). The production of high-N stainless steels of the ferritic class and the martensitic class is more challenging, because of the low solubility of N in ferrite (this is why N>0.08wt% is already considered high-N in the ferritic and martensitic classes by Speidel and Speidel [3]). Alloying with N is also very beneficial to these two classes of stainless steels. For example, the corrosion resistance and wear resistance of the N-containing AISI420 martensitic steel are much better than those of conventional AISI420 [4]. Nitrogen can hinder grain growth during an autensitisation treatment [5]. Precipitation of nitrides also enhances the tribological behaviour of the ferritic class [6].

Joseph Ki Leuk Lai, Kin Ho Lo and Chan Hung Shek

It is worth mentioning that although some people defined high-N martensitic steels as those whose N contents exceeded a meagre 0.08wt% [2], recent research efforts have been able to produce very high-N martensitic steels using a variety of techniques like powder metallurgy (2.9wt%) [7], gaseous injection of N into the melt (0.19wt%) [4] and gas nitriding (0.42wt%) [8].

2. PROBLEMS ASSOCIATED WITH NICKEL

Nickel is a very precious element ($15000/tonne in the year 2005 [9]) whose use as an alloying element for stainless steels is widespread. Substantial savings can be made by doing away with or reducing the use of Ni in the fabrication of stainless steels [10]. Economy aside, reducing the use of nickel also contributes to environmental protection because this element accounts for a significant percentage of the energy used in the fabrication of stainless steels, as calculated very recently by Johnson *et al.* [11].

One of the main problems associated with the use of Ni in stainless steels lies in its notoriety for causing allergy in both humans [12] and animals [13]. In fact, Ni and Ni-alloys are suspected of being carcinogenic to humans [14]. Consequently, legislations and directives aiming at restricting the use of Ni-containing parts in the human body are in place in the USA and some European countries [12, 15] (for instance, the European Directive 94/27/EC of 30-June-1994 [15]), although stainless steels used nowadays for medical and surgical purposes still contain Ni (like 316L [16]) [17-20]. It has to be noted that even those stainless steels listed in the ASTM standards contain Ni [21]. Because of the potential hazards that Ni could brings to humans, a significant amount of research on stainless steels has been devoted to developing the low-Ni [22] and Ni-free varieties [10, 15, 23, 24].

3. SUBSTITUTES FOR Ni-CONTAINING STAINLESS STEELS

A large chunk of usage of Ni-free/ultra-low-Ni stainless steels is in medicine [21, 23-26] and new steels are readily available in the market (like the BioDur 108 stainless steel [27]). The most commonly used austenitising substitutes for Ni are manganese, carbon and nitrogen [26]. Among these elements, carbon is the least welcome because of its sensitising effect. Manganese is favoured because of its austenitising ability and its high negative interaction parameter with nitrogen, which is conducive to the pickup of nitrogen [28]. Usually, both Mn and N are used for the following reasons: firstly, Mn alone cannot achieve complete austenite stabilisation for high Cr contents (>12%) [29, 30]. Secondly, attaining complete austenite stabilisation using a high nitrogen content (>0.2%) in the Fe-Cr-N system is not desirable because of the large positive interaction between Ni and N that may lead to a high tendency for nitride precipitation [28]. Also, high N contents may necessitate expensive high-pressure alloying techniques or time-consuming solid-state nitriding processes. Thirdly, Ni-free, high-N stainless steels undergo a ductile-to-brittle transition. The Ductile-To-Brittle Transition Temperature (DBTT) is directly proportional to %N as $DBTT(°K) = 300(\%N) - 30$ [15] and so N must not be too high.

These CrMnN steels in general possess resistance to corrosive wear [31]. For instance, the Hadfield steel has very good resistance to wear, but its corrosion resistance is not impressive. Gavriljuk *et al.* [26] have been able to fabricate a corrosion-resistant CrMnCN steel that is analogous to its Hadfield counterpart. The CrMnN steel produced by Mills and Knutsen [32] even tribologically outperforms the Hadfield steel in a corrosive environment. Additionally, CrMnN steels may exhibit TRIP-like behaviour and other properties of these steel are also satisfactory. These topics have been studied exhaustively by a number of workers [33-35]. Increasing Mn, while at the same time lowering Ni, has been found to be conducive to the resistance to stress corrosion cracking in a boiling $MgCl_2$ solution [36, 37]. This is because Mn selectively dissolves and the likelihood of localised corrosion attack is reduced as a consequence. A rule of thumb is that for every unit of Ni replaced, the requisite amount of Mn will be about two units. This rule forms that basis for the composition of the AISI200 series steels.

It has to be noted that the Mn content in CrMnN stainless steels may reach very high values (up to 19.0mass% in the Hadfield analogue, for instance [26]). And high manganese steels are known to be prone to cryogenic intergranular fracture [38]. Too much nitrogen is also believed to detract from the cryogenic toughness and impact properties of austenitic stainless steels [39-41]. To address the problem of low cryogenic toughness, Fu *et al.* [42] have recently shown that electroslag melting (ESR) and alloying with Cr and Mo may be beneficial.

Although manganese is widely used as a substitute for Ni, its use may detract from fabricability. To get around this problem, Niinomi *et al.* [43] have devised a new, ingenious way to fabricate Ni-Free austenitic stainless steel wires. These authors used Fe-24Cr-2Mo ferritic stainless steel as the starting material. After shaping the raw material into a wire form, the wires are heated in a nitrogen atmosphere, thereby realising austenitisation. Because of the shallow depths that are penetrable by N atoms, this process is limited to making small parts only [43]. Since products in the forms of wire and thin plate are commonly fabricated in the steel-making industry, the behaviour of the absorption of N into austenitic stainless steel wires and plates has been studied by Tsuchiyama *et al.* [44] recently.

4. CONSTITUTION DIAGRAMS FOR LOW-Ni/Ni-FREE STAINLESS STEELS

Traditionally, the effects of different alloying elements on phase balance have been quantified by using the nickel and chromium equivalent numbers. Several formulae for these two equivalent numbers, as proposed by Schaeffler [45], Long and DeLong [46] and Hull [47], have been in use for a very long time. A very good chronological review of the development of the various expressions of Cr_{eq} and Ni_{eq} and the different constitution diagrams is available in an early review by Olson [48].

While the effects of major alloying elements like Cr and Ni have been well-studied, the effect of Mn is still under investigation, because **1.** its role on phase stability is still controversial, **2.** the Mn-Cr-(N) stainless steels have been gaining importance in recent years. **3.** although a lot of early researchers had studied this subject area and put forth a number of constitution diagrams, they might not be applicable to high-N [49] and Mn-substituted stainless steels [50]. For example, Onozuka [50] has suggested that the austenite-stabilising ability of Mn might be overrated if the Schaeffler and Long-DeLong constitution diagrams are used for low-Ni, high-Mn austenitic stainless steels.

As far as the phase-stabilising effect of Mn is concerned, contradicting results have been reported by different workers. In early studies, Mn was both regarded as a ferritiser [51, 52] and an austenitiser [44, 45, 53]. Hull [46], however, suggested that the role of Mn on phase stability is dependent upon its own content. It acts as an austenitiser at low contents, while being a ferritiser at high contents [46].

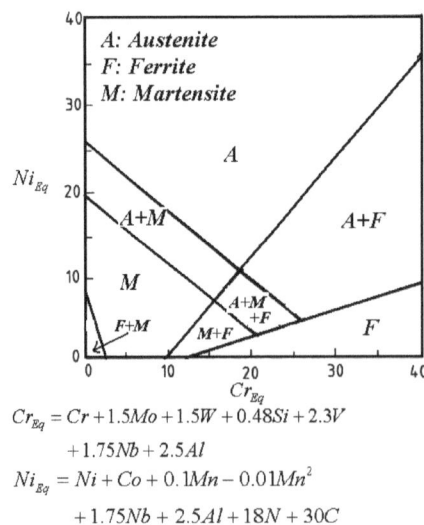

$$Cr_{Eq} = Cr + 1.5Mo + 1.5W + 0.48Si + 2.3V$$
$$+ 1.75Nb + 2.5Al$$
$$Ni_{Eq} = Ni + Co + 0.1Mn - 0.01Mn^2$$
$$+ 1.75Nb + 2.5Al + 18N + 30C$$

Figure 1: A rough replication of the modified Schaeffler diagram for high-N stainless steels [Drawn with reference to Balachandran G, Bhatia ML, Ballal NB, *et al.* Some theoretical aspects of designing nickel free high nitrogen austenitic stainless steels. ISIJ Int 2001; 41(9): 1018-27].

Recent studies [34, 49-51] have once again confirmed the complex role of Mn on phase stability. Kemp *et al.* [34] have conducted a study on a range of Cr-Mn stainless steels and found that Mn, at least at low nitrogen and carbon levels, is an austenitiser. Onozuka *et al.* [50], while also holding the view that Mn tends to stabilise the austenite phase, have suggested that the austenite-stabilising ability of Mn might be overestimated if the Schaeffler and Long-DeLong constitution diagrams are used for low-Ni, high-Mn austenitic stainless steels. For this type of stainless steels, they concluded that it would be more appropriate to use the Cr and Ni equivalent numbers proposed by Hull. The Schaeffler diagram, modified by Hull, has been adopted by many for the design of N-containing/Ni-free stainless steels [15, 52]. However, it has been pointed out that the Hull-modified Schaeffler diagram should be used with caution, especially for Cr between 12 and 18wt% [52]. A rough replication of the modified Schaeffler diagram for high-nitrogen stainless steels is drawn by the authors with reference to the work by Balachandran *et al.* (Fig. **1**) [52].

There is a requirement for ensuring a fully austenitic microstructure, which is is $Ni_{eq} > Cr_{eq} - 8$. The minimum nitrogen level for ensuring a fully austenitic microstructure calculated by the use of the Hull-modified Schaeffler approach is, in general, overestimated [52]. Balachandran *et al.* [52] have proposed an empirical formula for computing the minimum nitrogen level for achieving a fully austenitic microstructure.

$$\%N_{min} = -0.88wt\%C + 0.046wt\%Cr - 0.0009wt\%Mn + 0.038wt\%Mo - 0.053wt\%Si + 0.082wt\%Ni$$
$$- 0.208wt\%Cu - 0.032wt\%W - 0.278$$
$$Cr:12 \sim 24wt\%, Mn:10 \sim 20wt\%, C:0.05 \sim 0.6wt\%, Mo:1 \sim 3wt\%, Si:0.1 \sim 1.5wt\%, Ni:0.05 \sim 0.1wt\%$$

Regarding the interplay between Mn and other alloying elements, Raghavan [50] has found that the phase-stabilising effect of Mn depends on the contents of other alloys, too (not just on its own content, as suggested by Hull [46]). In his study, Raghavan [50] has also reported that Mn becomes a ferritiser if the content of Cr is high. The composition-dependent role of Mn on phase stability has been supported by a several other workers (Chen *et al.* [51], Schino *et al.* [53] and Mililitsky [54]). For Fe-Cr-Mn-C austenitic steels, Klueh *et al.* [55] have developed a modified Schaeffler diagram for phase prediction (Fig. **2**). The original Schaeffler diagram has to be modified because it was developed for high-Ni stainless steels. Other inadequacies of the original Schaeffler diagram as applied to the Fe-Cr-Mn-C system have been discussed in detail by these authors.

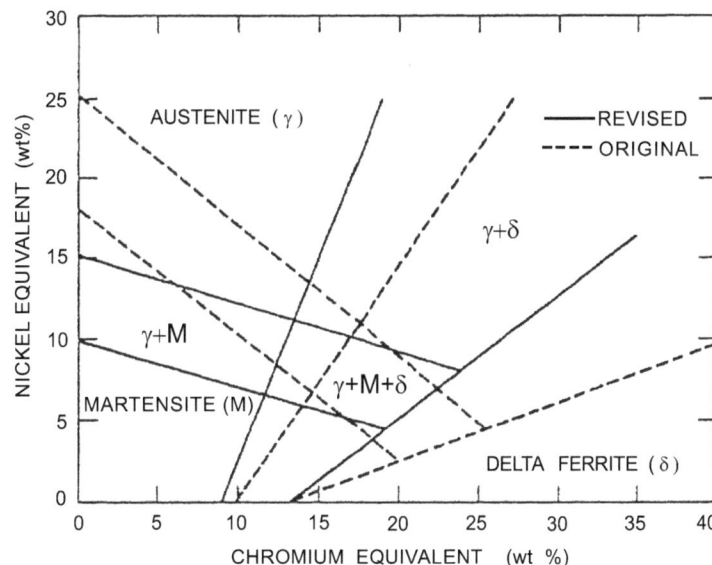

Figure 2: Constitution diagram for high-manganese stainless steels (the boundaries delineating different phases in the original Schaeffler diagram are shown as dashed lines) [Klueh RL, Maziasz PJ, Lee EH. Manganese as an austenite stabilizer in Fe-Cr-Mn-C steels. Mater Sci Eng A 1988; 102(1): 115-24. With permission for reproduction from Elsevier].

For the prediction of ferrite content in N-containing cast duplex stainless steels, Glownia *et al.* [49] have recently examined the applicability of the formula proposed by Hull [46]. The conclusion was that the formula did not seem to give a very reliable prediction.

5. HIGH-Ni, HIGH-N STAINLESS STEELS

Although high-N stainless steels were, as mentioned, mainly developed with the aim of replacing (at least partially) Ni-containing stainless steels, some research has focused on the development of high-Ni (10~45wt%), high-N stainless steels, which have been shown to possess remarkable properties [1]. For the Fe-Cr-Ni system, it is possible to achieve an entirely austenitic microstructure over a range of compositions at high temperatures (*e.g.*, at 1000°C) [3].

In the fully austenitic composition range, it is possible to accommodate high levels of N and Ni in solution. Experiments showed that up to 1wt% of N may be possible in a fully austenitic microstructure of 31wt% Ni at 1250°C. Rapid quenching may retain the fully austenitic microstructure at room temperature. The experimental high-N, high Ni stainless steels have been shown to possess high yield and ultimate strengths, high work-hardening potential, high resistance against pitting and crevice corrosion. In the experimental steels, 1wt% is equivalent to 20wt% of Cr as far as critical pitting and crevice corrosion temperatures are concerned. The stress corrosion resistance of the high-Ni, high N stainless steel in 22% NaCl solution at 105°C is also superior to types 304, 316, 321 and 347 austenitic stainless steels. While the latter steels failed transgranularly, the high-Ni, high-N stainless steels was found to be completely immune.

6. METHODS FOR INTRODUCING HIGH LEVELS OF N TO STAINLESS STEELS

Owing to the advantages of N alloying, a huge amount of research effort has been directed at introducing raised levels of nitrogen into stainless steels and a variety of techniques are in use. N alloying may be performed for steels either in the liquid state or in the solid state. Techniques employable for alloying N to steels in the molten state include the use of an induction furnace or an electric arc furnace, Argon Oxygen Decarburisation (AOD), gas purging of molten metal, pressure electroslag remelting, arc slag melting, plasma arc melting and high-pressure melting with HIP [2].

A review by Balachandran [2] in a recent monograph and an earlier one by Berns [56] on high nitrogen stainless steels have documented the different high-N steels produced by various workers using these techniques. The following are research efforts that were not included in the review by Balachandran. Nitrogen gas pressurised electroslag remelting (P-ESR) can produce steels of N contents up to about 1.1 mass% [57]. The production of ultra-high-N austenitic stainless steels has been shown to be possible by using hot isostatic pressure melting (up to 200MPa) and N as the pressurising gas [58, 59].

Tsuchiyama *et al.* [60] have successfully produced austenitic stainless steels of N contents exceeding 1 mass% by using nitrogen gas absorption in the solid state, thus avoiding the potential problem of blow holes during solidification.

In order to introduce even more nitrogen into steels, a variety of techniques have been proposed in several research efforts. Introducing N into steels by using powder metallurgy is frequently employed [59]. Ti-alloyed, high-N 316 steel has been successfully produced by Lee and Hendry [61] using powder metallurgy. Cisneros *et al.* [62], using this technique, have shown that up to about 5wt% of nitrogen may be incorporated into stainless steels powders. Rawers *et al.* [63, 64] have shown that mechanical alloying and powder injection moulding [65] in a nitrogen gas atmosphere are effective ways to accomplish raised levels of nitrogen in steel powders. Mechanical alloying produces elastic stress fields and nanosized grains (and hence a profusion of boundaries) in the powder particles, which serve as preferential sites of accommodation for nitrogen atoms.

As regards the methods for manufacturing Ni-Free, high-N austenitic stainless steels, those proposed by Carney [66], Hull [46] and Hsiao [67] are often used. These methods have recently been reviewed by Balachadran *et al.*

[52]. These authors have also compared the N contents of a range of Ni-free, fully austenitic stainless steels with the theoretical minimum N contents as predicted by the three methods. They found inconsistencies in their comparison and have proposed, by using multi-regression analysis of data in the literature, an expression which gives the minimum N content for obtaining a fully austenitic matrix [52].

In the experimental front, Ustinovshikov *et al.* [68, 69] have lately performed very detailed TEM characterisation of the High-Temperature Microstructures of a series of Fe-18Cr-N (N = 0.6-1.3%) alloys. These authors have constructed a diagram which shows the constituent phases of the Fe-18Cr-N (N = 0.6-1.3%) alloy system after quenching from high temperatures (Fig. **3**). This diagram is a handy aid for setting the heat treatment condition to ensure a fully austenitic microstructure upon quenching.

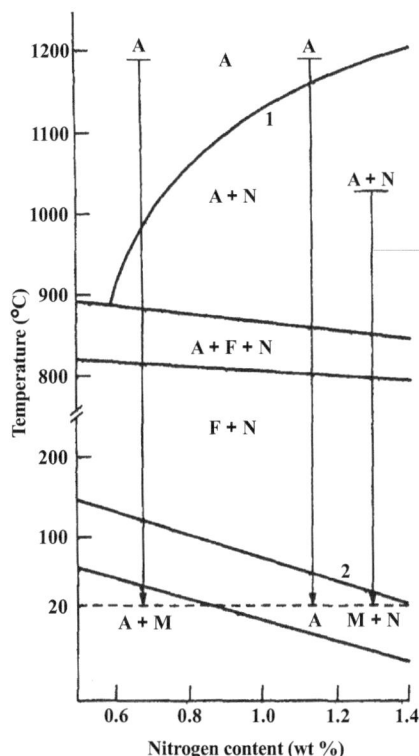

Figure 3: A diagram for predicting the phases of Fe-18Cr-N (N = 0.6-1.3%) alloys after quenching from high temperatures [Ustinovshikov Y, Ruts A, Bannykh O, *et al.* The microstructure of Fe-18%Cr alloys with high N contents. Acta Mater 1996; 44(3): 1119-25. With permission for reproduction from Elsevier].

While the afore-mentioned techniques may involve sophisticated and pricey equipment, duplex stainless steels and martensitic stainless steels containing elevated amounts of nitrogen have been fabricated successfully under ordinary atmospheric conditions [4, 10]. Balachandran *et al.* [70] have recently demonstrated that it is possible to retain the nitrogen of steels made by using other processing routes (like a conventional induction furnace and an arc furnace) through the use of conventional electroslag remelting without the application of any pressurised nitrogen gas. The strength of this method is that the porosity and inhomogeneity in high-N steels made by using conventional induction furnace can be successfully eliminated by a conventional ESR processing, but without the application of high pressure N gas over the melt (thereby saving the use of manpower and complex equipment).

For the design of high-N duplex stainless steels, a study by Weber and Uggowitzer [71] has provided some guidelines. These authors have discovered that N affects the partitioning of Cr and Mo between the ferrite phase and austenite phase. At high N contents, Mo will still preferentially segregate to ferrite, but Cr tends to partition equally in both ferrite and austenite. The reduction of Cr in ferrite inevitably adversely affects its PREN, making it even more susceptible to pitting attack compared with the austenite phase. Because the total amount of Mo plus Cr must be properly controlled in order to avoid the formation of intermetallics,

Weber and Uggowitzer [71], armed with their findings, have proposed that in the design of high-N superduplex stainless steels, the content of Mo should be increased, whereas that of Cr should be decreased.

The decrease in ferrite content in duplex stainless steels due to N addition has a pronounced effect on properties [72]. For instance, the decrease of the volume fraction of ferrite, which is vulnerable to hydrogen embrittlement and hence tends to cleave transgranularly [73-75], enhances corrosion-fatigue resistance [76].

One of the problems associated with the production of high-N stainless steels is bubbling. When the level of N exceeds about 0.2mass%, small, round pores show up on the surface of steel slabs. The pores are thought to be gas bubbles, which are generated by solute segregations in the spacing between dendrite arms during solidification [77]. Ridolfi and Tassa [77] have recently suggested that higher levels of ferritising elements in the melt will aggravate bubbling, on the grounds that N segregation from solid to liquid is aided by a ferritic microstructure.

The microstructural evolutions of a range of Ni-free, high-N austenitic stainless steels after different thermomechanical treatments have been closely examined by Balachandran *et al.* [78]. These authors have provided insights into the selection of processing parameters for obtaining the desired microstructure for this class of stainless steels.

7. EFFECTS OF N ALLOYING ON PROPERTIES

7.1. Effects of N on Strengths

Very high strength is attainable with N alloying. The strengthening mechanisms of N alloying are attributed to: **1**. interaction/pinning between dislocations and interstitial N atoms [79] because of electrostatic attraction [2] and **2**. formation of dislocation-N complexes that may exert a dragging effect on dislocations [80]. The thermal dependences of these two mechanisms have been analysed both qualitatively and quantitatively by Rawers and Grujicic [79]. The big lattice distortion associated with interstitial N also brings about strengthening [2, 12]. Short range ordering involving substitutional and interstitial elements (like Cr-N) may contribute to strengthening in N-containing stainless steels, too [81-83].

Schino and Kenny [84] have recently studied the grain size effect of high-N austenitic stainless steels, these authors have shown that the Hall-Petch relation is applicable (down to at least a grain size of 2.5μm). In the same study, a noteworthy result is that the high-N steel is stronger than its low-N counterpart at the same grain size. This is understandable because N has been found to increase the values of σ_0 and k in the Hall-Petch relation ($\sigma_y = \sigma_0 + k \cdot \sqrt{d}$) [85]. σ_0 increases linearly with the lattice distortion caused by nitrogen that is in solid solution [1]. At high contents, the effect of N on k is particularly strong [3], in as much as N binds well to dislocations. Therefore, the effect of grain size for high-N stainless steels must be taken into consideration.

High-N stainless steels, in addition to high strengths, also show high work-hardening rates [86, 87], which is ascribed to the presence of mechanical twins by Muller *et al.* [86]. Kubota *et al.* [87], however, have attributed the high rate of work hardening to the dislocation-pinning effect exerted by the Lomer-Cottrell reaction induced by the interactions among dislocations that are operative in two different intersecting slip planes. High-N austenitic stainless steels also possess better mechanical properties [53] and high-temperature properties [88] compared with their conventional counterparts. Merits of N-bearing austenitic stainless steels have recently been succinctly reviewed by Speidel [89].

Research efforts aiming at relating the mechanical properties of stainless steels to their compositions have been undertaken by several authors [90, 91] and most of these studies were devoted to austenitic stainless steels. Recently, Sieurin *et al.* [92] have proposed empirical expressions that were developed by using both austenitic stainless steels and duplex stainless steels. The effect of grain size was also taken into account by Sieurin *et al.* [92]. For Fe-Cr-Mn-N alloys, Balachandran *et al.* [52] have, by using regression analysis of a large set of available date in the literature, come up with sets of expressions that may be used to predict their yield strength and tensile strength. The effects of grain size, cold work and composition for steels in the as-solution-annealed condition are considered in these expressions. The expressions for yield strengths are listed below for reference purposes [52].

Yield stress as a function of composition in wt%

$$\sigma_y(MPa) = 3.03\%C + 335.60\%N + 5.08\%Mn + 8.41\%Cr - 9.39\%Si - 172.71\%P - 94.95\%S$$
$$- 21.94\%Mo + 15.19\%Ni - 94.25\%Cu + 3188.33\%V + 176.62$$

Yield stress as a function of composition in wt% and grain size

$$\sigma_y(MPa) = 257.69\%C + 479.49\%N + 5.74\%Mn + 1.46\%Cr - 92.58\%Si - 449.09\%P - 7354.36\%S$$
$$- 1.07(\text{Grain Size}) + 235.17$$

Yield stress as a function of composition in wt% and degree of cold work

$$\sigma_y(MPa) = 952.39\%C + 655.23\%N + 34.28\%Mn - 1.56\%Cr - 7.41\%Si - 736.13\%P + 7829.12\%S$$
$$+ 9.95\%Mo + 22.36(\%\text{Cold Work}) - 535.48$$

7.2. Effects of N on Fatigue

The addition of N to austenitic stainless steels will change stacking fault energy and hence the dislocation structure will also be changed, leading to changes in fatigue properties. With the increasing popularity of high-N stainless steels, a lot of recent research works are on the effects of N on fatigue properties.

Fatigue life is thought to be enhanced by N additions on the grounds of the planarity of dislocation structures [93] and retardation of Dynamic Strain Ageing (DSA) [94, 95] (Although retardation of DSA has been attributed to the lowering Cr diffusivity by nitrogen [96], Sourmail [97] has pointed out in a recent review that N may possibly enhance the diffusion of substitutional elements).

Planar dislocation arrangement due to N alloying is thought to be one of the reasons for enhanced fatigue property [93]. Maeng and Kim [98] have recently compared the dislocation structures at the fatigue crack tips of 316L and 316LN. These authors have concluded that nitrogen makes it difficult for dislocations in the plastic zone around crack tips to cross slip and thus fatigue crack growth is retarded in 316LN. It is to be noted that planar dislocation glide mode in the austenite phase of duplex stainless steels due to N alloying has been considered to be one of the contributors to improved fatigue resistance and higher fatigue strength, too [99-101]. The importance of planar dislocation arrangements to fatigue life enhancement can be further appreciated in the recent work by Rho and Nam [102]. These authors have studied the effect of N alloying on 304 steel. It has been found that if the level of N added to 304 is not sufficient to attain planar dislocation arrangements, then fatigue life will be shortened. This is because N alloying may degrade ductility [103], which will then adversely affect low-cycle fatigue life [102].

Besides austenitic stainless steels, the effects of N on N-containing duplex stainless steels have also been extensively investigated. Magnin and Lardon [104] stated that ferrite-like behaviour prevails at strain amplitudes exceeding 10^{-3}, whereas austenite-like behaviour dominates below this value for a low-N duplex stainless steel. The validity of Magnin's result as regards strain amplitudes $\Delta\varepsilon_p/2 < 10^{-3}$ has been questioned by Akdut [105]. A later study by Mateo *et al.* [106], which encompassed a wider range of strain amplitudes $(2.9\times10^{-5} < \Delta\varepsilon_p/2 < 6.6\times10^{-3})$, revealed that the cyclic stress-strain curve and the Coffin-Manson curve are in fact composed of three regimes (I-III). The duplex stainless steel used by Mateo *et al.* [106] exhibited an austenite-like behaviour in regime I $(\Delta\varepsilon_p/2 < 10^{-4})$, an austenite-plus-ferrite-like behaviour in regime II $(10^{-4} < \Delta\varepsilon_p/2 < 6.0\times10^{-4})$ and a ferrite-like behaviour in region III $(\Delta\varepsilon_p/2 > 6.0\times10^{-4})$. Mateo *et al.* suggested [106] and later on confirmed [99] that the three-regime behaviour should in general be applicable to duplex stainless steels, but the ranges of each regime hinges on the relative fractions of ferrite and austenite and chemical composition. A more recent study by Akdut [105] has actually shown that the fatigue behaviour of duplex stainless steels of higher N contents (from 0.15-0.35wt% N) do exhibit three regimes whose ranges are dependent on N content. It is to be noted that the content of nitrogen affects the phase balance of ferrite and austenite.

A recent study by Jencus *et al.* [107] using neutron diffraction has concluded that softening in low-cycle fatigue of the N-containing SAF2507 duplex stainless steel is due entirely to austenite. Armas *et al.* [108] have also concluded that at $\Delta\varepsilon_p / 2 = 6.02 \times 10^{-3}$, cyclic softening of SAF2507 in low-cyclic fatigue has to do with the unpinning of dislocations from N atoms in the austenite phase, while the hardening at higher strain amplitudes is due to dislocation locks that form at the intersections of two [111] slip planes, with subsequent formation of cellular structures. Polak *et al.* [101] have also observed that the formation of a cellular structure is suppressed by N. Therefore, nitrogen alloying is in general believed to better fatigue resistance of stainless steels [103, 109, 110], insofar as N reduces the tendency for dislocation cross slip and favours planar dislocation slip mode, promoting slip reversibility and giving rise to earlier cyclic softening [109, 110].

Nonetheless, Akdut [105] could not unambiguously discern the beneficial effect of N on the fatigue life of an N-containing duplex stainless steels. Mathew *et al.* [111] have also recently found that the beneficial effect of N towards fatigue life is not monotonic. Above a certain level, fatigue life is no more enhanced with further addition of N [111].

However, for corrosion fatigue (in a NaCl solution, for example), the beneficial effect of N alloying in duplex stainless steels is obvious [112], in as much as N can reduce the volume fraction of the ferrite phase, which succumbs to hydrogen embrittlement [73, 75, 112, 113]. For austenitic stainless steels, pitting-induced fatigue commonly lays waste to stainless steel components [114, 115]. Corrosion pits may aid crack initiation [116, 117] and there exists a vicious interplay between cyclic stress and pitting [116]. Rokhlin *et al.* [118] have also revealed a clear relation between the depth of corrosion pit and fatigue life. Because N strongly enhances pitting corrosion, N alloying, without a doubt, may improve the corrosion fatigue behaviour of austenitic stainless steels.

As regards the benefits brought about in N-containing stainless steels by grain refinement, several recent studies have found that the benefits may not be significant [95, 119] (while grain refinement improves a lot the fatigue properties of ordinary austenitic stainless steels [120]). This is because the generation of slip bands, which may act as crack initiators, is expedited by nitrogen [119]

7.3. Effects of N on Ductile-to-Brittle Transition

It was mentioned in the previous subsection that when the N content was increased, the cryogenic toughness of high-N austenitic stainless steels became a problem. As a matter of fact, high-N austenitic stainless steels are very peculiar as far as toughness is concerned, so much so that a number of recent works are devoted to this issue.

Metals possessing the fcc lattice usually exhibit good ductility without brittle fracture. However, austenitic stainless steels having high nitrogen and/or manganese contents are exceptions [15, 121]. There is a ductile-to-brittle transition temperature (DBTT) below which these austenitic stainless steels fracture in a brittle manner [122-127]. As a matter of fact, as the level of N increases, so will the ductile-to-brittle transition temperature (DBTT) [15] and brittleness [128].

It has been established that N significantly increases the concentration of conduction electrons, *i.e.*, the metallic component of interatomic bonds, in austenitic steels, resulting in good ductility at low temperatures. Nevertheless, when the N content exceeds about 1.5 at%, the covalent nature of interatomic bonds becomes prevalent and brittle fracture ensues, as observed in high-N austenitic steels. Mn is also believed to enhance the covalent nature of interatomic bonds.

A variety of deformation microstructures (stacking faults, dislocation networks, slip bands, twins and martensite) at cryogenic temperatures may be found on the [111] planes in austenitic stainless steels containing high contents of Mn and N [129, 130]. The fracture mechanism of high-N austenitic stainless steels has been variously attributed to these deformation microstructures by different authors.

Tobler and Meyn [122] have attributed the low-temperature, cleavage-like fracture of a Fe-18Cr-3Ni-13Mn-0.37N austenitic stainless steel to slip bands (slip-band cracking). They believed that in a suitably oriented grain, narrow shear (slip) bands would form and that activation of out-of-plane slip would be difficult. Also, they stated that shear would be restricted to the original plane and so shear strain would accumulate in the narrow shear (slip) band. Accumulation of dislocation-induced defects would eventually lead to cleavage-like fracture. These authors called the fracture cleavage-like because no river patterns, which are typical of classical cleavage facets for ferritic steels, could be discerned in their specimens. Ishizaka *et al.* [131], however, thought that the main culprit of the cleavage-like cracking of N-bearing 18Cr-18Mn austenitic steels should be the ε martensite. More recently, Muller *et al.* [132] have implicated microtwins and the inapplicability of the twin-twin penetration mechanism in the low-temperature brittle fracture of N-bearing austenitic stainless steels.

The twin-twin penetration mechanism may be understood with the disclination model. According to this model [133], when a secondary twin meets a primary twin, the twinning dislocations of the former shall form a dislocation wall having the properties of a disclination dipole [134-137]. In the stress field of this dipole, the coordinated motion of dislocations can transfer the shear strain of the secondary twin across the primary one, thereby leading to twin-twin penetration. This twin-twin penetration mechanism may explain deformation behaviour in the ductile regime [133].

Muller *et al.* [132], however, have proposed that at low temperatures, the twin-twin penetration mechanism breaks down in N(Mn)-bearing austenitic stainless steels, in as much as twins get thinner and thinner and the dislocation density decreases as temperature drops to low values [138]. As a result, the stress levels associated with the disclination dipole of very thin twins (of the order of a few nanometres) will be so low that nucleation of dislocations is impossible, and so the twin-twin penetration mechanism will be suppressed. Additionally, blunting of growing cracks gets more difficult [138]. Muller *et al.* [86, 132, 138] have further proposed that a lowering of the nickel content and an increase in the content of N (or Mn), as seen in low-Ni, N(Mn)-bearing austenitic stainless steels, also have the effect of decreasing twin spacing, which may render the twin-twin penetration mechanism inoperable. These authors have instead proposed that the intersections of microtwins in N(Mn)-bearing austenitic stainless steels will build up high internal stresses and so can act as crack initiators. These microcracks on different planes then cause micro-cleavage cracking.

Tomata *et al.* [139], however, have disputed the Muller model and the role played by the ε martensite, as suggested by Ishizaka *et al.* [131]. Instead of attributing the low-temperature, transgranular, brittle fracture to cleavage fracture, these authors have proposed that slipping off of active [111] slip planes with a high density of dislocations, speculated to be caused by the short range ordering of nitrogen, is to blame.

In the above-mentioned investigations, the fracture was observed to be transgranular. Nevertheless, intergranular cracking of austenitic steels containing high Mn(N) has also been reported [126].

7.4. Effects of N on Corrosion

In order to economise on the use of precious nickel, Mn has been used as a substitute alloying element. While Mn is deleterious to general and localised corrosion properties [140], the addition of N may offset its harmful effects [141].

With the increasing use of high-N stainless steels, the effects of N on corrosion have been the subject to very intensive research. Results on the effects of N on general corrosion are not consistent. For example, positive results were obtained by Sakamoto *et al.* [142], but Zagorski and Doraczynska [143], and Lu *et al.* [144] have reported negatively on N. Ahmed and Fathy [145] have recently studied the effect of partial replacement of Ni by N on the resistance to general corrosion of an austenitic stainless steel. It has been found that the resistance to general corrosion in Fe_3Cl of the steel decreases after N exceeds a certain level.

However, for localised corrosion phenomena, *i.e.*, pitting corrosion and crevice corrosion, the beneficial role of nitrogen is clearly identified in the bulk of the literature. For example, an early study by Speidel [1] suggested that the pitting potential increases linearly with N content. In fact, the positive effect of N is believed to be much more dramatic than that of Cr.

Nonetheless, in a recent review article by Mudali and Ningshen [146], the authors have pointed out that although there is a general consensus that N alloying improves corrosion resistance, there still exists no universally accepted theory with regard to the underlying mechanisms despite several hypotheses that have been propounded.

It is generally believed that reduction of the ferrite content in duplex stainless steels and austenitic weld metals by nitrogen will reduce pitting, since the amount of vulnerable sites like ferrite/austenite grain boundaries are decreased.

For single-phased austenitic stainless steels, several hypotheses have been put forth to explain the beneficial effects of N on localised corrosion. Some of these hypotheses are summarised below:

- The formation of NH_4^+ (or NH_3) by nitrogen in pits/crevices consumes protons, thereby leading to less severe reduction in the pH value [147-153] and promoting repassivation ($[N] + 4H^+ + 3e^- \rightarrow NH_4^+$). In sulphate solutions, it has been propounded that NH_4^+ may form a protective, passive ammonium sulphate layer [154]. Ammonium has also been proposed to react with free chlorine in chlorinated water, forming less oxidising species [155]. A lot of workers have confirmed the existence of NH_4^+ in a variety of media and NH_4^+ has been found to be present in passive, active and transpassive potentials [156]. Additionally, rate of generation of NH_4^+ increases proportionately with N content [147]

- The formation of corrosion inhibiting nitrates (NO_2^-, NO_3^-) [151, 157-160].

- Prevention of anion attack because of the enrichment of nitrogen at the film/substrate interface during passivation. Some authors have suggested that the formation of a stable protective layer of a Cr, Mo, Ni-rich nitride (Ni2Mo3N) is responsible for the protection [161-163], but Olefjord and Wergrelius [149] did not seem to endorse this notion. Alternatively, a layer of an intermetallic phase beneath the passive film has been suggested to be the reason [164]. Olsson [165], besides finding N accumulation at the oxide/substrate interface, also detected N at the top of the passive layer.

- The repulsive action between Cl^- (and other deleterious ions) and negatively charged nitrogen ($N^{\delta-}$) that accumulates at the passive layer [152, 166, 167]. The repulsion of Cl^- gives rise to fast repassivation of pits [145]. It has been suggested that $N^{\delta-}$ may change into ammonium [145, 164].

- Incorporation of nitrogen into the surface passive film [168-172]

- Nitrogen is also believed to accumulate preferentially to active sites such as kinks and steps, thereby hindering dissolution [173].

These mechanisms are variously used by different authors to explain their experimental findings. For instance, while a lot of workers favour the mechanism involving the formation of ammonia/ammonium in pits and crevices [174], Misawa and Tanabe [160], by using in-situ Raman spectroscopy, have stated that nitrate ions, instead of ammonia/ammonium ions, were observed at the precursor regions of pits. Also, self-repair of the passive film involves its absorption of nitrate [160].

A recent study by Mudali *et al.* [175] has discovered that too much nitrogen may in fact be harmful to cold-worked austenitic stainless steels. When the amount of N increases, the deformation substructures get finer, with an attendant increase in dislocation densities within them. The sheer number of dislocations in these deformation substructures provides suitable initiation sites for corrosion. Furthermore, the shearing of Cr-N clusters leads to Cr-rich clusters. These clusters and the relatively Cr-depleted adjacent matrix form microgalvanic cells, resulting in enhanced corrosion.

Besides pitting resistance, N can also improve crevice corrosion resistance [176, 177], because of the reduction of N, instead of O, at the crevice and the increase of pH value in the crevice due to the formation of ammonium ions [176].

For the commonly used N-bearing austenitic stainless steels 304 and 316, Mudali *et al.* [177] have established their time-temperature-sensitisation-pitting diagrams. These diagrams facilitate the use of these steels.

REFERENCES

[1]   Speidel MO. Properties and Applications of High Nitrogen Steels. In: Proceedings of the 1st International Conference on High Nitrogen Steels, (Foct J and Hendry A, EDS), The Institute of Metals, Lond, 1989, pp.92-96.

[2]   Balachandran G. Developments in the manufacture of high nitrogen stainless steels. In: High Nitrogen Steels and Stainless Steels-Manufacturing, Properties and Applications, Alpha Science International, Pangbourne, UK, 2004; pp.40-93.

[3]   Speidel H, Speidel MO. High nitrogen high nickel austenitic stainless steels. In: High Nitrogen Steels and Stainless Steels – Manufacturing, Properties and Applications (U.K.Mudali and B.Raj, eds). Alpha Sci. 2004; pp. 231-42.

[4]   Horovitz MB, Beneduce F, Garbogini A, *et al.* Nitrogen bearing martensitic stainless steels: microstructure and properties. ISIJ Int 1996; 36(7): 840-5.

[5]   Leda H. Nitrogen in martensitic stainless steels. J Mater Process Technol 1995; 53(1-2): 263-72.

[6]   Kliauga AM, Polh M, Laffke DK. A comparison of the friction and reciprocating wear behaviour between an austenitic (X2CrNiMo17 13 2) and a ferritic (X1CrNiMoNb 28 4 2) stainless steel after nitrogen ion implantation. Surf Coat Technol 1998; 102(3): 237-44.

[7]   Toro A, Falleiros NA, Rodriguez D, *et al.* P/M processing routes for high nitrogen martensitic stainless steels. Trans Indian Inst Metal 2002; 55(5): 481-7.

[8]   Toro A, Misiolek WZ, Tschiptschin AP. Correlations between microstructure and surface properties in a high nitrogen martensitic stainless steel. Acta Mater 2003; 51(12): 3363-74.

[9]   News and Updates in: J Materials (JOM), 58 (2006) p.6.

[10]   Wang J, Uggowitzer PJ, Magdowski R, *et al.* Nickel free duplex stainless steels. Scr Mater 1999; 40(1): 123-9.

[11]   Johnson J, Reck BK, Wang T, *et al.* The energy benefit of stainless steel recycling. Energy Policy 2008; 36(1): 181-92.

[12]   Menzel J, Kirschner W, Stein G. High Nitrogen Containing Ni-free Austenitic Steels for Medical Applications. ISIJ Int 1996; 36(7): 893-900.

[13]   Kimura T. Contact hypersensitivity to stainless steel cages (Chromium metal) in hairless descendants of Mexican hairless dogs. Environ Toxicol 2007; 22(2): 176-84.

[14]   IARC Monographs on the Evaluation of Carcinogenic Risks to Humans: Surgical Implants and Other Foreign Bodies. Lyon. Vol.74, 1999, p.65.

[15]   Uggowitzer PJ, Magdowski R, Speidel MO. Nickel free high nitrogen austenitic steels. ISIJ Int 1996; 36(7): 901-8.

[16]   Dahlke MG, Dabrowski JR, Dabrowski B. Characteristic of the porous 316 stainless steel for the friction element of prosthetic joint. Wear 2007; 263(Spec Iss): 10239.

[17]   Ciagada A, Rondelli G, Vicentini B, *et al.* Duplex stainless steels for osteosynthesis devices. J Biomed Mater Res 1989; 23(9): 1087-95.

[18]   Tang Y, Katsuma S, Fujimoto S, *et al.* Electrochemical study of Type 304 and 316L stainless steels in simulated body fluids and cell cultures. Acta Biomater 2006; 2(6): 709-15.

[19]   Muller R, Abke J, Schnell E, *et al.* Surface engineering of stainless steel materials by covalent collagen immobilization to improve implant biocompatibility. Biomater 2005; 26(34): 6962-72.

[20]   Filho AI, Rollo JMDA, Silva RV, *et al.* Alternative process to manufacture austenitic-ferritic stainless steel wires. Mater Lett 2005; 59(10): 1192-4.

[21]   Sumita M, Hanawa T, Teoh SH. Development of nitrogen-containing nickel-free austenitic stainless steels for metallic biomaterials—review. Mater Sci Eng C 2004; 24(6-8): 753-60.

[22]   Sieurin H, Sandstrom R, Westin EM. Fracture toughness of the lean duplex stainless steel LDX 2101. Metall Mater Trans A 2006; 37(10): 2975-81.

[23]   Fini M, Aldini NN, Torricelli P, *et al.* A new austenitic stainless steel with negligible nickel content: an *in vitro* and *in vivo* comparative investigation. Biomater 2003; 24(27): 4929-39.

[24]   Montanaro L, Cervellati M, Campoccia D, *et al.* Promising *in vitro* performances of a new nickel-free stainless steel. J Materials: Mater Med 2006; 17(3): 267-75.

[25]   Reclaru L, Ziegenhagen R, Eschler PY, *et al.* Comparative corrosion study of "Ni-free" austenitic stainless steels in view of medical applications. Acta Biomater 2006; 2(4): 433-44.

[26]   Gavriljuk VG, Tyshchenko AI, Razumov ON, *et al.* Corrosion-resistant analogue of Hadfield steel. Mater Sci Eng A 2006; 420(1-2): 47-54.

[27]   Walter MJ. Stainless steel for medical implants. Adv Mater & Process 2006; 164(4): 22-4.

[28]   Kunze J. Nitrogen and Carbon in Iron and Steel-Thermodynamics. Akademie Verlag Berlin. 1990. p.133.

[29]   Monypenny JHG. Stainless Iron and Steel (3rd edition). Vo.2. Chapman and Hall. Lond. (1951); pp.125-44.

[30]   Lula R. Manganese Stainless Steel. The Manganese Centre. France (1986); pp.29-31.

[31]   Bi HY, Jiang XX, Li SZ. The corrosive wear behavior of Cr–Mn–N series casting stainless steel. Wear 1999; 225-29: 1043-9.

[32]   Mills DJ, Knutsen RD. An investigation of the tribological behaviour of a high-nitrogen Cr---Mn austenitic stainless steel. Wear 1998; 215(1-2): 83-90.

[33]   Fu W, Zheng Y, He X. Resistance of a high nitrogen austenitic steel to cavitation erosion. Wear 2001; 249(9): 788-91.

[34]   Kemp M, Bennekom AV, Robinson FPA. Evaluation of the corrosion and mechanical properties of a range of experimental Cr---Mn stainless steels. Mater Sci Eng A 1995; 199(2): 183-94.

[35]   Lelyveld CDV, Bennekom AV. An evaluation of two experimental duplex Cr-Mn-N stainless steels after thermo-mechanical processing. Mater Sci Eng A 1996; 205(1-2): 229-38.

[36]   Devasenapathi A, Asawa M. Effect of high Mn on SCC behaviour of an austenitic stainless steel in 42% boiling $MgCl_2$ solution. J Mater Sci Lett 1997; 16(16): 1363-5.

[37]   Devasenapathi A, Ramakrishna GS, Raja VS. A comparative study of stress-corrosion cracking behavior of 3 manganese-containing stainless steels in boiling $MgCl_2$ solution. J Mater Sci Lett 1995; 14(18): 1254-5.

[38]   Miura R, Nakajima H. 32Mn-7Cr Austenitic steel for cryogenic applications. In: Advances in Cryogenic Engineering (Clark AF and Reed RP. EDS), Plenum Press, New York, Vol.32, 1986; pp.245-52.

[39]   Yoo O, Oh YJ, Lee BS, Nam SW. The effect of the carbon and nitrogen contents on the fracture toughness of Type 347 austenitic stainless steels. Mater Sci Eng A 2005; 405(1-2): 147-57.

[40]   Yuan Z, Dai Q, Cheng X, *et al.* Impact properties of high-nitrogen austenitic stainless steels. Mater Sci Eng A 2008; 475(1-2): 202-6.

[41]   Nakamura N, Takaki S. Structural Control of Stainless Steel by Nitrogen Absorption in Solid State. ISIJ Int 1996; 36(7) 922-6.

[42]   Fu R, Zheng Y, Ren Y. Mechanical properties of 32Mn-7Cr-0.6Mo-0.3N austenitic steel for cryogenic applications. J Mater Eng Perform 2001; 10(4): 456-9.

[43]   Niinomi M, Hanawa T, Narushima T. Japanese research and development on metallic biomedical, dental, and healthcare materials. J Mater (JOM) 2005; 57(4): 18-25.

[44]   Tsuchiyama T, Fukumaru T, Egashira M, *et al.* Calculation of Nitrogen Absorption into Austenitic Stainless Steel Plate and Wire. ISIJ Int 2004; 44(6): 1121-3.

[45]   Schaeffler AL. Constitution diagram for stainless steel weld metal. Metal Prog 1949; 56: 680-8.

[46]   Long CJ, DeLong WT. Ferrite content of austenitic stainless steel weld metal. Weld J 1973; 52(7): 281s-97s.

[47]   Hull FC. Delta ferrite and martensite formation in stainless steels. Weld J 1973; 52(5): 193s-203s.

[48]   Olson DL. Prediction of austenitic weld metal microstructure and properties. Weld J 1985; 64(10): 281s-95s.

[49]   Glownia J, Kalandyk B, Hubner K. Delta ferrite predictions for cast duplex steels with high nitrogen content. Mater Charact 2001; 47(2): 149-55.

[50]   Onozuka M, Saida T, Hirai S, *et al.* Low-activation Mn-Cr austenitic stainless steel with further reduced content of long-lived radioactive elements. J Nucl Mater 1998; 255(2-3); 128-38.

[51]   Szumachowski ER, Kotecki DJ. Effect of manganese on stainless steel weld metal ferrite. Weld J 1984; 63(5): 156s-61s.

[52] Ritter AM, Henry MF, Savage WF. High-temperature phase chemistries and solidification mode prediction in nitrogen strengthened austenitic stainless steels. Scr Metall Mater 1984; 15(7): 1339-51.

[53] Hammar O, Svensson U. Solidification and Casting of Metals. The Metal Soc, Lond, 1979, p.401.

[50] Raghavan V. Effect of manganese on the stability of austenite in Fe-Cr-Ni alloys. Metall Mater Trans 1995; 26(22): 237-46.

[51] Chen SR, Davies HA, Rainforth WM. Austenite phase formation in rapidly solidified Fe–Cr–Mn–C steels. Acta Mater 1999; 47(8): 4555-69.

[52] Balachandran G, Bhatia ML, Ballal NB, *et al.* Some theoretical aspects of designing nickel free high nitrogen austenitic stainless steels. ISIJ Int 2001; 41(9): 1018-27.

[53] Schino AD, Mecozzi MG, Barteri M, *et al.* Development of high nitrogen, low nickel, 18%Cr austenitic stainless steels. J Mater Sci 2000; 35(19): 4803-08.

[54] Milititsky M, DeCooman BC, Speer JG, *et al.* Room-temperature aging of manganese-alloyed high nitrogen duplex stainless steels. Metall Mater Trans A 2006; 37(7): 2117-123.

[55] Klueh RL, Maziasz PJ, Lee EH. Manganese as an austenite stabilizer in Fe-Cr-Mn-C steels. Mater Sci Eng A 1988; 102(1): 115-24.

[56] Berns H. Manufacture and application of high nitrogen steels. ISIJ Int 1996; 36(7): 909-4.

[57] Sagara M, Katada Y, Kodama T. Localized Corrosion Behavior of High Nitrogen-bearing Austenitic Stainless Steels in Seawater Environment. ISIJ Int 2003; 43(5): 714-9.

[58] Rawers JC, Asai G, aDunning JS. Change in mechanical properties and microstructure of 201 stainless steel with increased nitrogen alloying. J Mater Res 1994; 9(12): 3160-9.

[59] Romu JJ, Tervo JJ, Hanninen HE, *et al.* Development of Properties of P/M Austenitic Stainless Steels by Nitrogen Infusion. ISIJ Int 1996; 36(7): 938-46.

[60] Tsuchiyama T, Ito H, Kataoka K, *et al.* Fabrication of ultrahigh nitrogen austenitic steels by nitrogen gas absorption into solid solution. Metall Mater Trans A 2003; 34(11): 2591-9.

[61] Lee SL, Hendry A. Dissolution and reprecipitation of nitrides in an austenitic stainless steel produced by powder metallurgy. ISIJ Int 1996; 36(7): 932-7.

[62] Cisneros MM, Lopez HF, Mancha H, *et al.* Processing of nanostructured high nitrogen stainless steel by mechanical alloying. Metal Mat Trans A 2005; 36(5): 1309-16.

[63] Rawers J, Govier D, Cook D. Mechanical Alloying of Nitrogen into Iron Powders. ISIJ Int 1996; 36(7): 958-61.

[64] Rawers J, Doan R. Mechanical processing of iron powders in reactive and nonreactive gas atmospheres. Metall Mater Trans A 1994; 25(2): 381-8.

[65] Rawers J, Croydon F, Krabbe R, *et al.* Tensile characteristics of nitrogen enhanced powder injection moulded 316L stainless steel. Powder Metall 1996; 39(2): 125-9.

[66] Carney DJ. Blast furnace & steel plant Dec 1955; p.1377.

[67] Hsiao CM, Dulis EJ. Phase Relationship in Austenitic Cr-Mn-C-N Stainless Steel. Trans Am Soc Metal 1958; 50: 773-802.

[68] Ustinovshikov Y, Ruts A, Bannykh O, *et al.* The microstructure of Fe-18%Cr alloys with high N contents. Acta Mater 1996; 44(3): 1119-25.

[69] Ustinovshikov Y, Ruts A, Bannykh O, *et al.* Microstructure and properties of the high-nitrogen Fe-Cr austenite. Mater Sci Eng A 1999; 262(1-2): 82-7.

[70] Balachandran G, Bhatia ML, Ballal NB, *et al.* Processing Nickel Free High Nitrogen Austenitic Stainless Steels through Conventional Electroslag Remelting Process. ISIJ Int 2000; 40(5): 478-83.

[71] Weber L, Uggowitzer PJ. Partitioning of chromium and molybdenum in super duplex stainless steels with respect to nitrogen and nickel content Mater Sci Eng A 1998; 242(1-2): 222-9.

[72] Park Y, Lee Z. The effect of nitrogen and heat treatment on the microstructure and tensile properties of 25Cr–7Ni–1.5Mo–3W–xN duplex stainless steel castings. Mater Sci Eng A 2001; 297(1-2): 78-84.

[73] Krishnan KN. Mechanism of corrosion fatigue in super duplex stainless steel in 3.5 percent NaCl solution. Int J Fract 1998; 88(3): 205-13.

[74] Marrow TJ, Hippsley CA, King JE. Effect of mean stress on hydrogen assisted fatigue crack propagation in duplex stainless steel. Acta Metall Mater 1991; 39(6): 1367-76.

[75] Marrow TJ, King JE. Microstructural and environmental effects on fatigue crack propagation in duplex stainless steels. Fatigue Fract Eng Mater Struct 1994; 17(7): 761-7.

[76] Tseng C, Liou H, Tsai W. The influence of nitrogen content on corrosion fatigue crack growth behavior of duplex stainless steel. Mater Sci Eng A 2003; 344(1-2): 190-200.

[77]   Ridolfi MR, Tassa O. Formation of nitrogen bubbles during the solidification of 16-18% Cr high nitrogen austenitic stainless steels. Intermetalics 2003; 11(11-12): 1335-8.

[78]   Balachandran G, Bhatia ML, Ballal NB, *et al.* Influence of Mechanical Processing and Heat Treatment on Microstructure Evolution in Nickel Free High Nitrogen Austenitic Stainless Steels. ISIJ Int 2000; 40(5): 491-500.

[79]   J.Rawers and M.Grujicic. Effects of metal composition and temperature on the yield strength of nitrogen strengthened stainless steels. Mater Sci Eng A 1996; 207(2): 188-94.

[80]   Owen W. Nitrogen strengthening of austenitic stainless steels at temperatures above 500 K, In: Proceedings of the Conference on High Nitrogen Steels 90 (eds Stein G. and Witulski H.), Aachen, Germany, October, 1990, Stahleisen, Dusseldorf, October 1990; p.42.

[81]   Park ID, Ahn SH, Nam KW. Solid solution strengthening behavior of 25Cr-20Ni austenite stainless steel. Key Eng Mater 2004; 261-263: 1209-14.

[82]   Byrnes MLG, Grujicic M, Owen WS. Nitrogen strengthening of a stable austenitic stainless steel. Acta Metall 1987; 35(7): 1853-62.

[83]   Saller G, Spiradek-Hahn K, Scheu C, *et al.* Microstructural evolution of Cr-Mn-N austenitic steels during cold work hardening. Mater Sci Eng A 2006; 427(1-2): 246-54.

[84]   Schino AD, Kenny JM. Grain refinement strengthening of a micro-crystalline high nitrogen austenitic stainless steel. Mater Lett 2003; 57(12): 1830-4.

[85]   Ikegami Y, Nemoto R. Effect of Thermo-mechanical Treatment on Mechanical Properties of High-nitrogen Containing Cr-Mn-Ni Austenitic Stainless Steels. ISIJ Int 1996; 36(7): 855-61.

[86]   Muller P, Solenthaler C, Uggowitzer P, *et al.* On the effect of nitrogen on the dislocation structure of austenitic stainless steel. Mater Sci Eng A 1993; 164(1-2): 164-9.

[87]   Kubota S, Xia Y, Tomota Y. Work-hardening Behavior and Evolution of Dislocation-microstructures in High-nitrogen Bearing Austenitic Steels. ISIJ Int 1998; 38(5): 474-81.

[88]   Schino AD, Kenny JM, Barteri A. High temperature resistance of a high nitrogen and low nickel austenitic stainless steel. J Mater Sci Lett 2003; 22(9): 691-3.

[89]   Speidel MO. New nitrogen-bearing austenitic stainless steels with high strength and ductility. Metal Sci Heat Treat 2005; 47(11-12): 489-93.

[90]   Pickering FB. The Metallurgical Evolution of Stainless Steel, Metals Park. OH USA. 1979.

[91]   Eliasson J, Sandstrom R. Proof strength values for austenitic stainless steels at elevated temperatures. Steel Res 2000; 71(6-7): 249-54.

[92]   Sieurin H, Zander J, Sandstrom R. Modelling solid solution hardening in stainless steels. Mater Sci Eng A 2006; 415(1-2): 66-71.

[93]   Kim DW, Chang J, Ryu W. Evaluation of the creep-fatigue damage mechanism of Type 316L and Type 316LN stainless steel. Int J Press Vessel Pip 2008; 85(6): 378-84.

[94]   Kim DW, Ryu W, Hong JH, *et al.* Effect of nitrogen on the dynamic strain ageing behaviour of type 316L stainless steel. J Mater Sci 1998; 33(3): 675-9.

[95]   Kim DW, Kim WG, Ryu W. Role of dynamic strain aging on low cycle fatigue and crack propagation of type 316L(N) stainless steel. Int J Fatigue 2003; 25(9-11): 1203-07.

[96]   Kim DW, Ryu W, Hong JH, *et al.* Effect of nitrogen on high temperature low cycle fatigue behaviors in type 316L stainless steel. J Nucl Mater 1998; 254(2-3): 226-33.

[97]   Sourmail T. Precipitation in creep resistant austenitic stainless steels. Mater Sci Technol 2001; 17(1): 1-14.

[98]   Maeng W, Kim W. Comparative study on the fatigue crack growth behavior of 316L and 316LN stainless steels: effect of microstructure of cyclic plastic strain zone at crack tip. J Nucl Mater 2000; 282(1): 32-9.

[99]   Mateo A, Girones A, Keichel J, *et al.* Cyclic deformation behaviour of superduplex stainless steels. Mater Sci Eng A 2001; 314(1-2): 176-85.

[100]   Li Y, Laird C. Cyclic response and dislocation structures of AISI316L stainless steel. 2. polycrystals fatigued at intermediate strain amplitude. Mater Sci Eng A 1994; 186(1-2): 87-103.

[101]   Polak J, Fardoun F, Degallaix S. Analysis of the hysteresis loop in stainless steels II. Austenitic-ferritic duplex steel and the effect of nitrogen. Mater Sci Eng A 2001; 297(1-2): 154-61.

[102]   Rho BS, Nam SW. Effects of nitrogen on low-cycle fatigue properties of type 304L austenitic stainless steels tested with and without tensile strain hold. J Nucl Mater 2002; 300(1): 65-72.

[103]   Vogt JB, Foct J, Regnard C, *et al.* Low temperature fatigue of 316L and 316LN austenitic stainless steels. Metall Trans A 1991; 22(10): 2385-92.

[104] Magnin T, Lardon JM. Cyclic deformation mechanisms of a 2-phase stainless steel in various environmental conditions. Mater Sci Eng A 1998; 104: 21-8.

[105] Akdut N. Phase morphology and fatigue lives of nitrogen alloyed duplex stainless steels. Int J Fatigue 1999; 21: S97-S103.

[106] Mateo A, LLanes L, Iturgoyen L, *et al.* Cyclic stress-strain response and dislocation substructure evolution of a ferrite-austenite stainless steel. Acta Mater 1996; 44(3): 1143-53.

[107] Jencus P, Polak J, Lukas P, *et al. In situ* neutron diffraction study of the low cycle fatigue of the alpha-gamma dupleik stainless steel. Phys B 2006; 385-386: 597-9.

[108] Armas IA, Marinelli MC, Herenu S, *et al.* On the cyclic softening behavior of SAF 2507 duplex stainless steel. Acta Mater 2006; 54(19): 5041-9.

[109] Vogt JB. Fatigue properties of high nitrogen steels. J Mater Process Technol 2001; 117(3): 364-9.

[110] Massol K, Vogt JB, Foct J. Fatigue behaviour of new duplex stainless steels upgraded by nitrogen alloying. ISIJ Int 2002; 42(3): 310-5.

[111] Mathew MD, Kim DW, Ryu W. A neural network model to predict low cycle fatigue life of nitrogen-alloyed 316L stainless steel. Mater Sci Eng A 2008; 474(1-2): 247-53.

[112] Tseng C, Liou H, Tsai W. The influence of nitrogen content on corrosion fatigue crack growth behavior of duplex stainless steel. Mater Sci Eng A 2003; 344(1-2): 190-200.

[113] Makhlouf K, Sidhom H, Triguia I, *et al.* Corrosion fatigue crack propagation of a duplex stainless steel X6 CrNiMoCu25-6 in air and in artificial sea water. Int J Fatigue 2006; 25(2): 167-79.

[114] M.Sivakumar and S.Rajeswari. Investigation on biomechanically induced fatigue of a stainless-steel orthopaedic implant device. J Mater Sci Lett 1993; 12(3): 145-8.

[115] Sivakumar M, Mudali UK, Rajeswari S. investigation of failures in stainless steel orthopaedic implant devices – pit induced fatigue cracks. J Mater Sci Lett 1995; 14(2): 148-51.

[116] Xie J, Alpas AT, Northwood DO. A mechanism for the crack initiation of corrosion fatigue of Type 316L stainless steel in Hank's solution. Mater Charact 2002; 48(4): 271-7.

[117] Qian YR, Cahoon JR. Crack initiation mechanisms for corrosion fatigue of austenitic stainless steel. Corrosion 1997; 53(2): 129-35.

[118] Rokhlin SI, Kim JY, Nagy H, *et al.* Effect of pitting corrosion on fatigue crack initiation and fatigue life. Eng Fract Mech 1999; 62(4-5): 425-44.

[119] Di Schino A, Barteri M, Kenny JM. Fatigue behavior of a high nitrogen austenitic stainless steel as a function of its grain size. J Mater Sci Lett 2003; 22(21): 1511-13.

[120] Schino AD, Kenny JM. Grain size dependence of the fatigue behaviour of a ultrafine-grained AISI 304 stainless steel. Mater Lett 2003; 57(21): 3182-5.

[121] Gavriljuk VG. Nitrogen in iron and steel. ISIJ Int 1996; 36(5): 738-45.

[122] Tobler RL, Meyn D. Cleavage like fracture along slip planes in Fe-18Cr-3Ni-0.37N austenitic stainless steel at liquid helium temperature. Metall Trans A 1988; 19(6): 1626-31.

[123] Tomato Y, Endo S. Cleavage-like fracture at low temperatures in an 18Mn-18Cr-0.5N austenitic steel. ISIJ Int 1990; 30(8): 656-62.

[124] Ilola RJ, Hanninen HE, Ullakko KM. Mechanical Properties of Austenitic High-nitrogen Cr-Ni and Cr-Mn Steels at Low Temperatures. ISIJ Int 1996; 36(7): 873-7.

[125] You RK, Kao P, Gan D. Mechanical properties of Fe-30Mn-10Al-1C-1Si alloy. Mater Sci Eng A 1989; 117: 141-8.

[126] Vogt JB, Messai A, Foct J. Cleavage fracture of austenite induced by nitrogen supersaturation. Scr Metall Mater 1994; 31(5): 549-54.

[127] Simmons JW. Overview: high-nitrogen alloying of stainless steels. Mater Sci Eng A 1995; 207(2): 159-69.

[128] Sumita M, Hanawa T, Teoh SH. Development of nitrogen-containing nickel-free austenitic stainless steels for metallic biomaterials - Review. Mater Sci Eng C 2004; 24(6-8): 753-60.

[129] Fu R, Qiu L, Wang T, *et al.* Cryogenic deformation microstructures of 32Mn–7Cr–1Mo–0.3N austenitic steels. Mater Charact 2005; 55(4-5): 355-61.

[130] Dai D, Yang R, Chen K. Deformation Behavior of Fe-Mn-Cr-N Austenitic Steel. Mater Charact 1999; 42(1): 21-6.

[131] Ishizaka J, Orita K, Terao K. The influence of chemical composition and prestrain on impact toughness transition behaviour of 18%Mn-18%Cr-N austenitic steels. Tetsu-to-Hagane 1992; 78(12): 1846-53.

[132] Muller P, Solenthaler C, Uggowitzer PJ, *et al.* Brittle fracture in austenitic steels. Acta Metall Mater 1994; 42(7): 2211-7.

[133] Muller P. Solenthaler C. Speidel M.O.y, The intersection of deformation twins in austenitic steel. In: M.H.Yoo, M.Wutting (Eds), Proceedings of Twinning in Advanced Materials, Pittsburgh, PA, 1994, p.483.

[134] Muller P, Romanov AE. Between dislocation and disclination models for twins. Scr Metall Mater 1994; 31(12): 1657-62.

[135] Muller P, Solenthaler S. A proper model of a deformation twin for twin-intersection problems. Philos Mag Lett 1994; 69(3): 111-3.

[136] Muller P, Solenthaler S. The shape of a blocked deformation twin. Philos Mag Lett 1994; 69(4): 171-5.

[137] Kamat SV, Hirth JP, Muller P. The effect of stress on the shape of a blocked deformation twin. Philos Mag Lett 1996; 73(3): 669-80.

[138] Muller P. On the ductile to brittle transition in austenitic steel. Mater Sci Eng A 1997; 234-236: 94-7.

[139] Tomota T, Xia Y, Inoue K. Mechanism of low temperature brittle fracture in high nitrogen bearing austenitic steels. Acta Mater 1998; 46(5): 1577-87.

[140] Jackson EMLEM, Paton R. Influence of manganese on the properties of a vanadium-bearing ferritic stainless steel. ISIJ Int 1995; 35(5): 557-63.

[141] Pettersson RFAJ. Sensitization behaviour and corrosion resistance of austenitic stainless steels alloyed with nitrogen and manganese. ISIJ Int 1996; 36(7): 818-24.

[142] Sakamoto T, Abo H, Okazaki T, *et al.* Alloys for the Eighties, Climax Molybdenum Company, Ann Arbor, 1980, p.269.

[143] Zagorski K, Doraczynska A. Potentiodynmic polarization behaviour of 2 17Cr-13Ni-2.5Mo austenitic steels with different N contents. Corros Sci 1976; (6): 405-10.

[144] Lu YC, Ives MB, Clayton CR. Synergism of alloying elements and pitting corrosion resistance of stainless steels. Corros Sci 1993; 35(1-4): 89-96

[145] Ahmed A, Fathy A. Effect of slag properties and nitrogen addition on behaviour of alloying elements during ESR of AISI M41 tool steel. Ironmak Steelmak 2005; 59(26): 3311-14.

[146] Mudali UK, Ningshen S. Corrosion properties of nitrogen bearing stainless steels. In: High Nitrogen Steels and Stainless Steels-Manufacturing Properties and Applications (eds U.Kamachi Mudali and Baldev Raj), Alpha Science International Ltd., Pangbourne, UK, 2004; pp.133-81.

[147] Palit GC, Kain V, Gidayar HS. Electrochemical investigation of pitting corrosion in nitrogen-bearing type316LN stainless steel. Corrosion 1993; 49(12): 977-91.

[148] Baba H, Kodama T, Katada Y. Role of nitrogen on the corrosion behavior of austenitic stainless steels. Corrosion Sci 2002; 44(10): 2393-407.

[149] Olefjord I, Wergrelius L. The influence of nitrogen on the passivation of stainless steels. Corros Sci 1996; 38(7): 1203-20.

[150] Azuma S, Miyuki H, Kudo T. Effect of alloying nitrogen on crevice corrosion of austenitic stainless steels. ISIJ Int 1996; 36(7): 793-8.

[151] Yashiro H, Hirayasu D, Kumagai N. Effect of nitrogen alloying on the pitting of type 310 stainless steel. ISIJ Int 2002; 42(12): 1477-82.

[152] Vehovar L, Vehovar A, Hukovic MM, *et al.* Investigations into the stress corrosion cracking of stainless steel alloyed with nitrogen. Mater Corros 2002; 53(5): 316-27.

[153] Tsai WT, Reynders B, Stratmann M, *et al.* The effect of applied potential on the stress-corrosion cracking behavior of high-nitrogen steels. Corros Sci 1993; 34(10): 1647-56.

[154] Clayton CR, Rosenzweig L, Oversluizen M, *et al.* The influence of nitrogen on the passivity of 18-8(0.24%N) stainless steels. J Electrochem Soc 1986; 133(8): C303-C303.

[155] Ives MB, Lu YC, Luo JL. Cathodic reactions involved in metallic corrosion in chlorinated saline environment. Corros Sci 1991; 32(1): 91-102.

[156] Clayton CR, Martin KG. Evidence of Anodic Segregation of. Nitrogen in High-Nitrogen Stainless Steels and Its Influence on. Passivity. In: Proceedings of the Conference on High Nitrogen Steels HNS88. Foct J., Hendry A. Eds, Inst Metal Lond, 1989; pp.256-60.

[157] Newman RC, Ajjawi MAA. A microelectrode study of the nitride effect on pitting of stainless steels. Corros Sci 1982; 26(12): 1057-63.

[158] Baba H, Katada Y. Effect of nitrogen on crevice corrosion in austenitic stainless steel. Corros Sci 2006; 48(9): 2510-24.

[159] Yashiro H, Hirayasu D, Kumagai N. Effect of nitrogen alloying on the pitting of type 310 stainless steel. ISIJ Int 2002; 42(12): 1477-82.

[160] Misawa T, Tanabe H. *In-situ* Observation of Dynamic Reacting Species at Pit Precursors of Nitrogen-bearing Austenitic Stainless Steels. ISIJ Int 1996; 35(7): 787-92.

[161] Lu Y, Bandy R, Clayton CR, *et al.* Surface enrichment of nitrogen during passivation of a highly resistant stainless steel. J Electrochem Soc 1983; 130(8): 1774-6.

[162] Clayton CR, Halada GP, Kearns JR. Passivity of high-nitrogen stainless alloys – the role of oxyanions and salt films. Mater Sci Eng A 1995; 198(1-2): 135-44.

[163] Olefjord I, Brox B, Jevelstam U. Surface composition of stainless steels during anodic dissolution and passivation studied by ESCA. J Electrochem Soc 1985; 132(12): 2854-61.

[164] Garbke HJ. The Role of Nitrogen in the Corrosion of Iron and Steels. ISIJ Int 1996; 36(7): 777-86.

[165] Olsson COA. The influence of nitrogen and molybdenum on passive films formed on the austenoferritic stainless steel 2205 studied by AES and XPS. Corros Sci 1995; 37(3): 467-79.

[166] Vanini AS, Audouard JP, Marcus P. The role of nitrogen in the passivity of austenitic stainless steels. Corros Sci 1994; 36(110): 1825-34.

[167] Lothongkum G, Wongpanya P, Morito S, *et al.* Effect of nitrogen on corrosion behavior of 28Cr-7Ni duplex and microduplex stainless steels in air-saturated 3.5 wt% NaCl solution. Corros Sci 2006; 48(1): 137-53.

[168] Truman JE, Coleman MJ, Pirt KR. Note on influence of nitrogen content on resistance to pitting corrosion of stainless steels. Br Corros J 1977; 12(4): 236-8.

[169] Song S, Song W, Fang Z. The improvement of passivity by ion-implantation. Corros Sci 1990; 31: 395-400.

[170] Ningshen S, Mudali UK, Amarendra G, *et al.* Hydrogen effects on the passive film formation and pitting susceptibility of nitrogen containing type 316L stainless steels. Corros Sci 2006; 48(5): 1106-21.

[171] Lim AS, Atrens A. ESCA studies of nitrogen-containing stainless steels. Appl Phys A: Solid Surf 1990; 51(5): 411-18.

[172] Mudali UK, Ningshen S, Dayal DK. Study of passive films of nitrogen-bearing austenitic stainless steels using electrochemical impedance spectroscopy. Bull Electrochem 1999; 15(2): 74-8.

[173] Newman RC, Shahrabi T. The effect of alloyed nitrogen or dissolved nitrate ions on the anodic behavior of austenitic stainless steel in hydrochloric acid. Corros Sci 1987; 27(8): 827-38.

[174] Latha G, Rajeswari S. Versatility of superaustenitic stainless steels in marine applications. J Mate Eng Perform 1996; 5(5): 577-82.

[175] Mudali UK, Shankar P, Ningshen S, *et al.* On the pitting corrosion resistance of nitrogen alloyed cold worked austenitic stainless steels. Corros Sci 2002; 44(10): 2183-98.

[176] Mudali UK, Dayal RK. Influence of nitrogen addition on the crevice corrosion resistance of nitrogen-bearing austenitic stainless steels. J Mater Sci 2000; 35(7): 1799-803.

[177] Mudali UK, Dayal RK, Gnanamoorthy JB, *et al.* Relationship between Pitting and Intergranular Corrosion of Nitrogen-bearing Austenitic Stainless Steels. ISIJ Int 1996; 36(7): 799-806.

CHAPTER 8

The Various Precipitate Phases in Stainless Steels

Abstract: This chapter gives an overview of the various precipitate phases that can form in the different classes of stainless steels. Besides the phases that are well-studied, a newly discovered phase, called the J phase, is also introduced. The conditions under which the different phases will form, the problems they cause, the methods for alleviating these problems, and the new methods for the detection of the various phases are presented.

Keywords: Sigma phase, chi phase, martensite, carbide, J phase, G phase, Z phase, electron domain, S phase, spinodal decomposition, 475°C embrittlement, austenite, ferrite, intermetallic, sensitization.

1. INTRODUCTION

This chapter is on the different precipitate phases of stainless steels. Some of these phases may form in all classes of stainless steels, whereas some occur only in certain classes. For instance, while the chromium-rich ferrite α_{Cr} only forms in the ferritic and martensitic classes, the $M_{23}C_6$ carbide can form in all types of stainless steels. Nowadays, with the availability of sophisticated techniques, the detection and characterisation of the various precipitates has become more reliable and effective.

2. AUSTENITE AND FERRITE

When a precipitate forms, the understanding of its orientation relations with the matrix is of fundamental interest, because these relations have implications on formation sequence and the resulting property changes (*e.g.*, the degree of strengthening depends on the coherency of the precipitate with the matrix).

In the solution-annealed state, the microstructure of Fe-Cr alloys is either bcc or fcc.

Table 1: Orientation relationships for a microstructure comprising the the bcc and the fcc lattices.

Discoverers	Orientation Relationships
Headley-Brooks [6]	$(111)_{fcc}//(110)_{bcc},[\bar{1}10]_{fcc}//[\bar{1}10]_{bcc}$

Kurdjumov-Sachs [1]	$(111)_{fcc}//(110)_{bcc},[\bar{1}10]_{fcc}//[\bar{1}11]_{bcc},[11\bar{2}]_{fcc}//[\bar{1}1\bar{2}]_{bcc}$
	Deviations given by Qiu and Zhang [13] are:
	$[0\bar{1}1]_{fcc}//[1\,\overline{1.02}\,1.05]_{bcc}$ with $1.2°$ away from $[1\bar{1}1]_{bcc}$
	$(111)_{fcc}//(0.011\,0.96)_{bcc}$ with $1.1°$ away from $(011)_{bcc}$

Nishiyama-Wasserman [2]	$(111)_{fcc}//(110)_{bcc},[\bar{1}01]_{fcc}//[001]_{bcc},[\bar{1}2\bar{1}]_{fcc}//[\bar{1}10]_{bcc}$
Bain [14]	$(010)_{fcc}//(010)_{bcc},[001]_{fcc}//[101]_{bcc},[100]_{fcc}//[10\bar{1}]_{bcc},[101]_{fcc}//[100]_{bcc},$ $[\bar{1}01]_{fcc}//[001]_{bcc}$
Greninger-Troiano [4]	$(111)_{fcc}$ deviates from $(110)_{bcc}$ by about $1°$ $<112>_{fcc}$ deviates from $[1\bar{1}0]_{bcc}$ by about $2°$
Pitsch [5]	$(001)_{fcc}//(101)_{bcc},[\bar{1}10]_{fcc}//[\bar{1}1\bar{1}]_{bcc},[110]_{fcc}//[12\bar{1}]_{bcc}$

3. α'_{Cr}

As regards the orientation relations for a duplex microstructure comprising the bcc and the fcc lattices, many have been discovered. These are the Bain [1], the Kurdjumov-Sachs (K-S) [2], the Nishiyama-Wasserman (N-W) [2, 3], the Greninger-Troiano [4] and Pitsch [5] relationships. These orientation relations are listed in Table **1**. Very recently, Headley and Brooks [6], on studying the solidified structure of 304L and 309S, have found a new set of orientation relations, which is also listed in Table **1**. The K-S and N-W relations are the most frequently observed [7, 8], but there are usually small deviations [9-12].

Exposure of ferritic, martensitic and duplex stainless steels to temperatures below about 550℃ cause spinodal decomposition [15, 16] (as a matter of fact, radiation may also induce spinodal decomposition [17]). It is generally accepted that 475°C-embrittlement is associated with the formation of the α_{Cr} phase, which forms either through nucleation-and-growth or spinodal decomposition. Vitek *et al.* [18], however, have shown that both mechanisms can occur at the same time.

The occurrence of spinodal decomposition can be clearly revealed by using the TEM, as the formation of regions that are respectively enriched with Fe and Cr will cause a characteristic mottled contrast. However, Kobayashi *et al.* [19] have suggested recently that in high-N stainless steels, the compositional fluctuation of the interstitial element N may also cause the characteristic mottled contrast. Since N tends to segregate to the Cr-rich α_{Cr} phase [20], it may be difficult to separate the contributions to the mottled contrast by Cr and N. The other microstructural feature accompanying spinodal decomposition is the crisscrossing of dislocations (Fig. **1**) [21, 22].

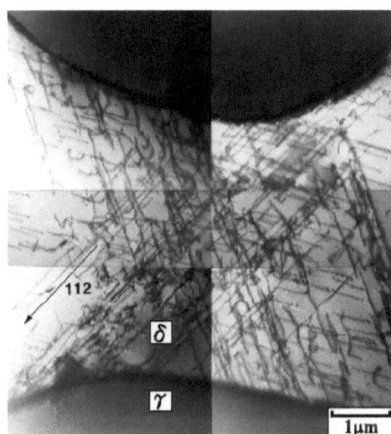

Figure 1: The criss-crossing of dislocations in the spinodally decomposed ferrite phase of a 2205 duplex stainless steel [Weng KL, Chen HR, Yang JR. The low-temperature aging embrittlement in a 2205 duplex stainless steel. MateR Sci Eng A 2004; 379(1-2): 119-32. With permission for reproduction from Elsevier].

Although both the ferritic class and the duplex class of stainless steels may undergo spinodal decomposition, Miller *et al.* [23] have shown that the spinodal decomposition of the ferrite phase of the duplex class proceeds more rapidly.

The austenite phase in duplex stainless steels was thought to be unaffected by spinodal decomposition for a long time, except for the precipitation of some carbides. A recent study by Smuk *et al.* [20], however, has shown that redistribution of substitutional elements does occur in the austenite phase during spinodal decomposition. May *et al.* [24] have also discovered a gradual enrichment of silicon in the surface oxide film of the austenite phase, which can aggravate the galvanic corrosion between the austenite and the ferrite phases.

Regarding the morphology of the spinodally decomposed ferrite, Shek *et al.* [25] have suggested that the α_{Fe} and the α_{Cr} domains have different morphologies. The α_{Cr} domains are thought to be embedded in the α_{Fe} matrix. Miller *et al.* [26], however, have propounded that the morphologies of α_{Fe} and α_{Cr} are temperature-dependent. Between 400 and 500°C, the two regions form an interconnected network, whereas isolated α_{Fe} and α_{Cr} regions will form at 550°C [26].

3.1. Methods of Detection of Spinodal Decomposition

Because of the severe degradation in mechanical and corrosion properties brought about by spinodal decomposition, a significant percentage of the research on spinodal decomposition has been devoted to its detection.

Acoustic methods that can measure the change in the longitudinal wave velocity may be used for this purpose, it has been observed that spinodal decomposition will cause a change in the lattice parameter of the ferrite phase [27], which in turn alters the elastic properties.

Changes in corrosion [28, 29] and electrical properties [30] are employable for the detection of spinodal decomposition, too. Both the DL-EPR test [28] and polarisation measurement [29] have been shown to be feasible tools for monitoring spinodal decomposition, as the formation of α_{Fe} and α_{Cr} will change the anodic and polarisation behaviours of the steel. As regards electrical property changes, the increase in the thermoelectric power (TEP) with the development of spinodal decomposition has been demonstrated to be a promising non-destructive monitoring technique [31].

Mechanically, both the small punch test [32] and the automated ball indentation method (ABI) [33] have been proved to be suitable methods for assessing the embrittlement caused by spinodal decomposition. The ABI method may also be used as an on-line monitoring tool [33].

Magnetic property changes have also been successfully utilised to monitor spinodal decomposition. The coercive force [34], the magnetic transition temperature [35, 36] (Curie temperature [37]) and the relative change in a.c. magnetic susceptibility [38] all change with the progress of spinodal decomposition.

4. THE G PHASE

The G phase (Fig. **2**) is a fcc silicide having and a nominal composition of $Ni_{16}Si_7Ti_6$ [39]. The substitutions of Ni and Ti by Cr, Fe, Mo, Mn, V, Hf, Ta, Zr and Nb are possible [39-41]. The composition of the G phase is thought to be temperature-dependent, but it is independent of time at a fixed temperature [42].

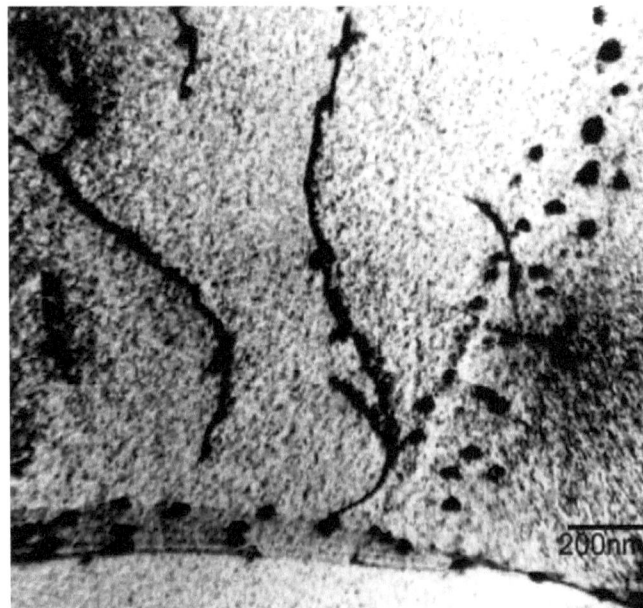

Figure 2: G phase in a duplex stainless steel aged at 475°C [May JE, De Sousa CAC, Kuri SE. Aspects of the anodic behaviour of duplex stainless steels aged for long periods at low temperatures. Corros Sci 2003; 45(7): 1395-403. With permission for reproduction from Elsevier]

The contribution of the G phase to hardening is a subject of some debate. Vitek *et al.* [18] have suggested that in the later stage of low temperature annealing, hardening will be mainly contributed by the G phase. Nonetheless, some researchers think that the hardening effect of the G phase is not significant [43, 44]. In order to isolate the embrittling effect of the G phase, Chung and Chopra [45] dissolved the α_{Cr} phase in spinodally decomposed steels by re-annealing them at a temperature that did not affect the G phase. It was found that both the microhardness and toughness were recovered in spite of the presence of the G phase.

Recently, a model describing the formation of the G phase has been formulated by Mateo *et al.* [46]. These authors have proposed that there exists a ferritic interdomain between the α_{Fe} and the α_{Cr} phases (Fig. **3**). The constituent elements of the G phase are continuously ejected into this interdomain by the α_{Fe} and the α_{Cr} phases. When the contents of the various elements reach the critical levels, the interdomain will transform into the G phase through slight atomic adjustments [46]. Experimentally, it has been observed by Leax *et al.* [42] that the G phase preferentially nucleates at the interface between the α_{Fe} and the α_{Cr} phases and that the formation of the G phase always lags behind spinodal decomposition. These observations corroborate with the model proposed by Matoe *et al.* [46].

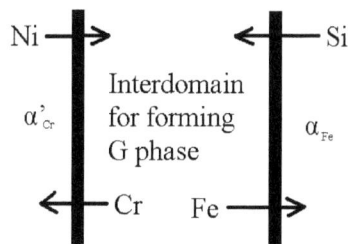

Figure 3: Schematic representation of the model for the formation of G phase as proposed by Mateo *et al.* [Figure drawn with reference to: Mateo A, Llanes L, Anglada, M, *et al.* Characterization of the intermetallic G-phase in an AISI 329 duplex stainless steel. J Mater Sci 1997; 32(17): 4533-40].

5. THE J PHASE AND E DOMAIN

Since stainless steels have been researched for such a long time, the different phases that may form in them are thought to be well understood. Additionally, the Fe-Cr phase diagram was considered well-established. However, a series of studies undertaken in about the last decade by Ustinovshikov *et al.* [47-50] on the Fe-rich portion the Fe-Cr system have revealed many interesting new features. A new phase has also been discovered by these authors [47-50]. This new phase occurs in the Fe-rich portion of the Fe-Cr system and was designated as the J phase by Ustinovshikov *et al.* [47-50]. The J phase is composed of alternating layers of pure Fe and pure Cr and it occurs in the Fe-rich side of the Fe-Cr system (Fig. **4**).

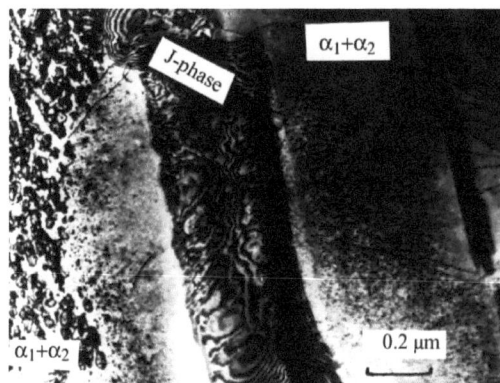

Figure 4: The J phase consists of alternating layers of pure Fe and pure Cr [Ustinovshikov Y, Pushkarev B. Alloys of the Fe–Cr system: the relations between phase transitions "order–disorder" and "ordering-separation". J Alloy Compd 2005; 389(1-2): 95-101. With permission for reproduction from Elsevier].

At high temperatures (up to the solidus temperature), Fe-Cr alloys are in general considered to be a homogeneous, fully disordered solid solution. Nevertheless, Ustinovshikov *et al.* [47] have found that even after solution-treatment at 1400℃, a disordered state will not exist when the Cr contents exceed 10%. In fact, as the temperature rises between 1200℃ and 1400℃, the microstructure becomes even more heterogeneous [49]. Only when the temperature reaches between 1400℃ and 1500℃ will a solid solution exist [49]. In this temperature range, the J phase can also form. Although the J phase is not an intermetallic compound, it was initially called the σ phase by Ustinovshikov *et al.* [47].

Besides the J phase, Cr-rich clusters have been found to exist in high-Cr samples in the as-water-quenched state, too. For Fe-Cr alloys containing 20~30% Cr, the as-water quenched state is a tweed-like structure [47]. In addition to the tweed-like structure and the J phase, Ustinovshikov *et al.* [48, 50] have also discovered the existence of the electronic domains (e domains) in the Fe-Cr system (Fig. **5**).

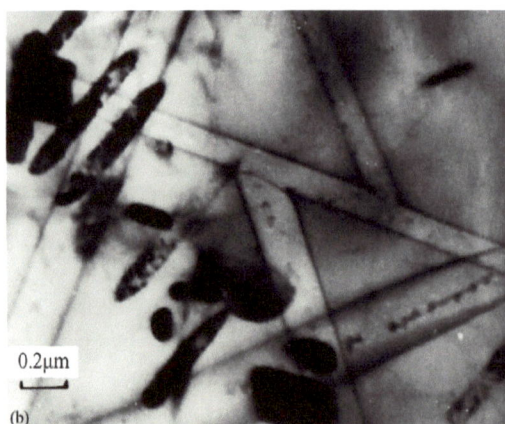

Figure 5: The e domains of a Fe–47Cr alloy that was solution-treated at 1200°C for 1 h, followed by water quenching and re-annealing at 700 8C for 1 h. The black precipitates are the J phase, the transparent laths are e domains [Ustinovshikov Y, Pushkarev B. Alloys of the Fe–Cr system: the relations between phase transitions "order–disorder" and "ordering-separation". J Alloy Compd 2005; 389(1-2): 95-101. With permission for reproduction from Elsevier].

Below about 550℃, the tweed-like structure and the J phase will decompose into ultrafine Fe-rich and Cr-rich phases [47]. For Fe-Cr alloys having high Cr contents, thermal annealing in the temperature regime where the intermetallic sigma phase precipitates will lead to the formation of the e domains [48, 50, 51]. In this sigma-precipitation temperature regime, the heterogeneous structure in high-Cr alloys that forms because of the chemical unmixing/separation process (with mixing energy Emix > 0) will transform into the sigma phase due to the tendency of ordering (with Emix < 0). The transformation of the electron configuration from Emix > 0 to Emix < 0 during the dissolution of the heterogeneous structure results in the formation of the e domains (Fig. **5**) [48, 50, 51]. The formation of the e domains precedes the precipitation of the sigma phase [48, 50].

6. THE R PHASE

Between 550°C and 700°C, the Mo-rich intermetallic R phase [52] may lead to rapid and significant loss of toughness [53]. In duplex stainless steels, the R phase may form both intergranularly and intragranularly in the ferrite phase. A recent TEM study by Redjaimia *et al.* [52] have revealed that the R phase contains a high number of planar faults that are parallel to the [1, 3, 9] and [11, 13, 23] lattice planes of the ferrite matrix.

There is disagreement on the precipitation sequence of the R phase and the intermetallic sigma phase. At 600°C, some authors believe that the formation of the sigma phase precedes the formation of the R phase [53, 54]. Kobayashi *et al.* [55] have also pointed out that the R phase appears only after prolonged ageing at 600°C. However, according to the precipitation-time-temperature (TTP) diagram constructed by Nilsson [56], the R phase forms earlier than the sigma phase.

7. THE R' PHASE

The as-solution-treated microstructure of maraging steels is fully martensitic. This initial martensitic microstructure is then subjected to thermal ageing between 400°C and 500°C to precipitate strengthening phases like Ni_3M (m = V, Mo, W and Nd) and the Laves phase. The duration of this strengthening ageing treatment must not be excessively long or else strength will decrease because of overageing.

The 1RK91 is a new maraging stainless steel (12Cr-9Ni-4Mo-(2Cu)) developed by the Sandvik company that is resistant to overageing [57-60]. Mo and Cu tend to encourage the formation of the R' phase. A nickel-rich, ordered phase, called the L phase, also forms in this new maraging stainless steel.

The resistance to overageing of 1RK91 may be attributed to a new, Mo-rich quasi-crystalline phase (the R' phase), which does not grow very much even after prolonged ageing at 475°C [57, 58]. The 1RK91 steel is very strong because quasicrystals are very hard. The R phase, Laves phase and the R' phase are thought to be intimately related because of they are closely related crystallographically [57].

8. THE S PHASE

A variety of surface treatment nitriding techniques [61], such as low-temperature plasma nitriding [62, 63], magnetron sputtering [64] and plasma immersion ion implantation [62, 65-67], have been utilised to enhance the surface properties of stainless steels.

If the temperature at which nitriding is conducted is above 500°C, then the precipitation of chromium nitrides will deteriorate corrosion resistance, albeit wear resistance will be improved. To get around this problem, nitriding may be performed below 450°C. When nitriding is conducted below this temperature, a thin layer saturated with nitrogen and carbon atoms will be obtained. This precipitate-free, interstitial-saturated layer possesses a good combination of mechanical, tribological and corrosion properties. This surface layer was first called the S phase by Ichii *et al.* [68]. Later researchers have variously named this surface layer as expanded or supersaturated austenite [62, 64, 65, 68], the *m* phase [69, 70], the ε ' phase [71], and the S' phase [72]. Here, this surface layer is termed the S phase, as this is the most frequently used in the literature.

The existence of the S phase enhances corrosion and wear properties substantially [73-75]. Magnetically, the S phase has been shown to possess some interesting features. It may either be paramagnetic [72, 76, 77] or ferromagnetic [72, 76, 78]. The ferromagnetism is believed to stem from lattice expansion caused by high stacking fault density [76] and N content [79].

8.1. Microstructure of the S Phase

X-ray diffraction characterisation of the S phase has revealed some interesting features. Firstly, the diffraction peaks pertaining to austenite shift to lower angles and broaden after nitriding [73, 75, 80], because of compressive residual stresses [81, 82] and the high density of stacking faults [82, 83]. Secondly, the change in the lattice spacing of the (200) crystallographic plane is bigger than those of the (111), (220), (311) and (222) planes [64, 67, 84-86].

One sticking point of the S phase is its microstructure. Some researchers [64, 67, 86] have suggested that the S phase to be face-centred-cubic with an expanded lattice parameter due to the presence of N. The more severe change in the lattice spacing of the (200) plane is attributed to the higher diffusion rate of nitrogen [86] and the higher residual stresses [64] associated with the (200) plane. However, researchers like Menthe and Rie [84] and Bacci *et al.* [87] have proposed that the S phase should possess a tetragonally distorted face-centred lattice. Marchev *et al.* [88], on the other hand, have stated that the S phase should have a body-centred-tetragonal lattice. These authors also pointed out that although a high number of stacking faults might cause the X-ray diffraction pattern of the S phase to deviate from the normal diffraction pattern of the fcc lattice, this is not very likely because such a deviation can also occur even when the N levels is not high

[88]. To make matters even more complicated, a triclinic lattice has been proposed by Fewell *et al.* [89]. Recent studies, nonetheless, have suggested that the S phase should have a face-centred lattice with a high number of stacking faults [85, 90].

Besides the austenitic class, the S phase can also form in the ferritic class [91-93], the precipitation-hardening class [94], the martensitic class [95], and the ferrite phase of the duplex class [96].Compared with the S phase of austenitic class, the S phase of the ferritic class is less stable against thermal annealing, because of the higher rate of diffusion of nitrogen in the bcc lattice [92].

8.2. The Carbon S Phase

In addition to nitriding, the S phase may also form in carburising [80, 93, 94-96] and nitrocarburising [93, 97, 98] (Fig. **6**). Several workers have undertaken to compare the nitrogen S phase and its carbon counterpart [80]. In general, the nitrogen S phase is harder and thinner than the carbon S phase.

Figure 6: The carbon S phase (the top layer) of a 316LVM austenitic stainless steel [Wu W, Li X, Chen J, *et al.* Design and characterization of an advanced duplex system based on carbon S-phase case and GiC coatings for 316LVM austenitic stainless steel. Surf Coat Technol 2009; 203(9): 1273-80. With permission for reproduction from Elsevier].

In nitrocarburising, the simultaneous introduction of nitrogen and carbon into austenitic stainless steels will result in two distinct layers of the S phase, with the nitrogen S phase being on top of the carbon S phase [93, 99]. Because N and C mutually enhance the diffusions of each other, a thick dual-layered structure of the S phase is obtainable in nitrocarburising [93]. The dual-layered structure may also be attained by carburising the steel first, and then nitriding it [100].

9. THE Z PHASE

The Z phase exists in different kinds of steels, especially in Nb-containing [101, 102] and N-containing austenitic stainless steels [101-105]. Even when the level of N is relatively low (like 0.06wt%), the Z phase may be able to form [106].

Although the lattice of the Z phase is in general thought to be the tetragonal, a cubic lattice has recently been discovered in the Cr(V,Nb)N and CrVN Z phases (but not for the CrNbN Z phase) in austenitic and martensitic creep-resistant steels containing 9-12wt% Cr [107]. In fact, the tetragonal and cubic lattices have been found to coexist in the same Z phase particle with the following orientation relations [107]:

$$(001)_{tetragonal} / /(001)_{cubic}$$
$$[010]_{tetragonal} / /[110]_{cubic}$$

10. THE χ (Chi) PHASE

The chi phase and the sigma phase are intermetallics that are found frequently in stainless steels. It is generally accepted that the chi phase forms before the sigma phase and it transforms to the sigma phase upon prolonged ageing [1108-110]. Fig. **7** shows the co-existence of the chi phase and the sigma phase. However, Redjaimia *et al.* [111] have proposed that the sigma phase forms ahead of the chi phase. Because carbon is believed to be soluble in the chi phase, it is also called $M_{18}C$ carbide.

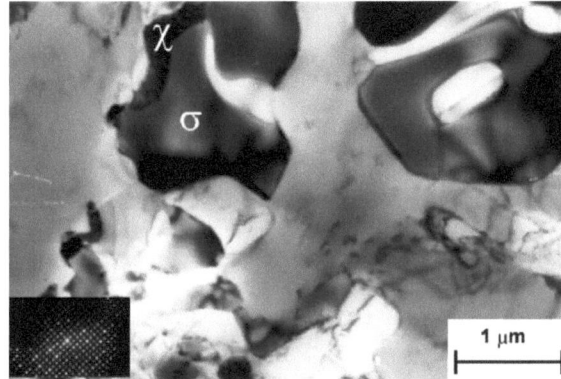

Figure 7: A TEM micrograph showing the intimate co-existence of the chi phase and the sigma phase [Michalska J, Sozanska M. Qualitative and quantitative analysis of σ and χ phases in 2205 duplex stainless steel. Mater Charact 2006; 56(4-5): 355-62. With permission for reproduction from Elsevier].

The chi phase is more enriched in Mo than is the sigma phase [108, 112] and so Mo tends to encourage chi formation. The replacement of Mo by W may improve stress corrosion cracking resistance [113] and W has a retarding effect on sigma phase formation.

In most cases, the precipitation of intermetallics leads to degradation of mechanical properties. However, several recent investigations have had differing views on the role of the chi phase and sigma phase. Chen *et al.* [108] have concluded that the impact toughness of duplex stainless steel 2205 will be greatly affected by the sigma phase. Nilsson *et al.* [114] have also found a close match between the curve representing an impact toughness of 27J and the curve representing the presence of 5% σ in a duplex stainless steel. But Calliari *et al.* [110] have argued that only a minute amount of the chi phase in 2205 is sufficient to degrade significantly the impact toughness and the degradation in impact toughness occurs well before the precipitation of the sigma phase.

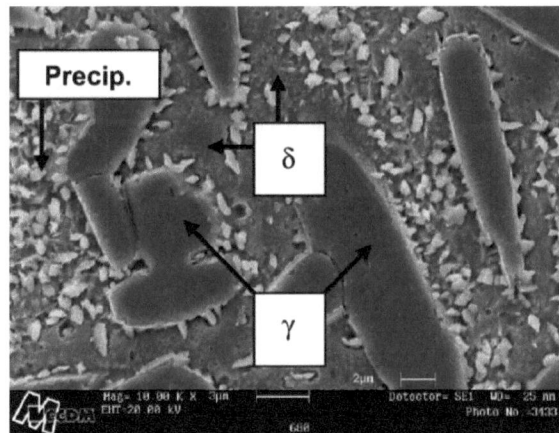

Figure 8: The unidentified precipitate (marked as Precip.) that forms in the ASTM A890 Gr 6A super duplex stainless steel [Martins M, Forti LRN. Effect of aging in impact properties of ASTM A890 grade 1C super duplex stainless steel. Mater Charact 2008; 59(2): 162-6. With permission for reproduction from Elsevier].

In this regard, the recent study by Martins *et al.* [115] is particularly noteworthy. These authors have shown that the impact property of the superduplex stainless steel, ASTM A890 Grade 1C, will get worse substantially after ageing between 580°C and 740°C. In this temperature range, the formation of the sigma phase in this steel was ruled out by the authors and they attributed the severe deterioration of impact property to an unidentified phase, which is rich in Fe and Cr, that forms intragranularly in the ferrite phase and at the ferrite/austenite boundaries (Fig. **8**). Energy-dispersive spectroscopy results conducted by the authors indicated that the Mo content of this mysterious phase was not particularly high. Insofar as the phases that may form in stainless steels are well documented, it is of fundamental interest to have a more in-depth characterisation of this superduplex stainless steel to identify the nature of this mysterious phase.

The chi phase is antiferromagnetic. Its Neel temperature is composition-dependent and may vary from 80K to 100K (Fig. **9**) [116, 117].

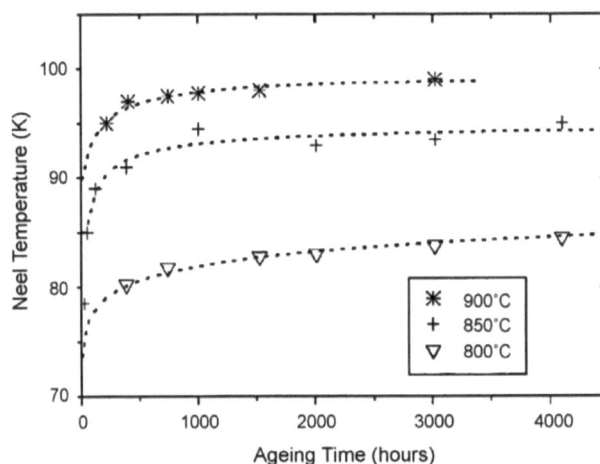

Figure 9: Variation of the Neel temperature of the chi phase with ageing time and ageing temperature [Shek CH, Lai MW, Lai JKL. Effect of thermal aging on the Neel temperature of a Fe-Cr-Ni-Mo alloy. J Mater Res 2002; 17(4): 879-883. With permission for reproduction from Materials Research Society].

11. THE Σ (SIGMA) PHASE

Because of its significant influence, the sigma phase has been researched for a very long time by a large number of workers. In spite of the very numerous past studies, research on this phase has continued apace.

11.1. Factors Affecting the Formation of the Sigma Phase

An early study by Sims [118] pointed out that the electron hole number (Nv) was a good predictor of sigma phase formation. When Nv exceeds 2.5, no sigma phase will form. The validity of using Nv was later verified in a 25Cr-20Ni austenitic stainless steels [119]. The diffusion of Cr is usually regarded as one of the controlling factors for sigma formation. Sasikala *et al.* [120], nevertheless, have pointed out that the diffusion of Mo could also be important. Vitek and David [121, 122], however, believe that the nucleation of the sigma phase is the limiting step for the ferrite-to-sigma transformation, while the enrichment of ferrite with Cr is not a sufficient condition.

An early work by Barcik [123] proposed that the grain size affects greatly the formation of the sigma phase in austenitic stainless steels. Recently, Schwind *et al.* [124] have concluded the grain shape will also influence sigma precipitation, although the effect of grain size is more important. Misorientations among grains also influence sigma formation. In duplex stainless steels, a higher crystallographic misorientation between the austenite phase and the ferrite phase facilitates sigma formation [125].

The precipitation sequence of the sigma phase has been re-examined recently and the suitability of using a single Johnson-Melh-Avrami (JMA) equation to describe sigma formation has been assessed by different authors. In studying the formation of the sigma phase in duplex stainless weld metals, Gill *et al.* [126, 127] have clearly identified the process to comprise two stages: the first stage is related to carbide precipitation and the second stage corresponds to sigma formation. Because the activation energies are quite different for the two stages, the JMA equations describing the two stages have different slopes. Nonetheless, Sasikala *et al.* [120] have stated that the difference between the activation energies of the two stages is not big. According to Sasikala *et al.* [120], the first stage of carbide formation involves grain boundary diffusion of Cr. Once carbide precipitation ends, the originally nearly continuous ferrite network will be broken up. Therefore, the second stage that corresponds to sigma formation will involve lattice diffusion of Mo. Since the activation energies for the boundary diffusion of Cr and lattice diffusion of Mo are almost the same, the JMA equations for both carbide precipitation and sigma formation should show no obvious discontinuity.

Although the $M_{23}C_6$ and the sigma phase are intimately related, it seems that no consensus has yet been reached regarding their interrelationship. The majority of researchers seem to consider $M_{23}C_6$ to be the precursor phase of sigma [128-130]. However, this view is disputed by some researchers [111]. For instance, Redjaimia *et al.* [111] have stated that in-situ transformation of $M_{23}C_6$ to the sigma phase is improbable, although they tend to be in contact. A more recent study on the precipitation sequence of a 316L steel below 500°C even concluded that the sigma phase forms before $M_{23}C_6$ [131]. This study has pointed out that the precursor phase of the sigma phase in this temperature regime is precipitates that are unidentifiable even by using high-resolution TEM. These unidentified precipitates are extremely small and they form on dislocations and grain boundaries.

11.2. Influence of the Sigma Phase on Properties

The role of the sigma phase on creep strength is a subject of a lot of debate. Some early literature reported negatively on the effects of the sigma phase on creep strength [132, 133]. Nonetheless, Li *et al.* [134] pointed out that if the sigma phase is finely dispersed intragranularly, then creep strength is enhanced. Recent studies by Shek *et al.* [135, 136] have found that when the distribution and morphology of the sigma phase are properly controlled through appropriate pre-treatments, the creep strength, ductility, yield strength and tensile strength of a 25Cr-8Ni duplex stainless steel will be improved.

It is believed that a fine and homogenous dispersion of ($\sigma + \gamma$) may increase tensile elongation [137]. Also, the sigma phase is believed to enhance the superplasticity of duplex stainless steels by retarding grain growth at high temperatures. However, in the study by Sagradi *et al.* [138], the sigma phase acts as cavity nucleation site during grain sliding and so is detrimental to the superplastic behaviour of duplex stainless steels.

While the effects of the sigma phase on mechanical properties of stainless steels are well researched, the properties of the sigma phase itself are less studied. A recent nanoindentation study by Ohmura *et al.* [139] has determined the hardness of the sigma phase to be about 17GPa at the peak load of 500 μN.

11.3. Detection of the Sigma Phase

The detection of sigma phase formation using magnetic methods is very convenient [36, 140]. In duplex stainless steels (and ferritic stainless steels), the only ferromagnetic phase at room temperature is ferrite. When the sigma phase forms, it forms at the expense of the ferrite. Therefore, sigma formation results in a decrease in the ferromagnetism of the steel. A recent study by Meszaros *et al.* [140] has found that the field strength corresponding to the maximum relative permeability is sensitive to sigma phase precipitation.

As regards the magnetic property of the sigma phase itself, it has been shown that the sigma phase of the Fe-Cr alloys (and a number of other alloys) undergoes a paramagnetic-to-ferromagnetic at cryogenic temperatures. The Curie temperature of the Fe-Cr sigma phase is about 60K [141]. For the sigma phase of the Fe-Cr-Ni-Mo alloys, the magnetic transition of its sigma phase occurs at about 77K [142, 143].

Recently, Lo [144, 145] has studied the effect of ageing temperature on the Curie temperature (Fig. **10**) of the sigma phase of the Fe-Cr-Ni-Mo duplex stainless steel. At a fixed temperature, the Curie temperature and the paramagnetic Curie temperature vary with ageing time at the early stage, but eventually, they level off to steady state values. The Curie temperature is also dependent on the ageing temperature, because the composition of the sigma phase is temperature-dependent [144, 145]. This cryogenic magnetic transition of the sigma phase has been utilised for hot-spot indication in a power plant [146].

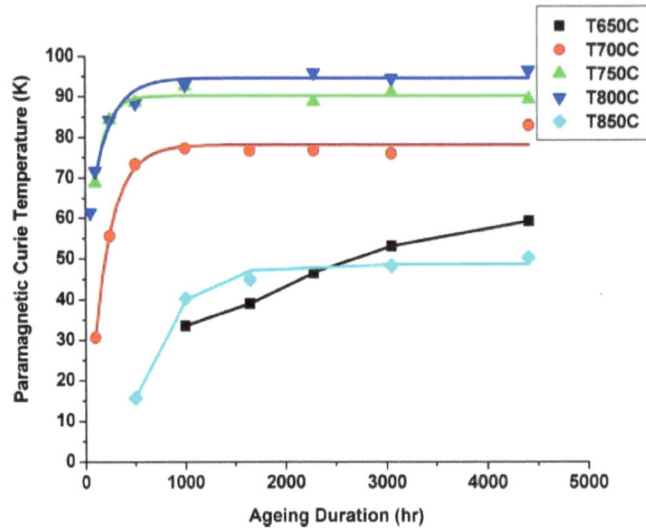

Figure 10: Variation of the Neel temperature of the chi phase with ageing time and ageing temperature [Lo KH, Shek CH, Lai JKL. Recent developments in stainless steels. Mater Sci Eng R Rep 2009; 65(4-6): 39-104. With permission for reproduction from Elsevier].

12. CARBIDES

There are many types of carbides in stainless steels. Among them, the $M_{23}C_6$ carbide is the most extensively studied. This section will cover the various of types of carbides and the many recent research findings on each of these carbides.

12.1. $M_{23}C_6$

This carbide has a fcc lattice. Many of the early studies on this carbide were on its TEM characterisation [247-149]. When carbides are observable under the TEM, they are already in a relatively well developed state. Studies on the early formation stages of $M_{23}C_6$ were not undertaken because of the unavailability of sophisticated equipment in the past. Lately, several studies have aimed at addressing this lack of information on the early formation stage of $M_{23}C_6$.

Below 500°C, Wasnik *et al.* [131], by using using DSC and high-resolution TEM, have discovered that the precipitation of $M_{23}C_6$ is preceded by the formation of unidentified, plate-like precipitates. These unidentified precipitates measure just a few atomic planes in size and form at dislocations and grain boundaries. In another recent study using positron annihilation spectroscopy, it has been revealed that the formation of carbides begins with the association of carbon atoms with vacancy clusters [150].

The interrelationship between $M_{23}C_6$ and grain boundary serration has been studied in detail. Hong and Nam [151] have discovered that grain boundary serration occurs before $M_{23}C_6$ precipitation and exerts a dramatic effect on its formation in austenitic stainless steels. Carbides that form on planar grain boundaries tend to be triangular in shape and are numerous, whereas those that form on serrated grain boundaries are usually planar or faceted, and are less numerous [151]. These authors have also observed that an increase in the misorientation favours the formation of triangular carbides [208]. The planar carbides is less damaging to creep-fatigue resistance than is their triangular counterparts [152-154]. Because planar carbide is less harmful, heat treatments that can generate them have been devised am these are documented by Kim *et al.* [153].

Addition of nitrogen has been known to retard the formation of $M_{23}C_6$ [155, 156]. Recent ab initio studies have shown that nitrogen is absolutely insoluble in $M_{23}C_6$ and this carbide is destabilised as soon as a small amount of C is replace by N [157].

12.2. MC

Besides nitrogen addition, the addition of stabilising elements, *i.e.*, strong carbide-forming alloying elements, such as V, Nb, Ti, Zr, Hf and Ta can also be used to reduce $M_{23}C_6$ precipitation [128]. When stabilising elements are added, MC carbides form in preference to $M_{23}C_6$ and so the severity of sensitisation is lowered. Compared with $M_{23}C_6$, MC carbides have a lower interfacial energy and so are less deleterious to mechanical properties like creep resistance [158].

In terms of the effectiveness in restricting $M_{23}C_6$ formation in ferritic stainless steels, Kuzucu *et al.* [159] have recently suggested the following order: V > Nb > Ti > Mo. For more effective stabilisation, two or more stabilising elements may be used at the same time [160, 161]. Other transition metals like Sc and Y have also been suggested to be useful for alleviating $M_{23}C_6$ formation in ferritic stainless steels [162]. It has to be stressed that although stabilising elements can mitigate the formation of $M_{23}C_6$, complete suppression of this carbide might be unachievable [156].

Besides lessening M23C6 formation, stabilising elements are also solid-solution strengtheners [163, 164]. The formation of fine carbide/nitride comprising stabilising elements [165, 166] and Nb-N clusters [167] also contribute to strengthening. Some of the stabilising elements (such as Nb) may form the Laves phase, *e.g.*, the Fe2Nb Laves phase [168]. When this Laves phase is fine, it will strengthen the steel substantially. However, this Laves phase is incoherent with the matrix and so it has a high proclivity to grow upon thermal annealing, thereby losing its strengthening capability [168]. While the use of stabilising elements may engender additional strengthening, it must be emphasised that excessive stabilisation may compromise ductility [160].

12.2.1. Effects MC on Properties

Because the extent of $M_{23}C_6$ formation is reduced, stabilisation usually results in improved corrosion resistance. However, the use of zirconium as a stabilising element has been found to compromise corrosion resistance [169]. In a cast 18-8 type stainless steel, Duan *et al.* [169] have found that zirconium ties up silicon, titanium, *etc.*, instead of carbon. As a consequence, pitting corrosion is degraded.

Quite often, the formation of fine MC carbides yields better mechanical properties. For example, in the new high-temperature cast stainless steel CF8C-Plus, which was developed by the Oak Ridge National Laboratories and Caterpillar®, strengthening is mainly achieved by the precipitation of nanosized NbC [170]. Fine MC carbides are also effective grain growth inhibitors [171].

While fine MC carbides are desirable, coarse MC carbides degrade properties [172-176]. Luckily, there are many simple methods for restricting the growth of MC carbides. The addition of nitrogen [177] and suitable amounts of plastic deformation [174] have been found to impede the growth of TiC. Keeping M and C below their solubility limit may also be helpful [155].

12.2.2. Stabilising Treatment Condition

It has been mentioned in the preceding paragraphs that the excessive use of stabilising elements may produce harmful effects. Therefore, it is highly desirable to add just the 'right amounts' of stabilising elements. For titanium, Steigerwald [178] has proposed the following expression $\%Ti = 5 \times (\%C + \%N)$. Pardo *et al.* [179], however, have found that this formula needs modification to reflect the higher importance of nitrogen. Since nitrogen has a very high affinity for titanium, titanium nitrides, instead of titanium carbides, tend to form. As a result, less titanium is available to tie up carbon, thereby expediting the formation of chromium carbides. Similar conclusion applies to other stabilising elements like niobium [180].

In order to reduce sensitisation between 600°C and 800°C, stabilised austenitic stainless steels are invariably subjected to a stabilising heat treatment at high temperatures to encourage the precipitation of MC carbide. Moura *et al.* [181] and Sousa [171] *et al.* have demonstrated that if the stabilising heat treatment temperature is too high, then the sensitisation-suppressing capability of stabilising elements will be reduced.

12.2.3. Relations Between MC/M(CN) and Other Phases

In the section on the sigma phase, it has been pointed out that the $M_{23}C_6$ carbide could be the precursor phase of the sigma phase. Hence, the retardation of $M_{23}C_6$ through stabilisation treatment may indirectly result in a reduction of sigma phase formation [128].

The main aim of the addition of stabilising elements, as mentioned, is to form MC, instead of $M_{23}C_6$. Nevertheless, the MC carbide may transform into the $M_{23}C_6$ carbide on long-term thermal ageing [182, 183], possibly because the stacking sequence in the NbC/austenite interface closely resembles the [111] γ planes, thereby aiding the formation of $M_{23}C_6$ [184].

Besides $M_{23}C_6$, the MC carbide/carbonitride may also transform into other phases. For instance, a recent study has suggested that the dissolution of Nb(CN) might encourage the formation the Z phase (Fig. 11) [180]. However, it has been proposed that NbC is more stable than the Z phase. In addition to the Z phase, Nb(CN) may also transform to the G phase upon prolonged ageing [186-189].

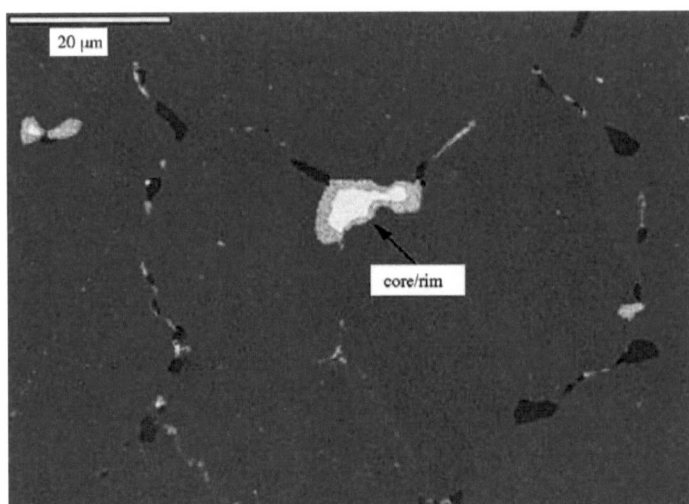

Figure 11: A precipitate that formed in an Nb-stabilised, type 347 austenitic stainless steel after ageing at 800 8C for 77,782h. The precipitate is composed of a rim and a core, with the former suggested to be the Z phase and the latter Nb(C, N) [Erneman J, Schwind M, Liu P, *et al.* Precipitation reactions caused by nitrogen uptake during service at high temperatures of a niobium stabilised austenitic steel. Acta Mater 2004; 52(14): 4337-50. With permission for reproduction from Elsevier].

12.3. M_6C, M_7C_3 and M_5C_2

The M_6C carbide has an fcc, diamond-type lattice. It is frequently found in Mo-containing austenitic stainless steels [169] and is believed to be able to accommodate some nitrogen [169, 190]. Superaustenitic stainless steels typically contain high levels of Mo and N, and both of these elements are believed to promote the formation of M_6C. Nb, especially when present in high levels, is also believed to encourage M_6C formation [155].

M_7C_3 is thought to possess a pseudo-hexagonal lattice [169] and is found only when the carbon levels are very high (*e.g.*, in carburisation [191, 192]). In austenitic stainless steels having a colossal supersaturation of carbon, the Hagg carbide M_5C_2 may also form.

12.4. Detection Methods for Carbides

Caballero *et al.* [193, 194] have demonstrated the feasibility of using Thermoelectric Power (TEP) measurement to monitor the dissolution and precipitation of carbides, because TEP is sensitive to the level of carbon dissolved in the iron matrix. In addition to carbon, TEP may also reflect the content of nitrogen that is dissolved in the matrix [195].

For martensitic stainless steels, the precipitation of $M_{23}C_6$ lowers the saturation magnetisation of the matrix, because of the removals of Cr and C from the matrix [196]. Such carbide-induced changes in magnetic properties are potentially utilisable for monitoring the tempering behaviour of martensitic stainless steels.

13. THE π PHASE

The π phase is a nitride that was first reported by Nilsson *et al.* [59]. This phase has been found to precipitate intragranularly in the ferrite phase of the 22Cr-3Mo-8Ni duplex stainless steel [58] and in a Mn-alloyed austenitic stainless steel [197].

14. THE η PHASE AND THE T PHASE

The η phase is rich in Si and it has a fcc lattice and the Fd3m structure. The Fd3m structure distinguishes the η phase from the G phase, the latter is of the Fm3m structure. Preliminary studies by Pettersson [197, 198] have concluded that the T phase possesses an orthorhombic structure.

15. THE τ PHASE

The τ phase is an intermetallic and may form upon thermal ageing between 550°C and 650°C. The τ phase possesses a needle-like morphology and an orthorhombic microstructure [199]. It forms in the ferrite phase and contains a lot of stacking faults. The orientation relations between the τ phase and the ferrite matrix are as follows:

$$(\bar{1}10)_\delta \, / /(100)_\tau$$
$$<001>_\delta \, // <001>_\tau$$
$$<110>_\delta \, // <010>_\tau$$

REFERENCES

[1]   Kurdjumov G, Sachs G. Uber den Mechanismus der Stahlhartung. Z Phys1964; 30: 325-43.

[2]   Nishiyama Z. Sci Rep Res Inst Tohoku Univ 1934; 23: 637-64.

[3]   Wasserman G. Arch. Eisnhuttenwes 1933; 16: 647-59.

[4]   Grehinger AB, Troiano AR. Trans.TMS-AIME 1949; 185: 590.

[5]   Pitsch W. The Martensite Transformation in thin foils of iron-nitrogen alloys. Philos Mag 1959; 4: 577-84.

[6]   Headley TJ, Brooks JA. A new bcc-fcc orientation relationship observed between ferrite and austenite in solidification structures of steels. Metall Mater Trans A 2002; 33(1): 5-15.

[7]   Xie G, Song M, Mitsuishi K, *et al.* Orientation of gamma to alpha transformation in Xe-implanted austenitic 304 stainless steel. J Nucl Mater 2000; 281(1): 80-3.

[8]   Lee KM, Cho HS, Choi DC. Effect of isothermal treatment of SAF 2205 duplex stainless steel on migration of delta/gamma interface boundary and growth of austenite. J Alloy Compd 1999; 285(1-2): 156-61.

[9]   Shek CH, Dong C, Lai JKL, *et al.* Early-stage Widmanstatten growth of the gamma phase in a duplex steel. Metall Mater Trans A 2000; 31(1): 15-9.

[10] Ameyama K, Weatherly GC, Aust KT. A study of grain-boundary nucleated Widmansttatan precipitates in a 2-phase stainless steel. Acta Metall Mater 1992; 40(8): 1835-46.

[11] Qiu D, Zhang WZ. A TEM study of the crystallography of austenite precipitates in a duplex stainless steel. Acta Mater 2007; 55(20): 6754-64.

[12] Jiao H, Aindow M, Pond RC. Precipitate orientation relationships and interfacial structures in duplex stainless steel Zeron-100. Philos Mag 2003; 83(16): 1867-87.

[13] Qiu D, Zhang WZ. A TEM study of the crystallography of austenite precipitates in a duplex stainless steel. Acta Mater 2007; 55: 6754-64.

[14] Bain E. The nature of martensite. Trans AIME 70; 1924: 25-46.

[15] DeNys T, Gielen PM. Spinodal decomposition in Fe-Cr system. Metall Trans 1971; 2(5): 1423-37.

[16] Nichol TJ, Datta A, Aggen G. Embrittlement of ferritic stainless steels. Metall Trans A 1980; 11(4): 573-85.

[17]   Miller MK, Stoller RE, Russell KF. Effect of neutron-irradiation on the spinodal decomposition of Fe-32% Cr model alloy. J Nucl Mater 1996; 230(3): 219-25.

[18]   Vitek JM, David SA, Alexander DJ, *et al.* Low temperature aging behavior of type-308 stainless steel weld metal. Acta Metall Mater 1991; 39(4): 503-16.

[19]   Kobayashi S, Nakai K, Ohmori Y. Decomposition Processes of δ-Ferrite during Continuous Heating in a 25Cr–7Ni–0.14N Stainless Steel. ISIJ Int 2000; 40(8): 802-8.

[20]   Smuk O, Jagodzinski Yu, Tarasenko O, *et al.* Internal friction study of decomposition kinetics of SAF 2507 type duplex stainless steel. Scr Mater 1999; 40(3): 321-6.

[21]   Weng KL, Chen HR, Yang JR. The low-temperature aging embrittlement in a 2205 duplex stainless steel. Mater Sci Eng A 2004; 379(1-2): 119-32.

[22]   Brenner SS, Miller MK, Soffa WA. Spinodal decomposition of Fe-32at%Cr at 470C. Scr Metall 1982; 16(7): 831-6.

[23]   Miller MK, Hyde JM, Cerezo A, *et al.* Comparison of low-temperature decomposition in Fe-Cr and duplex stainless steels. Appl Surf Sci 1995; 87/88(1-4): 323-8.

[24]   May JE, Sousa CAC, Kuri SE. Aspects of the anodic behaviour of duplex stainless steels aged for long periods at low temperatures. Corros Sci 2003; 45(7): 1395-403.

[25]   Shek CH, Shao YZ, Wong KW, *et al.* Spatial fractal characteristic of spinodal decomposition in Fe-Cr-Ni duplex stainless steel. Scr Mater 1997; 37(4): 529-33.

[26]   Miller MK, Anderson IM, Bentley J, *et al.* Phase separation in the Fe-Cr-Ni system. Appl Surf Sci 1996; 94/95: 391-7.

[27]   Tane M, Ichitsubo T, Ogi H, *et al.* Elastic property of aged duplex stainless steel. Scr Mater 2003; 48(3): 229-34.

[28]   Park CJ, Kwon HS. Effects of aging at 475°C on corrosion properties of tungsten-containing duplex stainless steels. Corros Sci 2002; 44(12): 2817-830.

[29]   Yi YS, Shoji T. Detection and evaluation of material degradation of thermally aged duplex stainless steels: Electrochemical polarization test and AFM surface analysis. J Nucl Mater 1996; 231(1-2): 20-8.

[30]   Maeda N, Goto T, Kamimura T, *et al.* Changes in electromagnetic properties during thermal aging of duplex stainless steel. Int J Press Vessel Pip 1997; 71(1): 7-12.

[31]   Kawaguchi Y, Yamanaka S. Mechanism of the change in thermoelectric power of cast duplex stainless steel due to thermal aging. J Alloy Compd 2002; 336(1-2): 301-14.

[32]   Cheon JS, Kim IS. Evaluation of thermal aging embrittlement in CF8 duplex stainless steel by small punch test. J Nucl Mater 2000; 278(1): 96-103.

[33]   Mathew MD, Lietzan LM, Murty KL, *et al.* Low temperature aging embrittlement of CF-8 stainless steel. Mater Sci Eng A 1996; 269(1-2): 186-92.

[34]   Pedrosa PDS, Teodosio JR, Tavares SSM, *et al.* Magnetic and mechanical hardening of Fe-based alloys. J Alloy Compd 2001; 329(1-2): L14-7.

[35]   Tavares SSM, Pedrosa PDS, Teodosio JR, *et al.* Magnetic properties of the UNS S39205 duplex stainless steel. J Alloy Compd 2003; 351(1-2): 283-8.

[36]   Tavares SSM, Silva MR, Neto JM. Magnetic property changes during embrittlement of a duplex stainless steel. J Alloy Compd 2000; 313: 168-73.

[37]   Silva MR, Tavares SSM, Fruchart D, *et al.* The use of thermomagnetic analysis for detection quantification of 475 degrees C embrittlement of duplex stainless steels. J Magn Magn Mater 2001; 226-30: 1103-5.

[38]   Shek CH. Growth of α' clusters and associated changes in magnetic properties in a duplex stainless steel. Phys Status Solid A 2001; 186(1): R7-9.

[39]   Miller MK, Bentley J. APFIM and AEM investigation of CF8 and CF8M primary coolant pipe steel. Mater Sci Technol 1990; 6(3): 285-92.

[40]   Spiegel FX, Bardos D, Beck PA. Ternary G and L filicides and Germanides of Transition Elements. Trans AIME 1963; 227: 575-9.

[41]   Vitek JM. G phase formation in aged type-308 stainless steel. Metall Trans A 1987; 18(10): 154-6.

[42]   Leax TR, Brenner SS, Spitznagel JA. Atom probe examination of thermally aged CF8M cast stainless steel. Metall Trans A 1992; 23(10): 2725-36.

[43]   Gonzalez JJ, Solana F, Sanchez L, *et al.* Low-temperature aging kinetics in cast duplex stainless steels: Experimental characterization. J Test Eval 1997; 25(2): 154-62.

[44]   Brown JE, Smith GDW. Atom probe studies of spinodal processes in duplex stainless steels and single phase and dual phase Fe-Cr-Ni alloys. Surf Sci 1991; 246(1-3): 285-91.

[45]   Chung HM, Chopra OK. Kinetics and mechanism of thermal aging embrittlement of duplex stainless steels. In: Proceedings of the 3rd International Symposium on Environmental Degradation of Materials in Nuclear Power Systems-Water Reactors, (Theus G.J., Weeks J.R., Eds), The Metal Soc, Warrendale, PA, 1988; pp.359.

[46]   Mateo A, Llanes L, Anglada,et M, *et al.* Characterization of the intermetallic G-phase in an AISI 329 duplex stainless steel. J Mater Sci 1997; 32(17): 4533-40.

[47]   Ustinovshikov Y, Shirobokova M, Pushkarev B. A structural study of the Fe-Cr system alloys. Acta Mater 1996; 44(12): 5021-32.

[48]   Ustinovshikov Y, Oushkarev B. Alloys of the Fe-Cr system: the relations between phase transitions "order-disorder" and "ordering-separation". J Alloy Compd 2005; 389(1-2): 95-101.

[49]   Ustinovshikov Y, Oushkarev B. Morphology of Fe-Cr alloys. Mater Sci Eng A 1998; 241(1-2): 159-68.

[50]   Ustinovshikov Y, Oushkarev B, Igumnov I. Fe-rich portion of the Fe-Cr phase diagram: electron microscopy study. J Mater Sci 2002; 37(10): 2031-42.

[51]   Ustinovshikov Y, Pushkarev B, Sapegina J. Conditions of existence of a disordered solid solution having chemical interactions between the atomic species. J Alloy Compd 2005; 394(1-2): 200-6.

[52]   Redjaimia A, Morniroli JP, Donnadieu P, *et al.* Microstructural and analytical study of heavily faulted Frank-Kasper R-phase precipitates in the ferrite of a duplex stainless steel. J Mater Sci 2002; 37(19): 4079-91.

[53]   Shimoide Y, Cui J, Kang C, *et al.* Effect of R Phase Formation on the Impact Toughness of a 25%Cr-7%Ni-3%Mo Duplex Stainless Steel. ISIJ Int 1999; 39(2): 191-4.

[54]   Cui J, Park I, Kang C, *et al.* Degradation of impact toughness due to formation of R phase in high nitrogen 25Cr-7Ni-Mo duplex stainless steels. ISIJ Int 2001; 41(2): 192-5.

[55]   Kobayashi S, Nakai K, Ohmori Y. Isothermal decomposition of delta-ferrite in a 25Cr-7Ni-0.14N stainless steel. Acta Mater 2001; 49(11): 1891-902.

[56]   Nilsson JO. Super duplex stainless steels. Mater Sci Technol 1992; 8(8): 685-700.

[57]   Liu P, Stigenberg AH, Nilsson JO. Quasi-crystalline and crystalline precipitation during isothermal tempering in a 12Cr-9Ni-4Mo maragin stainless steel. Acta Metall Mater 1995; 43(7): 2881-90.

[58]   Hattestrand M, Nilsson JO, Stiller K, *et al.* Precipitation hardening in a 12%Cr-9%Ni-4%Mo-2%Cu stainless steel. Acta Mater 2004; 52(4): 1023-37.

[59]   Nilsson JO, Stigenberg AH, Liu P. Isothermal formation of quasi-crystalline precipitates and their effect on strength in a 12Cr-9Ni-4Mo maraging stainless steel. Metall Mater Trans A 1994; 25(10): 2225-33.

[60]   Sha W. Thermodynamic calculations for precipitation in maraging steels. Mater Sci Technol 2000; 16(11-12): 1434-6.

[61]   Mudali UK, Khatak HS, Raj B, *et al.* Surface alloying of nitrogen to improve corrosion resistance of steels and stainless steels. Mater Manuf Process 2004; 19(1): 61-73.

[62]   Samandi M, Shedden SA, Smith DI, *et al.* Microstructure, corrosion and tribological behavior of plasma immersion ion-implanted austenitic stainless steel. Surf Coat Technol 1993; 59(1-3): 261-6.

[63]   Gontijo LC, Machado R, Kuri SE, *et al.* Corrosion resistance of the layers formed on the surface of plasma-nitrided AISI 304L steel. Thin Solid Films 2006; 515(3): 1093-6.

[64]   Saker A, Leroy Ch, Michel M, *et al.* Properties of sputtered stainless steel nitrogen coatings and structure analogy with low-temperature plasma nitrided layers of austenitic steels. Mater Sci Eng A 1991; 140: 702-8.

[65]   Williamsons DL, Ozturk O, Glick S. Microstructure of ultrahigh dose nitrogen-implanted iron and stainless steel. Nucl Instrum Method Phys Res B 1991; 59-60; 731-41.

[66]   Wei R, Vajo JJ, Matossian JN, *et al.* A comparative study of beam ion implantation, plasma ion implantation and nitriding of AISI 304 stainless steel. Surf Coat Technol 1996; 83(1-3): 235-42.

[67]   Collins GA, Hutchings R, Short KT, *et al.* Nitriding of austenitic stainless steel by plasma immersion ion-implantation. Surf Coat Technol 1995; 74-75(1-3) 417-24.

[68]   Lei MK, Zhang ZL. Plasma source ion nitriding – a new low-temperature, low-pressure nitriding approach. J Vac Sci Technol A 1995; 13(6): 2986-90.

[69]   Marchev K, Hidalgo R, Landis M, *et al.* The metastable m phase layer on ion-nitrided austenitic stainless steels Part 2: crystal structure and observation of its two-directional orientational anisotropy. Surf Coat Technol 1999; 112(1-3): 67-70.

[70]   Meletis EI, Singh V, Jiang JC. On the single phase formed during low-temperature plasma nitriding of austenitic stainless steels. J Mater Sci Lett 2002; 21(15): 1171-4.

[71]   Angelini E, Burdese A, DeBenedetti B. Ion nitriding of austenitic stainless steels. Metall Sci Technol 1988; 6(2): 33-9.

[72]   Gontijo LC, Machado R, Miola EJ, *et al.* Study of the S phase formed on plasma-nitrided AISI 316L stainless steel. Mater Sci Eng A 2006; 431(1-2): 315-21.

[73]   Menthe E, Bulak A, Olfe J, *et al.* Improvement of the mechanical properties of austenitic stainless steel after plasma nitriding. Surf Coat Technol 2000; 133-34: 259-63.

[74]   Zhu XM, Lei MK. Fitting corrosion resistance of high nitrogen f.c.c. phase in plasma source ion nitrided austenitic stainless steel. Surf Coat Technol 2000; 131(1-3): 400-3.

[75]   Fossati A, Borgioli F, Galvanetto E, *et al.* Glow-discharge nitriding of AISI 316L austenitic stainless steel: influence of treatment time. Surf Coat Technol 2006; 200(11): 3511-7.

[76]   Blawert C, Mordike BL, Jiraskova Y, *et al.* Structure and composition of expanded austenite produced by nitrogen plasma immersion ion implantation of stainless steels X6CrNiTi1810 and X2CrNiMoN2253. Surf Coat Technol 1999; 116-119: 189-98.

[77]   Ozturk O, Williamson DL. Phase and composition depth distribution analyses of low-energy, high-flux N-implanted stainless steel. J Appl Phys 1995; 77(8): 3839-50.

[78]   Williamson DL, Davis JA, Wilbur PJ. Effect of austenitic stainless steel composition on low-energy, high-flux, nitrogen ion beam processing. Surf Coat Technol 1998; 103-104: 178-84.

[79]   Williamson DL, Davis JA, Wilbur PJ, *et al.* Relative roles of ion energy, ion flux, and sample temperature in low-energy nitrogen ion implantation of Fe-Cr-Ni stainless steel. Nucl Instrum Method Phys Res B 1997; 127-128: 930-4.

[80]   Thaiwatthana S, Li XY, Dong H, *et al.* Comparison studies on properties of nitrogen and carbon S phase on low temperature plasma alloyed AISI 316 stainless steel. Surf Eng 2002; 18(6): 433-7.

[81]   Mandl S, Rauschenbach B. Anisotropic strain in nitrided austenitic stainless steel. J Appl Phys 2000; 88(6): 3323-9.

[82]   Christiansen T, Somers MAJ. On the crystallographic structure of S-phase. Scr Mater 2004; 50(1): 35-7.

[83]   Sun Y, Li XY, Bell T. X-ray diffraction characterisation of low temperature plasma nitrided austenitic stainless steels. J Mater Sci 1999; 34(19): 4793-803.

[84]   Menthe E, Rie KT. Further investigation of the structure and properties of austenitic stainless steel after plasma nitriding. Surf Coat Technol 1999; 116-119: 199-204.

[85]   Borgioli F, Fossati A, Galvanetto E, *et al.* Glow discharge nitriding of AISI 316L austenitic stainless steel: Influence of treatment pressure. Surf Coat Technol 2006; 200(18-19): 5505-13.

[86]   Williamson DL, Ozturk O, Wei R, *et al.* Metastable phase formation and enhanced diffusion in fcc alloys under high-dose, high-flux nitrogen implantation at high and low ion energies. Surf Coat Technol 1994; 65(1-3): 15-23.

[87]   Bacci T, Borgioli F, Galvanetto E, *et al.* Glow-discharge nitriding of sintered stainless steels. Surf Coat Technol 2001; 139(2-3): 251-6.

[88]   Marchev K, Landis M, Vallerio R, *et al.* The m phase layer on ion nitrided austenitic stainless steel (III): an epitaxial relationship between the m phase and the gamma parent phase and a review of structural identifications of this phase. Surf Coat Technol 1999; 116-119: 184-8.

[89]   Fewell MP, Mitchell DRG, Priest JM, *et al.* The nature of expanded austenite. Surf Coat Technol (1-3): 300-6.

[90]   Xu XL, Wang L, Yu ZW, *et al.* Microstructural characterization of plasma nitrided austenitic stainless steel. Surf Coat Technol 2000; 132(2-3): 270-4.

[91]   Kliauga AM, Pohl M, Robert CC, *et al.* Phase transformations in a super ferritic stainless steel containing 28% Cr after nitrogen ion implantation. J Mater Sci 1999; 34(16): 4065-73.

[93]   Jiraskova Y, Blawert C, Schneeweiss O. Thermal stability of stainless steel surfaces nitrided by plasma immersion ion implantation. Phys Status Solid A 1999; 175(2): 537-48.

[93]   Blawert C, Mordike BL, Collins GA, *et al.* Characterisation of duplex layer structures produced by simultaneous implantation of nitrogen and carbon into austenitic stainless steel X5CrNi189. Surf Coat Technol 2000; 128-29: 219-25.

[94]   Cheng Z, Li CX, Dong H, *et al.* Design and characterisation of an advanced duplex system based on carbon S-phase case and GiC coatings for 316LVM austenitic stainless steel. Surf Coat Technol 2009; 203(9): 1273-80.

[95]   Li XY, Thaiwatthana S, Dong H, *et al.* Comparison studies on properties of nitrogen and carbon S phase on low temperature plasma alloyed AISI 316 stainless steel. Surf Eng 2002; 18(6): 433-7.

[96]   Sun Y, Li X, Bell T. Low temperature plasma carburising of austenitic stainless steels for improved wear and corrosion resistance. Surf Eng 1999; 15(1): 49-54.

[97]   Cheng Z, Li CX, Dong H, *et al.* Low temperature plasma nitrocarburisig of AISI 316 austenitic stainless steel. Surf Coat Technol 2005; 191(2-3): 195-200.

[98]   Nie X, Tsotsos C, Wilson A, *et al.* Characteristics of a plasma electrolytic nitrocarburising treatment for stainless steels. Surf Coat Technol 2001; 139(2-3): 135-42.

[99]   Sun Y. Enhancement in corrosion resistance of austenitic stainless steels by surface alloying with nitrogen and carbon. Mater Lett 2005; 59(27): 3410-3.

[100] Christiansen T, Somers MAJ. Low temperature gaseous nitriding and carburising of stainless steel. Surf Eng 2005; 21(5-6): 445-55.

[101] Giordani EJ, Jorge Jr. AM, Balancin O. Proportion of recovery and recrystallization during interpass times at high temperatures on a Nb- and N-bearing austenitic stainless steel biomaterial. Scr Mater 2006; 55(8): 734-6.

[102] Ritter AM, Henry MF. Phase transformations during aging of a nitrogen-strengthened austenitic stainless steel. Metall Trans A 1985; 16(10): 1759-71.

[103] Mataya MC, Perkins CA, Thompson SW, *et al.* Flow stress and microstructural evolution during hot working of alloy 22Cr-13Ni-5Mn-0.3N austenitic stainless steel. Metall Mater Trans A 1996; 27(5): 1251-66.

[104] Giordani EJ, Guimaraes VA, Pinto TB, *et al.* Effect of precipitates on the corrosion-fatigue crack initiation of ISO 5832-9 stainless steel biomaterial. Int J Fatigue 2004; 26(10): 1129-36.

[105] Ornhagen, C, Nilsson JO, Vannevik H. Characterization of a nitrogen-rich austenitic stainless steel used for osteosynthesis devices. J Biomed Mater Res 1996; 31(1): 97-103.

[106] Erneman J, Schwind M, Liu P, *et al.* Precipitation reactions caused by nitrogen uptake during service at high temperatures of a niobium stabilised austenitic stainless steel. Acta Mater 2004; 52(14): 4337-50.

[107] Danielsen HK, Hald J, Grumsen FB, *et al.* On the crystal structure of Z-phase Cr(V,Nb)N. Metall Mater Trans A 2006; 37(9): 2633-40.

[108] Chen TH, Weng KL, Yang JR. The effect of high-temperature exposure on the microstructural stability and toughness property in a 2205 duplex stainless steel. Mater Sci Eng A 2002; 338(1-2): 259-70.

[109] Dobranszky J, Szabo PJ, Berecz T, *et al.* Energy-dispersive spectroscopy and electron backscatter diffraction analysis of isothermally aged SAF 2507 type superduplex stainless steel. Spectrochim Acta Part B 2004; 59(10-11): 1781-8.

[110] Calliari I, Zanesco M, Ramous E. Influence of isothermal aging on secondary phases precipitation and toughness of a duplex stainless steel SAF 2205. J Mater Sci 2006; 41(22): 7643-9.

[111] Redjaimia A, Metauer G, Gantois M. Decomposition of delta ferrite in a Fe-22Cr-5Ni-3Mo-0.03C duplex stainless steel. In: Charles J and Bernhardsson S, Editors, Proceedings of Duplex Stainless Steels *91* Beaune, France (1991); pp. 119–126.

[112] Michalska J, Sozanska M. Qualitative and quantitative analysis of sigma and chi phases in 2205 duplex stainless steel. Mater Charact 2006; 56(4-5): 355-62.

[113] Kim KY, Zhang PQ, Ha TH, *et al.* Electrochemical and stress corrosion properties of duplex stainless steels modified with tungsten addition. Corrosion 1998; 54(11): 910-21.

[114] Nilsson JO, Kangas P, Karlsson T, *et al.* Mechanical properties, microstructural stability and kinetics of sigma-phase formation in 29Cr-6Ni-2Mo-0.38N superduplex stainless steel. Metall Mater Trans A 2000; 31(1): 35-45.

[115] Martins M, Rodrigues L, Forti N. Effect of aging on impact properties of ASTM A890 Grade 1C super duplex stainless steel. Mater Charact 2008; 59(2): 162-6.

[116] Shek CH, Lai MW, Lai JKL. Effect of thermal aging on the Neel temperature of a Fe-Cr-Ni-Mo alloy. J Mater Res 2002; 17(4): 879-83.

[117] Lai JKL, Shek CH, Shao YZ, *et al.* Magnetic properties of thermal-aged 316 stainless steel ands its precipitated phases. Mater Sci Eng A 2004; 379(1-2): 308-12.

[118] Sims CT. The Superalloys. John Wiley & Sons. New York. 1972.

[119] Yamane T, Suzuki K, Minamino Y. Sigma phase precipitation in 25Cr-20Ni austenitic stainless steels substituting cobalt for iron. J Mater Sci Lett 1985; 4(3): 296-8.

[120] Sasikala G, Ray SK, Mannan SL. Kinetics of transformation of delta ferrite during creep in a type 316(N) stainless steel weld metal. Mater Sci Eng A 2003; 359(1-2): 86-90.

[121] Vitek JM, David SA. The sigma phase transformation in austenitic stainless steels. Weld J 1986; 65(4): 106-11.

[122] David SA, Vitek JM, Alexander DJ. Embrittlement of austenitic stainless steel welds. J Nondestruct Eval 1996; 15(3-4): 129-36.

[123] Barcik J. Mechansim of sigma phase precipitation in Cr-Ni austenitic stainless steels. Mater Sci Technol 1988; 4(1): 5-15.

[124] Schwind M, Kallqvist J, Nilsson JO, *et al.* sigma-phase precipitation in stabilized austenitic stainless steels. Acta Mater 2000; 48(10): 2473-81.

[125] Sato YS, Kokawa H. Preferential precipitation site of sigma phase in duplex stainless steel weld metal. Scripta Mater 1999; 40(6): 659-63.

[126] Gill TPS, Sharkar V, G.Pujar M, *et al.* Effect of composition on the transformation of delta ferrite to sigma in type316 stainless steel weld metals. Scripta Metall Mater 1995; 32(10): 1595-600.

[127] Gill TPS, Shankar V, Vijayalakshmi M, *et al.* Influence of carbon on the transformation kinetics of delta ferrite in type316 stainless steel weld metals. Scr Metall 1992; 27(3): 313-8.

[128] Guan K, Xu X, Xu H, *et al.* Effect of aging at 700 degrees C on precipitation and toughness of AISI 321 and AISI 347 austenitic stainless steel welds. Nucl Eng Des 2005; 235(23): 2485-94.

[129] Liu F, Hwang TH, Nam SW. The effect of post weld heat treatment on the creep-fatigue behavior of gas tungsten arc welded 308L stainless steel. Mater Sci Eng A 2006; 427(1-2): 35-47.

[130] Chen TH, Yang JR. Effects of solution treatment and continuous cooling on sigma-phase precipitation in a 2205 duplex stainless steel. Mater Sci Eng A 2001; 311(1-2): 28-41.

[131] Wasnik DN, Dey GK, Kain V, *et al.* Precipitation stages in a 316L austenitic stainless steel. Scr Mater 2003; 49(2): 135-41.

[132] Lai JKL, Wickens A. Effect of intergranular particle size and spacing on creep ductility of type316 stainless steel. Acta Metall 1979; 13(12): 1197-8.

[133] McMahon Jr CJ. On the mechanisms of creep damage in type316 stainless steel. Scr Metall 1985; 19(6): 733-7.

[134] Li DJ, Gao Y, Tan JL, *et al.* Effect of delta phase on the creep properties of Cr25Ni20 stainless steel. Scr Metall 1989; 23(8): 1319-21.

[135] Shek CH, Li DJ, Wong KW, *et al.* Creep properties of aged duplex stainless steels containing sigma phase. Mater Sci Eng A 1999; 266(1-2): 30-6.

[136] Shek CH, Wong KW, Lai JKL, *et al.* Hot tensile properties of 25Cr-8Ni duplex stainless steel containing cellular ($\sigma+\gamma_2$) structure after various thermal treatments. Mater Sci Eng A 1997; 231(1-2): 42-7.

[137] Han YS, Hong SH. The effects of thermo-mechanical treatments on superplasticity of Fe-24Cr-7Ni-3Mo-0.14N duplex stainless steel. Scr Mater 1997; 36(5): 557-63.

[138] Sagradi M, Sagradi DP, Medrano RE. The effect of the microstructure on the superplasticity of a duplex stainless steel. Acta Mater 1998; 46(11): 3857-62.

[139] Ohmura T, Tsuzaki K, Sawada K, *et al.* Inhomogeneous nano-mechanical properties in the multi-phase microstructure of long-term aged type 316 stainless steel. J Mater Res 2006; 21(5): 1229-36.

[140] Meszaros I, Szabo PJ. Complex magnetic and microstructural investigation of duplex stainless steel. NDT & E Int 2005; 38(7): 517-21.

[141] Gich M, Shafranovsky EA, Roig A, *et al.* Aerosol nanoparticles in the Fe1-xCrx system: Room-temperature stabilization of the sigma phase and sigma ->alpha-phase transformation. J Appl Phys 2005; 98(2): Article Number: 024303-1.

[142] Lai JKL, Wong KW, Li DJ. Magnetic properties of Feroplug material at sub-ambient temperature. Acta Mater 1996; 44(2): 567-71.

[143] Wong KW. Ph.D Thesis. Ageing characteristics and microstructural studies of duplex stainless steels. City Univ Hong Kong, Hong Kong. 1996.

[144] Lo KH. Ph.D Thesis. Transformation and magnetic behaviour of duplex stainless steels. City Univ Hong Kong. Hong Kong. 2007.

[145] K.H.Lo, C.H.Shek and J.K.L.Lai. Recent developments in stainless steels. Mater Sci Eng R 2009; 65(4-6): 39-104.

[146] Lai JKL, Shek CH, Wong KW. A novel technique to detect hot spots in high temperature boilers. Sens Actuat A 2001; 95(10): 51-4.

[147] Southwick PD, Honeycombe RWK. Precipitation of $M_{23}C_6$ at austenite ferrite interfaces in duplex stainless steel. Metal Sci 1982; 16(10): 475-81.

[148] Howell PR, Bee JV, Honeycombe RWK. Crystallography of the austenite-ferrite-carbide transformation in Fe-Cr-C alloy. Metall Trans A 1979; 10(9): 1213-22.

[149] Ricks RA, Howell PR, Honeycombe RWK. Effect on Ni on the decomposition of austenite in Fe-Cu alloys. Metall Trans A 1979; 10(8): 1049-58.

[150] Terada M, Saiki M, Costa I, *et al.* Microstructure and intergranular corrosion of the austenitic stainless steel 1.4970. J Nucl Mater 2006; 358(1): 40-6.

[151] Hong HU, Nam SW. The occurrence of grain boundary serration and its effect on the M23C6 carbide characteristics in an AISI 316 stainless steel. Mater Sci Eng A 2002; 332(1-2): 255-61.

[152] Hong HU, Rho BS, Nam SW. Correlation of the M23C6 precipitation morphology with grain boundary characteristics in austenitic stainless steel. Mater Sci Eng A 2001; 318(1-2): 285-92.

[153] Kim KJ, Hong HU, Min KS, *et al.* Correlation between the carbide morphology and cavity nucleation in an austenitic stainless steels under creep-fatigue. Mater Sci Eng A 2004; 387-389: 531-5.

[154] Hong HU, Nam SW. Improvement of creep-fatigue life by the modification of carbide characteristics through grain boundary serration in an AISI 304 stainless steel. J Mater Sci 2003; 38(7): 1535-42.

[155] Sourmail T. Precipitation in creep resistant austenitic stainless steels. Mater Sci Technol 2007; 17(1): 1-14.

[156] Padilha AF, Rios PR. Decomposition of Austenite in Austenitic Stainless Steels. ISIJ Int 2002; 42(4): 325-37.

[157] Gavriljuk VG, Berns H. High Nitrogen Steels. Berlin. Springer-Verlag. 1999.

[158] Min KS, Nam SW. Correlation between characteristics of grain boundary carbides and creep-fatigue properties in AISI 321 stainless steel. J Nucl Mater 2003; 322(2-3): 91-7.

[159] Kuzucu V, Aksoy M, Korkut MH. The effect of strong carbide-forming elements such as Mo, Ti, V and Nb on the microstructure of ferritic stainless steel. J Mater Process Technol 1998; 82(1-3): 165-71.

[160] Dowling NJE, Kim H, Kim JN, *et al.* Corrosion and toughness of experimental and commercial super ferritic stainless steels. Corrosion 1999; 55(8): 743-55.

[161] Cavazos JL. Characterization of precipitates formed in a ferritic stainless steel stabilized with Zr and Ti additions. Mater Charact 2006; 56(2): 96-101.

[162] Ogwu AA, Davis TJ. Improving the sensitisation resistance of ferritic stainless steels. Scripta Mater 1997; 37(3): 259-63.

[163] Fujita N, Ohmura K, Kikuchi M, *et al.* Effect of Nb on high-temperature properties for ferritic stainless steel. Scripta Mater 1996; 35(6): 705-10.

[164] Fujita N, Ohmura K, Yamamoto A. Changes of microstructures and high temperature properties during high temperature service of Niobium added ferritic stainless steels. Mater Sci Eng A 2003; 351(1-2): 272-81.

[165] Pilloni G, Quadrini E, Spigarelli S. Interpretation of the role of forest dislocations and precipitates in high-temperature creep in a Nb-stabilised austenitic stainless steel. Mater Sci Eng A 2000; 279(1-2): 52-60.

[166] Tendo M, Tadokoro Y, Suetsugu K, *et al.* Effects of Nitrogen, Niobium and Molybdenum on Strengthening of Austenitic Stainless Steel Produced by Thermo-Mechanical Control Process. ISIJ Int 2001; 41(3): 262-7.

[167] Kallqvist J, Andren HO. Microanalysis of a stabilised austenitic stainless steel after long term ageing. Mater Sci Eng A 1999; 270(1): 27-32.

[168] Sim GM, Ahn JC, Hong SC, *et al.* Effect of Nb precipitate coarsening on the high temperature strength in Nb containing ferritic stainless steels. Mater Sci Eng A 2005; 396(1-2): 159-65.

[169] Duan H, Yan X, Wei B, *et al.* Influence of trace alloying elements on corrosive resistance of cast stainless steel. J Iron Steel Res Int 2005; 12(5): 52-7.

[170] Shingledecker JP, Maziasz PJ, Evans ND, *et al.* Creep behavior of a new cast austenitic alloy. Int J Press Vessel Pip 2007; 84(1-2): 21-8.

[171] Sousa RC, Filho JCC, Tanaka AA, *et al.* Effects of solution heat treatment on grain growth and degree of sensitization of AISI 321 austenitic stainless steel. J Mater Sci 2006; 41(8): 2381-6.

[172] Yoo O, Oh Y, Lee B, *et al.* The effect of the carbon and nitrogen contents on the fracture toughness of Type 347 austenitic stainless steels. Mater Sci Eng A 2005; 405(1-2): 147-57.

[173] Yoon J, Yoon E, Lee B. Correlation of chemistry, microstructure and ductile fracture behaviours of niobium-stabilized austenitic stainless steel at elevated temperature. Scr Mater 2007; 57(1): 25-8.

[174] Reytier MC, Allais L, Caes C, *et al.* Mechanisms of stress relief cracking in titanium stabilised austenitic stainless steel. J Nucl Mater 2003; 323(1): 123-37.

[175] Sandhya R, Rao KBS, Mannan SL. Creep-fatigue interaction behaviour of a 15Cr-15Ni, Ti modified austenitic stainless steel as a function of Ti/C ratio and microstructure. Mater Sci Eng A 2005; 392(1-2): 326-34.

[176] Lee BS, Oh YJ, Yoon JH, *et al.* J-R fracture properties of SA508-1a ferritic steels and SA312-TP347 austenitic steels for pressurized water reactor's (PWR) primary coolant piping. Nucl Eng Des 2000; 199(1-2): 113-23.

[177] Gustafson A. Coarsening of TiC in austenitic stainless steel - experiments and simulations in comparison. Mater Sci Eng A 2000; 287(1): 52-8.

[178] Steigerwald R. Metals Handbook. Vol.13, Metals Park (OH), ASM Int, 1990; p.123.

[179] Pardo A, Merino MC, Coy AE, *et al.* Influence of Ti, C and N concentration on the intergranular corrosion behaviour of AISI 316Ti and 321 stainless steels. Acta Mater 2007; 55(7): 2239-51.

[180] Erneman J, Schwind M, Liu P, *et al.* Precipitation reactions caused by nitrogen uptake during service at high temperatures of a niobium stabilised austenitic steel. Acta Mater 2004; 52(14): 4337-50.

[181] Moura V, Kina AY, Tavares SSM, *et al.* Influence of stabilization heat treatments on microstructure, hardness and intergranular corrosion resistance of the AISI 321 stainless steel. J Mater Sci 2008; 43(2): 536-40.

[182] Howell PR, Nilsson JO, Dunlop GL. Effect of creep deformation on stability of intergranular carbide dispersion in an austenitic stainless steel. J Mater Sci 1978; 13(9): 2022-8.

[183] Thorvaldsson T, Dunlop GL. Precipitation reactions in Ti-stabilized austenitic stainless steel. Metal Sci 1980; 14(11): 513-8.

[184] Sasmal B. Mechanism of the formation of $M_{23}C_6$ plates around undissolved NbC particles in a stabilized austenitic stainless steel. J Mater Sci 1997; 32(20): 5439-44.

[185] Ayer R, Klein CF, Marzinsky CN. Instabilities in stabilized austenitic stainless steels. Metall Tran A 1992; 23(9): 2455-67.

[186] Powell DJ, Pilkington R, Miller DA. The precipitation characteristics of 20%Cr25%NiNb stabilized stainless steel. Acta Metall 1988; 36(3): 713-24.

[187] Ecob RC, Lobb RC, Kohler VL. The formation of G phase in 20/25Nb stainless steel AGR fuel cladding alloy and its effect on creep properties. J Mater Sci 1987; 22(8): 2867-80.

[188] Almeida LH, Ribeiro AF, May IL. Microstructural characterization of modified 25Cr–35Ni centrifugally cast steel furnace tubes. Mater Charact 2003; 49(3): 219-29.

[189] Piekarski B. Effect of Nb and Ti additions on microstructure, and identification of precipitates in stabilized Ni-Cr cast austenitic steels. Mater Charact 2001; 47(3-4): 181-6.

[190] Minami Y, Kimura H, Tanimura M. In: New Developments in Stainless Steel Technology (ed R.Lula), Metals Park, OH, Am Soc Metals, 1985; pp.231.

[191] Cao Y, Ernst F, Michal GM. Colossal carbon supersaturation in austenitic stainless steels carburized at low temperature. Acta Mater 2003; 51(14): 4171-81.

[192] Ernst F, Cao Y, Michal GM. Carbides in low-temperature-carburized stainless steels. Acta Mater 2004; 52(6): 1469-77.

[193] Caballero FG, Capdevila C, Alvarez LF, *et al.* Thermoelectric power studies on a martensitic stainless steel. Scr Mater 2004; 50(7): 1061-6.

[194] Caballero FG, Junceda AG, Capdevila C, *et al.* Precipitation of M23C6 carbide: thermoelectric power measurements. Scr Mater 2005; 52(6): 501-5.

[195] Lasseigne AN, Olson DL, Kleebe HJ, *et al.* Microstructural assessment of nitrogen-strengthened austenitic stainless-steel welds using thermoelectric power. Metall Mater Trans A 2005; 36(11): 3031-9.

[196] Tavares SSM, Fruchart D, Miraglia S, *et al.* Magnetic properties of an AISI 420 martensitic stainless steel. J Alloy Compd 2007; 312(1-2): 307-14.

[197] Pettersson RFAJ. Phase transformations in a manganese-alloyed austenitic stainless steel. Scr Metall Mater 1994; 30(9): 1233-8.

[198] Pettersson RFAJ. Precipitation in a nitrogen-alloyed stainless steel at 850°C. Scr Metall Mater 1993; 28(11): 1399-403.

[199] Kolts J, Sridhar N, Zeller MV. In: Proceedings of the Symposium on Metallography and Corrosion, ASM-NACE-IMS, Calgary, 1983.

CHAPTER 9

Problems in Stainless Steels: Recent Advances and Discoveries

Abstract: Although possessing superior properties, stainless steels are still afflicted by a host of problems, especially when they are exposed to elevated temperatures. Problems covered in this chapter are: hydrogen embrittlement, sensitisation, metal dusting, pitting corrosion, microbiologically-induced corrosion, high-temperature oxidation, stress-corrosion cracking, fatigue, creep, and dynamic strain ageing. The emphases of this chapter are on the recent findings on the underlying mechanisms of these problems, the main factors affecting them, new methods for their alleviation/elimination and detection, and modelling.

Keywords: Hydrogen embrittlement, martensite, sensitisation, metal dusting, fatigue, creep, dynamic strain ageing, high-temperature oxidation, pitting, stress corrosion cracking, microbiologically-induced corrosion, corrosion.

1. HYDROGEN EMBRITTLEMENT (HE)

1.1. HE of Austenitic Stainless Steels

The activation enthalpy for hydrogen migration in austenitic stainless steels is higher than those of nickel and α-iron (0.52-0.57eV [1]). However, austenitic stainless steels still suffer from hydrogen embrittlement (HE) (for instance, in cathodic charging). Although the HE of austenitic stainless steels has been known for a long time, it seems that only until recently has the most probable underlying mechanism(s) of HE been discovered.

1.1.1. Mechanisms of HE of Austenitic Stainless Steels

Over the years, many mechanisms have been put forth to explain the HE of austenitic stainless steels. The representative ones are:

- Hydrogen-induced localised plasticity (HELP) [2-4].

- Hydrogen-induced decohesion [5, 6].

- Formation of H-induced ε martensite or pseudo-hydride [7-10].

- Lowering of stacking fault energy by hydrogen [11-15].

- The formation of bcc α ' martensite [16, 17].

- Trapping/accumulation of hydrogen at the interfaces between matrix and precipitates [18, 19].

- The transport of hydrogen to defects/precipitates/crack tips and the interactions between hydrogen and dislocations [20, 21].

Among these mechanisms, a lot of recent research has pointed to the importance of HELP. For nickel, Robertson and Birnbaum [22] asserted that the culprit causing HE is HELP. Later on, HELP was also blamed for causing HE in austenitic stainless steels by Rozenak *et al.* [23]. Lately, Gavriljuk *et al.* [24] and Nibur *et al.* [25] have confirmed the importance of HELP in the HE of austenitic stainless steels. According to Gavriljuk *et al.* [24], hydrogen atoms enhance the metallic character (*i.e.*, an increase in free electrons) of the interatomic bonds in austenitic stainless steels, which then enhance plasticity and toughness because dislocation mobility is increased. The hydrogen-enhanced dislocation mobility has been observed by Ferreira *et al.* [26] and Robertson [27]. In addition to austenitic stainless steels, HELP and the blistering associated with hydrogen pressure build-up have been utilised to explain the HE in the ferrite phase of duplex steels [12].

Besides pointing out the importance of HELP, Gavriljuk *et al.* [28] have also concluded that it is very unlikely for H-induced decohesion to cause HE in austenitic stainless steels. This is because the increase of

Joseph Ki Leuk Lai, Kin Ho Lo and Chan Hung Shek

the metallic character of the interatomic bonds after hydrogen-charging [28, 29] is at odds with the H-induced decohesion mechanism.

The formation of H-induced ε martensite, ε hydride (and the fcc hydride γ * [10], which is also denoted as γ_H by some authors [30]) is thought to cause HE because their habit plane coincides with the active $(111)\gamma$ planes on which brittle fracture occurs [31].

Nonetheless, in the studies by Gavriljuk *et al.* [28, 32], the correlation between the ε martensite and HE cannot be established. In these studies, these authors increased the amounts of the ε martensite in their samples through the addition of silicon. However, the resistance of the samples to HE was not reduced. In this regard, Ulmer and Altstetter [30] also concluded that the hydride γ * might not contribute to ductility loss.

As to the role of hydrogen in causing the formation of the ε martensite/ε hydride, different views have been held out by different workers. Some researchers argued that hydrogen could lower the stacking fault energy (SFE) [13, 15], leading to easier formation of the ε martensite, which would compromise ductility [10]. It has to be noted that the role of the ε martensite in causing HE has been disputed as just mentioned. Glowacka and Swiatnicki [13] have observed that the stacking fault density of the austenite phase will increase after H-charging, which implies that hydrogen lowers the SFE. Regarding the role of the SFE in HE, it is believed that the formation of the Lomer-Cottrell locks, which are associated with the decrease of the SFE by hydrogen, is responsible [11]. Also, some authors believe that since the SFE is decreased by hydrogen, the deformation mode of austenite will change from cross slip to planar slip, with an attendant ductility loss [12]. However, in the study conducted by Hardie and Zheng [14], it was found that the amount of the ε martensite in a 316 steel was directly proportional to the SFE. This seems to be in contradiction to the view that hydrogen lowers the SFE.

In addition to the afore-mentioned mechanisms, other mechanisms have been proposed to explain HE. It is well established that the ε martensite may transform to the bcc α ' martensite [33]. This α ' martensite has also been implicated in causing HE [16, 17, 34-39], as hydrogen may induce its cracking [40]. Nevertheless, it must be noted that α ' does not form in the 310 steel, yet it still suffers from H-induced embrittlement.

Trapping of H at the interfaces between the matrix and second phase particles has also been suspected of causing HE in austenitic stainless steels [18, 19]. In precipitation-hardening stainless steels, this mechanism is considered to be very important [41, 42].

Long range redistribution of hydrogen to defects and precipitates through mobile dislocation during deformation is regarded by some researchers as the main mechanism of HE [20, 21]. Experimental supports for this view are: **1**. at 77K, the diffusion of hydrogen is almost impossible and so HE does not occur, and **2**. HE decreases with increasing strain rate. However, the need for long range transport of H *via* dislocations has been disputed by Harvey *et al.* [43], because these authors observed embrittlement in impact test (very high strain rates) at 77K. In addition to the above-mentioned mechanism, the tensile stress induced by hydrogen entry into the steel is also thought to enhance HE [44].

1.2. Hydrogen Embrittlement of Ferritic and Duplex Stainless Steels

Ferritic-austenitic duplex stainless steels are vulnerable to HE, especially the ferrite phase [45-49]. This is because the diffusion rate of hydrogen in the bcc ferrite is much higher than that in the fcc austenite [50-53]. In a cathodically-charged duplex stainless steel, it has been observed that hydrogen promotes transgranular fracture of the ferrite phase, which then induces microcracks in the austenite phase [54].The deleterious effect of ferrite has also been confirmed in a recent study by Luppo *et al.* [12]. These authors have concluded that both the amount and morphology of the ferrite phase will greatly affect the susceptibility of the duplex stainless steel to HE.

For duplex stainless steels, Zakroczymski *et al.* [48] have established that the following two ratios are good H-induced embrittlement indexes:

HE index 1: (the time to failure in the presence of H)/(the time to failure in air).

HE index 2: (the reduction in area in the presence of H)/(the reduction in area in air).

2. SENSITISATION

The most accepted theory of sensitisation is the Cr-depletion theory [55]. According to this theory, if the fraction of Cr-deficient boundaries (Cr<11wt%) reaches the critical value, then the steel is embrittled [56]. In austenitic stainless steels, both $M_{23}C_6$ and the sigma phase may cause intergranular Cr depletion [57, 58]. In a recent study, Nagae [59] has also demonstrated that grain boundary Cr depletion may occur because of stress-enhanced diffusion through dislocations.

Although sensitisation almost always deteriorates the properties of stainless steels, Arioka *et al.* [60, 61] have recently shown that sensitised austenitic stainless steels may be more resistant to intergranular stress corrosion cracking. This is because the grain boundary carbides may hinder grain boundary sliding.

2.1. Factors Affecting Sensitisation

2.1.1. Steel Chemistry

The effective chromium content (Cr^{eff}) provides a convenient, quantitative way to predict the propensity of stainless steels to sensitisation [62, 63].

$$Cr^{eff} = Cr + 1.45Mo - 0.19Ni - 100C + 0.13Mn - 0.22Si - 0.51Al - 0.20Co$$
$$+ 0.01Cu + 0.61Ti + 0.34V - 0.22W + 9.2N$$

Cr^{eff} is indicative of the chromium activity at the interface between the carbide and the matrix [63]. A higher Cr^{eff} means less severe grain boundary Cr-depletion, which means it will take a longer time for sensitisation to occur [63].

One way to alleviate sensitisation is to reduce the carbon content [64]. Trillo and Murr [65] have established that there exists a threshold grain boundary free energy (and a corresponding threshold carbon content) for sensitisation to occur at a given temperature.

2.1.2. Grain Size Effect

As to the effect of grain size, it has been established that as the grain size decreases, the time required for desensitisation becomes shorter [66, 67]. When the grain sizes decrease to a certain value, the time difference between sensitisation and desensitisation will be negligible. This is because **1**. Cr diffusion becomes highly efficient due to the profusion of grain boundaries, and **2**. the distances for Cr to diffuse from grain interiors to the grain boundary regions are very small. Because of these two reasons, any Cr-depleted regions that occur as a result of carbide precipitation will be rapidly replenished with Cr atoms.

2.1.3. Effects of Deformation and Deformation-induced Microstructures

Singh *et al.* [68], by using a wide range of degrees of prior deformation, testing temperature and testing duration, have concluded that prior deformation tends to accelerate sensitisation. Nevertheless, these authors have found that the degree of sensitisation and the amount of cold work are not directly proportional. That is, the relationship between the degree of plastic deformation and sensitisation is not monotonic [68, 69]. When the degree of deformation is very high, the diffusion of Cr atoms are expedited and so desensitisation may occur pretty fast.

For stable austenitic stainless steels (such as the 316 steel), deformation-induced slip bands and their intersections are the prime sites for carbide precipitation [70]. In these stable steels, intragranular carbide precipitation reduces intergranular carbide precipitation by taking up some of the carbon [71, 72].

For metastable austenitic stainless steels (such as the 304 steel), transgranular precipitation of carbides is facilitated by deformation-induced martensite and martensite lath boundaries [73-75]. For thermally aged (at 670°C) 304 containing deformation-induced martensite, Alvarez *et al.* [76] and Trillo *et al.* [77] have found that it is the fine-grained regions created by pseudo-recrystallisation that promote rapid sensitisation-desensitisation and carbide precipitation. Advani *et al.* [78] have concluded that the deformation-induced martensite will transform to clustered zones comprising thermal lath martensite and very fine austenite upon thermal ageing. It is these clustered zones that are the potent sites for transgranular carbide formation (although these clustered sites are not necessarily required).

2.2. Methods of Evaluation/Detection for Sensitisation

The Huey test, the Strauss test, and the electrochemical potentiokinetic reactivation (EPR) method are quite often used to assess the susceptibility of stainless steels to carbide precipitation. Among these methods, the EPR method is more favoured because it is quantitative, non-destructive and more efficient compared with the Huey and Strauss tests [55, 79].

For the EPR tests, Kain *et al.* [80] have recently proposed that the actual covered grain boundary area of Cr-depleted regions be incorporated into the conventional EPR test parameters, as this will raise the test sensitivity. The double-loop EPR (DLEPR) test has been demonstrated by many researchers to be employable for categorising the various sensitised microstructures (ditch, dual and step) of austenitic stainless steels [81, 82].

In spite of its advantages, the EPR method has been pointed out by Shaikh *et al.* [79] to be inadequate for quantifying the degree of sensitisation. For instance, two compositionally different steels exhibiting the same degree of sensitisation could have different reactivation ratios in the DLEPR test. Because of these limitations, these authors have suggested that the eddy current test to be more suitable for quantitative non-destructive evaluation of sensitisation [79].

Alteration of the grain boundary composition due to solute segregation and depletion may bring about changes in magnetic properties. For this reason, magnetic non-destructive evaluation techniques have been used for the detection of sensitisation [83, 84]. However, the applicability of this magnetic detection technique depends on the stability of austenite. The criteria on steel composition for using the magnetic technique have been set out by Kamada *et al.* [83]. A very recent work by Huang *et al.* [85] has demonstrated the feasibility of using atomic force microscopy for monitoring sensitisation.

3. METAL DUSTING

Metal dusting is a high-temperature corrosion process that occurs between 400°C and 900°C. During metal dusting, metals are saturated with carbon and they subsequently disintegrate into metal particles and graphite. This corrosion process usually occurs when hydrocarbons is present or in a strong carburising atmosphere (carbon activity ac>>1). The exact mechanisms, which may act synergistically, of metal dusting are still not well understood and are material-dependent. For stainless steels, new metal dusting mechanisms have been propounded recently [86].

Broadly speaking, there are four mechanisms of metal dusting (denoted as Types I – IV here in accordance with the nomenclature suggested by Szakalos [86]). The Type I mechanism is mainly applicable to pure iron and low alloy steels. This mechanism has lately been improved by Grabke [87, 88]. In this mechanism, metal dusting is considered to be associated with the decomposition of metastable carbides (like cementite (Fe3C) and some Hagg carbide (Fe5C2)) into graphite and iron [89]. The Type II mechanism was mainly developed for nickel and Ni-based alloys [88, 90]. In this mechanism, graphite forms into and inside the carbon-saturated metal matrix. As a consequence, small fragments of the metal matrix are surrounded by graphite. Additionally, graphite is present along grain boundaries [91]. While carbon is implicated in the Types I and II mechanisms, Schmid *et al.* [92] have taken into account the effect of oxygen and proposed the Type III mechanism. In this mechanism, metal dusting is viewed as a combination of carbide formation and the subsequent selective oxidation of Cr-rich carbides. The end products of these processes are spinel oxide and the Fe3C particles, the latter will subsequently decompose according to the Type I mechanism. Very recently, Szakalos [86] has proposed the Type IV mechanism for ferritic stainless steels. In this mechanism, fragments of ferrite coming off the metal matrix act as a catalyst and dissolve carbon. Graphite then forms as a consequence. This fragmentation process goes on and the particles shrink in size. Eventually, the ferrite particles will be so small that solid carbon can only form on their surfaces, resulting in the formation of carbon nanotubes. Note that the ferrite particles are regarded as undergoing the Type II mechanism. Szakalos [86] considers the Type IV mechanism to be a secondary mechanism because it just involves further disintegration of the corrosion products (the ferrite fragments) generated by the other metal-dusting mechanisms. Here, it is worth mentioning that in the austenitic stainless steel 304, Lin and Tsai [93] have also found carbon filaments on the surface of this steel after the metal-dusting process (Fig. **1**).

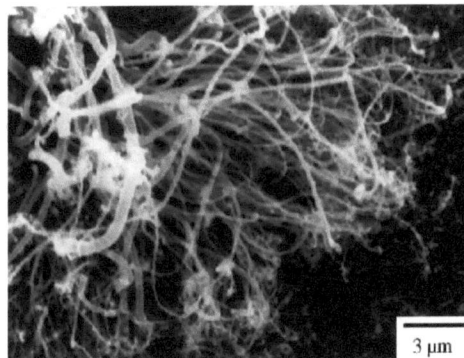

Figure 1: Nanosized carbon filaments that formed on an AISI304 that had been pre-carburised and subsequently exposed to CO/CO$_2$ mixed gases at 700°C for 30 h [Lin C, Tsai W. Nano-sized carbon filament formation during metal dusting of stainless steel. Mater Chem Phys 2003; 82(3): 929-36].

The applicability of these theories to stainless steels has been studied by several groups. Stevens *et al.* [94] have shown that the Type II mechanism for Ni-based alloys should be more relevant for Fe-Cr-Ni alloys. By exposing the 304L steel to a synthetic gas of low water vapour content (25%CO + 3%H2O + H2) at 650°C for 1000h, Szakalos *et al.* [95] have concluded that the type I mechanism would be unlikely to be operative in commercial austenitic stainless steels, whereas the Type II and III mechanisms will act simultaneously and additively. Szakalos [100] has modelled the metal dusting of ferritic stainless steels by using the Types I, II and IV mechanisms. The same author has modelled the metal dusting of austenitic stainless steels by using the Types II, III and IV mechanisms [86].

4. PITTING

4.1. Effects of Impurities

The most accepted view on the cause of pitting corrosion is impurities. Although a large number of studies have been devoted to this topic, controversies still exist. One of the controversies is on the role of manganese sulphide (MnS) in causing pitting corrosion.

4.1.1. Role of MnS

It is generally accepted that the presence of MnS in stainless steels is undesirable [96-98]. Even for the highly alloyed stainless steels, MnS is still believed to be able to compromise pitting resistance [99]. Although the harmful effect of MnS has been well recognised, the exact pitting mechanism(s) associated with this inclusion has not been unambiguously identified yet.

Some researchers believe that it is the dissolution of MnS and the attendant formation of corrosive dissolution products (like elemental sulphur [100, 101], thiosulphates [102], H2S [103], HS⁻ [104]) that are responsible for pit formation [105]. Since MnS does not passivate, its presence will disrupt the continuity of the surface passive film. So, pitting may begin in the MnS/matrix interfaces [106-109].

Williams and Zhu [110] have suggested that Cr and Fe will replace some of the Mn in the sulphide during solidification. Therefore, the sulphide is less stable and a Cr-depleted zone exists around it. Pits will form as a result [110]. The role of the Cr-depleted zone in causing pitting, however, has been a subject of debate and a series of recent studies on MnS has focused on the existence/non-existence of Cr-depleted regions in its immediate vicinity. Ryan *et al.* [111] have stated that Cr-depleted regions do exist around some of the MnS particles in their samples (but not all [112]) and these authors have proposed a mechanism on the basis of these Cr-depleted regions. Using the focused ion beam and secondary ion mass spectroscopy (FIB/SIMIS) (which were also the techniques used by Ryan *et al.* [111]), Meng *et al.* [113, 114], however, have been unable to detect any Cr-depleted zones (even in the nanometre scale). Schmuki *et al.* [115] used the scanning Auger microscopy to examine a large number of MnS particles with a lateral resolution of about 20nm, but they also failed to find any Cr-depleted regions [115].

4.1.1.1. Methods for Suppressing the Formation of MnS

The addition of rare earth elements in suitable amounts may suppress the formation of MnS considerably [116-118]. High temperature annealing (at 1000°C) has also been found to alleviate the detrimental effect of MnS, as this annealing may transform MnS into $(Cr,Mn)_2(O,S)_3$ [119]. The addition of proper amounts of titanium and carbon to ferritic stainless steels have been found to suppress the formation of MnS by producing Ti4C2S2 instead during solidification.

4.2. Effects of Deformation

Although deformation bands [120] and the interface between austenite and deformation-induced martensite [121] are preferred sites for pit formation, the effect of deformation on corrosion is very complex. In fact, the severity of corrosion attack [122, 123] and pit initiation frequency [124] may not be directly proportional to the degree of cold work [122, 123]. The time for pit nucleation has also been reported to be affected or unaffected by cold working. For example, while some workers reported that cold working did not seem to affect the time for pit nucleation [125], others have reported a shorter nucleation time [126]. Different results have also been reported regarding the relation between pit density and the degree of cold work [127, 128].

4.3. PREN and MARC

The pitting resistance equivalent number (PREN) is frequently used to indicate the susceptibility to pitting corrosion. For example, Merello *et al.* [129] have found that the pitting potential of a low Ni, high Mn-N DSS correlates exponentially with its PREN16. As regards the expressions of PREN, there are several versions and the most widely used ones are the PREN16 and the PREN30 [130].

$$PREN16 = \%Cr + 3.3\%Mo + 16\%N$$

$$PREN30 = \%Cr + 3.3\%Mo + 30\%N$$

For N, while a lot of researchers prefer the PREN16 [131], some opt for the PREN30 in order to underscore the importance of nitrogen in retarding pitting. Besides these values, 12, 20, 25 and 32 have also been used as the coefficients of N by various authors [132, 133]. Insofar as N and Mo may act synergistically [130, 134-136], Pettersson [130] has proposed the following new PREN formula, which also takes into consideration the decreasing solubility of N when its content is high:

$$PRE = \%Cr + 3.3\%Mo + 51\%N + 6\%Mo\%N - 1.6(\%N)^2$$

In addition to the elements in the PREN expressions, other elements will also affect pitting resistance. Cu is thought to improve pitting resistance [137-139] and it may act synergistically and beneficially with Sn [139]. Nb, when it is incorporated into the surface protective oxide, also improves pitting resistance [140]. Si, while being beneficial to corrosion resistance of austenitic stainless steels, may degrade pitting resistance if it is present in sufficient amounts to form ferrite [141]. Lately, Rondelli *et al.* [142], Hanninen

[143], and Park and Kwon [144] have incorporated the effects of elements like P, S, Mn and W into the PREN expressions. It has to be noted that PREN may give misleading results if these elements are not taken into account [131].

When using the PREN for duplex stainless steels, the bulk composition must not be used [145]. The compositions of the ferrite phase and the austenite phase must be used individually in the PREN expressions [145-148]. The weaker of these two phases will cause pitting.

The Measure of Alloying for Resistance for Corrosion (MARC) is another parameter that has been found to be indicative of the susceptibility to pitting corrosion and crevice corrosion.

4.4. Methods of Assessment and Non-destructive Evaluation for Pitting

The cyclic polarisation test may be used to assess the vulnerability of stainless steels to localised corrosion attack and Ravichandran *et al.* [149] have been able to correlate the area of the hysteresis loop in this test to pitting susceptibility.

The use of acoustic emission (AE) for monitoring of pitting corrosion has got a long history [150, 151] (although some researchers have disputed the suitability of this technique [152]). In general, AE signals are thought to be related to the evolution and friction caused by hydrogen bubbles in the pits [150-154], changes in stress on the surface [155], and rupture of oxide or covers of pits [156]. Recently, the number of pits or the pitted area have been found to be quantitatively correlated to the number of AE activities [157].

Besides pitting, AE can also detect crevice corrosion [158]. As matter of fact, Kim *et al.* [158] have reported in a recent study that the acoustic parameters generated by pitting and crevice corrosions are similar. In this respect, electrochemical noise (EN) measurement has been found to be capable of detecting and discerning pitting corrosion and crevice corrosion [159].

Optically, holographic interferometry has been demonstrated to be viable for the monitoring of pit initiation [160]. This technique can potentially be used to build a 3D interferometric microscope [160].

5. MICROBIOLOGICALLY-INDUCED CORROSION (MIC)

In microbiologically-induced corrosion (MIC), microorganisms/bacteria attach themselves to the steel surface (the formation of biofilms) and they subsequently change the composition of the underlying material. MIC may manifest itself in the form of pitting [161, 162] and selective dissolution of ferrite or austenite in duplex stainless steels [163, 164]. High-N stainless steels, which are used frequently nowadays in medicine, are pretty susceptible to MIC, especially when nitrifying bacteria are present [165]. For stainless steels that are exposed to seawater, the biofilms may induce a shift of the corrosion potential to the noble direction [166, 167]. Sensitised stainless steels are particular vulnerable [168], because of the presence of Cr-depleted zones [169].

Recent studies have demonstrated that the addition of Cu and Ag may be beneficial to antibacterial property [161, 162, 165, 170]. It has to be noted that Cu must precipitate out in the surface oxide layer in order to be antiseptic (such as the precipitation of ε-Cu in the oxide layer in a Cu-modified SUS304 steel [171]). Some elements (such as niobium), which may not possess the antibacterial ability in their own right, can indirectly enhance the antibacterial property by encouraging the uniform precipitation Cu-contaning precipitates [170].

6. HIGH-TEMPERATURE OXIDATION

Stainless steels are have good high-temperature oxidation resistance in dry oxidising environments, because of the presence of the Cr-rich surface oxide α-(Cr$_x$Fe$_{1-x}$)$_2$O$_3$. However, the chromium oxide gradually loses its protective capability if a wet atmosphere is present above 600°C. This is because chromia tends to react with water vapour to form a volatile product, *i.e.*, $\frac{1}{2}Cr_2O_{3(S)} + H_2O_{(g)} + \frac{3}{4}O_{2(g)} \rightarrow CrO_2(OH)_{2(g)}$. If chromium cannot diffuse fast enough to the surface to replenish the oxide with Cr, then α -(CrxFe1-x)2O3 becomes Fe-rich and it loses protectiveness. In this regard, grain refinement may be a good way to ameliorate breakaway oxidation, as Cr may diffuse efficiently due to the profusion of grain boundaries [172]. Ferritic stainless steels are less vulnerable than austenitic stainless steels, because of the higher diffusivity of Cr in the bcc microstructure.

The benefits of using Si in lessening breakaway oxidation are well recognised [173-177]. Compared with Cr2O3, it is believed that SiO2 is more resistant to oxidation and hot corrosion above 1000℃ [178]. The high temperature oxidation resistance of SiO2 is enhanced because a continuous vitreous silica layer, which acts as a diffusion barrier, will form between the metal and oxide interface. Additonally, silica may act as the nucleation site of a protective chromia layer. Although Si may mitigate breakaway oxidation, this element encourages sigma phase formation [179]. Aluminium has been found to improve the surface properties and oxidation resistance of stainless steels [247-249]. Nitrogen benefits high-temperature oxidation resistance [180, 181], because of the formation of a protective nitride layer on the surface [180].

7. STRESS CORROSION CRACKING: THE NISHIMURA CORROSION-ELONGATION TEST

As regards stress corrosion cracking (SCC), the series of recent work undertaken by Nishimura [182-192] is particularly noteworthy. Nishimura has shown that the corrosion-elongation test may be used conveniently to predict SCC of stainless steels (both the austenitic class [182, 183, 185-191] and the ferritic class [184]).

In the corrosion-elongation test, samples are pulled isothermally in a corrosive medium. The corrosion-elongation curve is composed of three stages (primary, secondary and tertiary). Several parameters such as the rate of elongation in the secondary stage (iss), the time that demarcates the secondary and the tertiary stages (tss), and the time to failure (tf) may be defined from the corrosion-elongation curve.

When SCC prevails, the logarithm of iss and the logarithm of tf exhibit a linear relationship, irrespective of the stress level, temperature and the type of anion in the corrosive medium [182, 183, 185-191]. Therefore, iss can be used to predict the time to failure tf. This parameter can also be used to tell whether there exists a threshold stress and the critical pH value that may lead to SCC. The parameter, tss/tf, can be used to indicate the occurrence of stress corrosion cracking, too.

8. FATIGUE

8.1. New Models for Fatigue

Commonly used models for fatigue life prediction (the Coffin-Mason relation and S-N curves, for example) are used on the premise that the behaviour of cyclically loaded materials will stabilise at the early stage of life span. However, Hong *et al.* [193, 194], while studying the cyclic stress response of a cold-worked 316 steel, have found that this premise may be invalid. These authors have suggested that the plastic strain energy density (W_P) should be a more suitable parameter for fatigue life prediction of the 316 steel, because this parameter stabilises at the early stage of life until almost up to 80% of the life span. For metastable austenitic stainless steels (the 316 is a stable steel against martensitic transformation), cumulative strain energy density is a very important parameter that affects the formation of deformation-induced martensite, which in turn critically affects fatigue life [195].

8.2. Factors Affecting the Fatigue Life of Stainless Steels

There are many factors that affect the fatigue behaviour of stainless steels, such as temperature, steel composition (which may affect dislocation structures and the formation of deformation-induced martensite), and loading conditions. The following subsections will report the recent results on the effects of the pernicious persistent slip bands, and deformation-induced martensite.

8.2.1. Persistent Slip Bands (PSBs)

Upon fatigue deformation, different types of dislocation structures may form, depending on the temperature, steel chemistry, loading conditions, strain amplitude, *etc.* Persistent slip bands (PSBs) are one of the dislocation structures that may form during fatigue. The formation of PSBs is unwelcome because when they impinge on grain boundaries, the steel may be cracked [196]. A recent study by Kaneko *et al.* [197] has given insights into some of the factors that influence PSB formation. Using electron channelling contrast imaging (ECCI), these authors have found in a Fe-19Ni-11Cr austenitic stainless steel that PSBs seem to preferentially form in grains having large diameters and that the Schmid factor does not seem to have a significant influence on the formation of PSBs [197]. But as the stress amplitude increases, even small grains of low Schmid factors may contain PSBs because of the influence of dislocation pile-ups in neighbouring grains.

From these results, it is clear that a reduction in grain size may prolong fatigue life. And this corroborates with a recent model developed by Hong *et al.* [198]. Using the concept of equilibrium crack density and dimensional analysis, these authors have revealed that the number of cycles to fracture (N_f) is inversely proportional to grain size (d), *i.e.*, $N_f \propto d^{-1}$, inasmuch as grain boundaries may hinder short cracks. From the study of Kaneko *et al.* [197], it seems that the restriction of PSB formation may also play a role in extending fatigue life. It is worth mentioning that in studying the fatigue damage of miniaturised microelectromechanical system (MEMS) devices made of austenitic stainless steel, Zhang *et al.* [199, 200] have found no PSBs in their samples.

8.2.2. Deformation-induced Martensite

There are contrasting views on the effects of martensitic transformation on fatigue life. While it has been reported that the brittle martensite aids crack propagation [201], some researchers have found that fatigue resistance is enhanced because of martensitic transformation [202, 203]. Fatigue resistance is improved because **1.** the martensitic transformation process absorbs energy [204], **2.** the martensite islands may block the cracks [198], and **3.** the stress field in the crack tip is perturbed by the volume expansion accompanying the martensitic transformation process [205, 206].

Miller *et al.* [207] have recently utilised magnetic force microscopy (MFM) to observe *in-situ* the martensitic transformation in the vicinity of fatigue cracks. These authors have observed that stringers of martensite will extend to about 3-4 μ m on both sides of the crack and these martensite stringers form at approximately 45° to the direction of crack growth. When two cracks come close to one another, the martensitic zone will spread over a much larger region (up to 15 μ m) [207].

For pre-existing martensite, it has been found that if its amount is below a certain critical value, the martensite that forms during fatigue loading will retard crack propagation. But above the critical value, crack propagation will be enhanced by the fatigue-induced martensite [208].

8.3. Methods for Detecting Fatigue

Because the propagation of fatigue cracks involves yielding at the edge of the plastic zone at crack tips, acoustic emission (AE) events will be generated during the process. For this reason, acoustic technique may be used to study fatigue crack propagation [209]. In fact, it has been determined that the count rate correlates well with the crack growth rate [209]. The Paris law is often used to describe fatigue crack growth. For steels, it has been found that there are two subregions in the Paris region [201-212]. AE has been found to be able to discern the transition between these two subregions [213]. Because the cyclic plastic zones are of different sizes for these two subregion, the extents of dislocation multiplication and rearrangement are also different. This is the reason why AE can differentiate between these two subregions [213].

The generation of ferromagnetic deformation-induced martensite (α') in austenitic stainless steels has been proved to be useful for fatigue assessment [214-216]. The magnetic flux associated with α', which is usable for both fatigue damage assessment and the estimation of crack length [217]. Besides martensite, cyclic deformations may also produce open-volume defects, such as vacancies, dislocations and microcracks. These defects have been shown to reduce magnetic moments and as a result, Superconducting Quantum Interference Devices (SQUID) may be utilised for fatigue damage detection [218].

As regards the detection of fatigue damage at the early stage, it has been demonstrated that the use of positron annihilation may be handy [219, 220]. The average positron lifetime has been found to be able to reveal the early stage of fatigue damage [219, 220].

9. CREEP

9.1. A New Alumina-based Creep Resistant Stainless Steel and Effects of Alloying

A lot of recent research on creep has focused on creep property enhancement through alloying. A new aluminium-containing steel produced by Yamamoto *et al.* [221-223] has been shown to possess very good high-temperature creep strength and oxidation resistance.

High creep strength is usually attained by alloying with Nb, Ti and V, followed by suitable heat treatments to precipitate nanosized carbides containing these alloying elements. For Nb, both stoichiometric addition [224] and understabilisation [225] have been found to engender good creep properties. Sourmail *et al.* [226], however, have established by using neural network calculations that the stabilisation ratios of both Ti and Nb for enhanced creep life are not the same for different time scales. For short periods, a stabilisation ratio of Nb close to unity is good, while stabilisation below stoichiometry should be adopted for long-term use [226]. Laha *et al.* [227] have also shown that too much Ti is undesirable as far as long-term creep strength is concerned, as Ti may accelerate sigma phase formation. Nonetheless, short-term creep strength is enhanced when Ti is present in high amounts [227].

The benefit associated with a high stabilisation ratio boils down to the fact that large Ti(C, N) may form even during solution treatment (at 1453K, e.g) [227]. Since there is a large misfit between these large Ti(C,N) and the matrix, a high densities of dislocation tangles around these carbides will be generated in the grain interiors during cooling [228]. Insofar as the precipitation of TiC is dislocation-controlled [229], the high dislocation densities will result in copious intragranular precipitation of fine, closely-spaced Ti(C, N) particles. These fine particles then give rise to good creep strength.

Another way to achieve good oxidation resistance is to generate an alumina-based protective layer to steels. Since the alumina-based layer is nearly impervious to attack in a moist environment between 600℃ and 850℃, it gives very good protection to the underlying steel [230]. Attempts aimed at producing Al_2O_3-forming austenitic stainless steels having high creep strength without the use of additional surface treatment has been made by Yamamoto *et al.* [221-223] and with success. The most notable feature of the new steel (Fe-20Ni-14Cr-2.5Al) is its relatively low aluminium content, which is achieved by reducing/eliminating the use of Ti and V. These two elements, when they are present simultaneously, may deteriorate the alumina layer. Nevertheless, the single use of either Ti or V in small quantities for strengthening purposes is allowable [231]. To compensate for the reduced levels of Ti and V in the new steel, more Nb must be used [223].

The effects of other less commonly used alloying elements have also been investigated recently. Yttrium alloying has been found to strengthen the creep resistance of heat-resistant cast stainless steels [232, 233]. The strengthening is attributable to the fragmentation of the detrimental chromium carbide networks. It has been discovered that yttrium carbides are the earliest to form during solidification. They then act as heterogeneous nucleation sites for chromium carbide and so the latter's distribution becomes more fragmentary. Alloying with B and Ce can also improve creep cavitation resistance [234-237]. Boron tends to segregate to grain boundary cavity surfaces [234, 236, 237] and so the nucleation and growth of cavities is hindered. The formation of borides also benefits creep strength [235]. Additionally, boron may stabilise fine grain boundary $M_{23}C_6$ and so grain boundary sliding is retarded [238]. Ce is believed to getter S and form Ce_2O_2S, thereby allowing more B to segregate to the grain boundary cavities [234, 236, 237]. The reduction of the segregation of S to grain boundary cavities will dampen grain growth [234, 236, 237].

9.2. Method of Detection of Creep

Small angle neutron scattering (SANS) has recently been used to study and quantify creep damage in austenitic stainless steels [239, 240]. It has been demonstrated that SANS has the potential of measuring the size distribution and volume fraction of cavities [239, 240]. Electromagnetic acoustic resonance (EMAR) may monitor the attenuation and velocity of shear waves. This technique has been demonstrated by Ohtani [241] to have the potential for remaining life prediction under creep loading conditions. Regardless of the applied stress levels, a peak in shear wave attenuation has been found to occur at about 60% (70%) of the lifetime when the cell structure dominates [241].

10. DYNAMIC STRAIN AGEING

When dynamic strain ageing (DSA) occurs, fatigue life is reduced [242-245] because the cyclic hardening rate is raised [244, 246], and crack initiation and propagation are facilitated [247]. Nevertheless, recent research findings have revealed some positive features of DSA. For example cold-work strengthening is enhanced in the presence of DSA.

In low cyclic fatigue, the following phenomena are indicative of DSA:

- Negative temperature dependence of the peak stress.

- Very high rate of cyclic hardening.

- High ratio of maximum stress to the stress at the first cycle with increasing temperature.

- Negative strain rate sensitivity.

During DSA, solute atoms are believed to intermittently pin and release dislocations. As a result, serrated yielding is often observed when DSA occurs. Nonetheless, some researchers have found in a cyclically loaded 316L steel that DSA occurs before serrated yielding becomes observable [248, 249].

At the beginning of this section, it was mentioned that DSA could degrade mechanical properties. However, Samuel *et al.* [250] have recently found that the ductile fracture toughness of a cold-worked 15Cr-15Ni Ti-modified stainless steel will be improved by DSA. Furthermore, cold working in the temperature regime where DSA occurs has been found to be more effective in strengthening austenitic stainless steels than by using traditional cold working [251]. The higher strengthening is believed to be due to a more uniform distribution of dislocations and their more efficient generation [251].

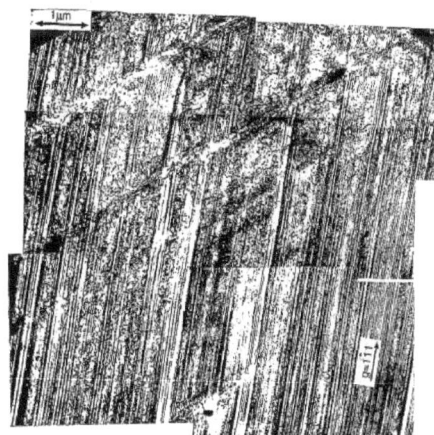

Figure 2: The generalised corduroy dislocation structure developed in a 316L austenitic stainless steel upon cyclic deformation at 400°C in a vacuum for 135000 cycles ($\varepsilon_{P/2}= 2 \times 10^{-3}$) [Gerland M, Alain R, Saadi BA, *et al.* Low cycle fatigue behaviour in vacuum of a 316L-type austenitic stainless steel between 20 and 600 C. Part II dislocation structure evolution and correlation with cyclic behaviour. Mater Sci Eng A 1997; 229(1-2): 68-86].

Between 200℃ and 400℃ and in a vacuum, it has also been observed by Alain *et al.* [252] that the fatigue life of 316L will be prolonged significantly alongside strong secondary hardening caused by DSA. This prolonging of fatigue life, however, disappears without a vacuum [252]. So, it seems that the environment does exert a big influence on fatigue life when DSA occurs. Besides this interesting observation, these authors have also discovered the corduroy dislocation structure (Fig. **2**), which is seldom observed. Alain *et al.* [252] attributed the secondary hardening in the 316L steel to the corduroy dislocation structure. Nevertheless, the corduroy dislocation structure is thought not to contribute solely to the concurrent increase in fatigue life [252].

10.1. Solute Atoms That Cause DSA

Different solute atoms (both substitutional and interstitial) can cause DSA. In duplex stainless steels, DSA is thought to take place in either the austenite phase or the ferrite phase.

Although chromium is considered to cause DSA in the Fe-Cr alloy systems (both for the fcc [253] and bcc [254] microstructures), its role is taken over when other alloying elements are added. For instance, Al [255, 256], Ti [257], Cu [245] and Mo and Ni [258] are also implicated for causing DSA in different stainless steels. Impurities like P and Si are also suggested to be able cause DSA [259].

Below 673K, interstitial solutes (C and N) are thought to cause DSA in austenitic stainless steels, whereas substitutional elements are regarded as being responsible between 673K and 873K [260]. Nitrogen, besides causing DSA, can widen the range of DSA occurrence to room temperature [261].

Between 300K and 773K, the DSA in the DIN1.4460 duplex stainless steel has been thought to be associated with carbon atoms in the ferrite phase [242]. However, Girones *et al.* [262] have proposed that the DSA in a superduplex stainless steel between 548K and 748K is associated with Cr in both the ferrite and the austenite phases, with the former being affected to a greater extent. It is interesting to note that above 673K, the DSA in single-phased austenitic stainless steels is thought to be associated with Cr [248, 263, 264].

REFERENCES

[1]   Quick NR, Johnson HH. Permeation and diffusion of hydrogen and deuterium in 310 stainless steel, 472L to 779K.Metall Trans A 1979; 10(1): 67-70.

[2]   Beachem CD. New model for hydrogen-assisted cracking (hydrogen embrittlement). Metall Trans 1972; 3(2): 437-45.

[3]   Lynch SP. Environmentally-assisted cracking – overview of evidence for an absorption-induced localized slip process. Acta Metall 1988; 36(10): 2639-61.

[4]   Birnbaum HK, Sofronis P. Hydrogen enhanced localized plasticity – a mechanism for hydrogen-related fracture. Mater Sci Eng A 1994; 176(1-2): 191-202.

[5]   Steigerwald EA, Schaller FW, Troiano AR. The role of stress in hydrogen induced delayed failure. Trans AIME 1960; 218: 832-41.

[6]   Briant CL, Feng HC, McMahon Jr CJ. Embrittlement of a 5% Ni high-strength steel by impurities and their effects on hydrogen-uinduced cracking. Metall Trans A 1978; 9(5): 625-33.

[7]   Szummer A, Janko A. hydride phases in austenitic stainless steels. Corros 1979; 35(10): 461-4.

[8]   Inoue A, Hosoya Y, Masumoto T. Effect of hydrogen on crack propagation behavior and microstructures around cracks in austenitic stainless steels. Trans Iron Steel Inst Japan. 1979; 19(3): 170-8.

[9]   Zheng W, Hardie D. The effect of hydrogen on the fracture of a commercial duplex stainless steel. Corros Sci 1991; 32(1): 23-36.

[10]   Narita N, Altstetter CJ, Birnbaun HK. Hydrogen related phase transformations in austenitic stainless steels. Metall Trans A 1982; 13(8): 1355-6.

[11]   Jani S, Marek M, Hochman RF, *et al.* A mechanistic study of transgranular stress corrosion cracking of type 304 stainless steel. Metall Trans A 1991; 22(6): 1453-61.

[12]   Luppo MI, Hazarabedian A, Garcia JO. Effects of delta ferrite on hydrogen embrittlement of austenitic stainless steel welds. Corros Sci 1999; 41(1): 87-103.

[13]   Glowacka A, Swiatnicki WA. Effect of hydrogen charging on the microstructure of duplex stainless steel. J Alloy Compd 2003; 356-357: 701-4.

[14]   Hardie D, Zheng W. Surface cracking and phase transformations associated with cathodic charging of austenitic stainless steel. Mater Sci Technol 1994; 10(9): 817-22.

[15]   Pontini AE, Hermida JD. X-ray diffraction measurement of the stacking fault energy reduction induced by hydrogen in an AISI 304 steel. Scr Mater 1997; 37(11): 1831-7.

[16]   Hanninen H, Hakarainen T. Effects of α' martensite in hydrogen embrittlement of a cathodically charged AISI type-304 austenitic stainless steel. Corros 1980; 36(1): 47-51.

[17]   Schuster G, Altstetter CJ. The effect of hydrogen on the creep-rupture properties of Fe-Ni alloys. Metall Trans A 1980; 11(10): 1657-64.

[18]   Thompson AW, Brooks JA. Hydrogen performance of precipitation-strengthened stainless steels based on A286. Metall Trans A 1975; 6(7): 1431-42.

[19]   Pound BG. The ingress of hydrogen into precipitation-hardened alloys A-286 and C17200. Corros Sci 2000; 42(7): 1269-81.

[20]   Hyzak JM, Rawl Jr DE, Louthan Jr MR. Hydrogen affected impact behavior of type 304 stainless steel. Scr Metall 1981; 15(8): 937-9.

[21]   Brass AM, Chene J. Hydrogen uptake in 316L stainless steel: Consequences on the tensile properties. Corros Sci 2006; 48(10): 3222-42.

[22]   Robertson IM, Birnbaum HK. An HVEM study of hydrogen effects on the deformation and fracture nickel. Acta Metall 1986; 34(3): 353-66.

[23]   Rozenak P, Robertson IM, Birnbaum HK. HVEM studies of the effects of hydrogen on the deformation and fracture of AISI type 316 austenitic stainless steel. Acta Metall Mater 1990; 38(11): 2031-40.

[24]   Shivanyuk VN, Foct J, Gavriljuk VG. Hydrogen-enhanced microplasticity of austenitic steels studied by means of internal friction. Mater Sci Eng A 2001; 300(1-2): 284-90.

[25]   Nibur KA, Bahr DF, Somerday BP. Hydrogen effects on dislocation activity in austenitic stainless steel. Acta Mater 2006; 54(10): 2677-84.

[26]   Ferreira PJ, Robertson IM, Birnbaum HK. Hydrogen effects on the interaction between dislocations. Acta Mater 1998; 46(5): 1749-57.

[27]   Robertson IM. The effect of hydrogen on dislocation dynamics. Eng Fract Mech 2001; 68(6): 671-92.

[28]   Gavriljuk VG, Shivanyuk VN, Foct J. Diagnostic experimental results on the hydrogen embrittlement of austenitic steels. Acta Mater 2003; 51(5): 1293-305.

[29]   Shivanyuk VN, Shanina BD, Tarasenko AV, *et al.* Effect of hydrogen on atomic bonds in austenitic stainless steel. Scr Mater 2001; 44(12): 2765-73.

[30]   Vogt JB. Hydrogen-induced phase transformation in a high nitrogen austenitic stainless steel. Corros Sci 1999; 41(3): 519-28.

[31]   Ulmer DG, Altstetter CJ. Hydrogen induced strain localization and failure of austenitic stainless steels at high hydrogen concentrations. Acta Metall Mater 1991; 39(6): 1237-48.

[32]   Shivanyuk VN, Foct J, Gavriljuk VG. On a role of hydrogen-induced epsilon-martensite in embrittlement of stable austenitic steel. Scr Mater 2003; 49(6): 601-6.

[33]   Yang Q, Qiao LJ, Chiovelli S, *et al.* Critical hydrogen charging conditions for martensite transformation and surface cracking in type 304 stainless steel. Scr Mater 1999; 40(11): 1209-14.

[34]   Buckley JR, Hardie D. Effect of low temperature on resistance of stable austenitic steel to embrittlement by hydrogen. Mater Sci Technol 1993; 9(3): 259-63.

[35]   Pan C, Su YJ, Chu WY, *et al.* Hydrogen embrittlement of weld metal of austenitic stainless steels. Corros Sci 2002; 44(9): 1983-92.

[36]   Hardie D, Xu J, Charles EA, *et al.* Hydrogen embrittlement of stainless steel overlay materials for hydrogenators. Corros Sci 2004; 46(12): 3089-100.

[37]   Pan C, Chu WY, Li ZB, *et al.* Hydrogen embrittlement induced by atomic hydrogen and hydrogen-induced martensites in type 304L stainless steel. Mater Sci Eng A 2003; 351(1-2): 293-8.

[38]   Han G, He J, Fukuyama S, *et al.* Effect of strain-induced martensite on hydrogen environment embrittlement of sensitized austenitic stainless steels at low temperatures. Acta Mater 1998; 46(13): 4559-70.

[39]   Chen H, Chu WY, Gao KW, *et al.* Effect of hydrogen-induced martensite on stress corrosion cracking of type 304 stainless steel in boiling $MgCl_2$. J Mater Sci Lett 2002; 21(17): 1337-8.

[40]   Narita N, Birnbaum H. On the role of phase transition in the hydrogen embrittlement of stainless steels. Scr Metall 1980; 14(12): 1355-8.

[41]   Srivatsan TS, Sudarshan TS. Hydroen effects on tensile behavior of a precipitation-hardened stainless steel. J Mater Sci Lett 1993; 12(20): 1600-2.

[42]   Chiang WC, Pu CC, Yu BL, *et al.* Hydrogen susceptibility of 17-4 PH stainless steel. Mater Lett 2003; 57(16-17): 2485-8.

[43]   Harvey II DP, Terrell JB, Sudarshan TS. Hydrogen effects on the ductile to brittle transition behavior of 21-6-9 stainless steel. J Mater Sci 1994; 29(20): 5485-90.

[44]   Zhang T, Chu WY, Gao KW, *et al.* Study of correlation between hydrogen-induced stress and hydrogen embrittlement. Mater Sci Eng A 2003; 347(1-2): 291-9.

[45]   Moraes FD, Bastian FL, Ponciano JA. Influence of dynamic straining on hydrogen embrittlement of UNS-G41300 and UNS-S31803 steels in a low H2S concentration environment. Corros Sci 2005; 47(6): 1325-35.

[46]   Chou SL, Tsai WT. Hydrogen embrittlement of duplex stainless steel in concentrated sodium chloride solution. Mater Chem Phys 1999; 60(2): 137-42.

[47]   Chou SL, Tsai WT. Effect of grain size on the hydrogen-assisted cracking in duplex stainless steels. Mater Sci Eng A 1999; 270(2): 219-24.

[48]   Zakroczymski T, Glowacka A, Swiatnicki W. Effect of hydrogen concentration on the embrittlement of a duplex stainless steel. Corros Sci 2005; 47(6): 1403-14.

[49]   Tsai WT, Chou SL. Environmentally assisted cracking behavior of duplex stainless steel in concentrated sodium chloride solution. Corros Sci 2000; 42(10): 1741-62.

[50]   Wei WY, Tzeng CH, Wu JK. Hydrogen transport and degradation of the duplex Fe-Mn-Al alloy. J Mater Sci Lett 1990; 9(11): 1357-8.

[51]   Chen SS, Wu TI, Wu JK. Effects of deformation on hydrogen degradation in a duplex stainless steel. J Mater Sci 2004; 39(1): 67-71.

[52]   Cocco VD, Franzese E, Iacoviello F, *et al.* 22Cr5Ni duplex and 25Cr7Ni superduplex stainless steel: Hydrogen influence on fatigue crack propagation resistance. Eng Fract Mech 2008; 75(3-4): 705-14.

[53]   Young MC, Chan SLI, Tsay LW, *et al.* Hydrogen-enhanced cracking of 2205 duplex stainless steel welds. Mater Chem Phys 2005; 91(1): 21-7.

[54]   Luu WC, Liu PW, Wu JK. Hydrogen transport and degradation of a commercial duplex stainless steel. Corros Sci 2002; 44(8): 1783-91.

[55]   Zahumensky P, Tuleja S, Orszagova J, *et al.* Corrosion resistance of 18Cr-12Ni-2,5Mo steel annealed at 500-1050C. Corros Sci 1999; 41(7): 1305-22.

[56]   Goudett MA, Scully JR. Distributions of Cr depletion levels in sensitized AISI304 stainless steel and its implications concerning intergranular corrosion phenomena. J Electrochem Soc 1993; 140(12): 3425-35.

[57]   Terada M, Escriba DM, Costa I, *et al.* Investigation on the intergranular corrosion resistance of the AISI 316L(N) stainless steel after long time creep testing at 600C. Mater Charact 2008; 59(6): 663-8.

[58]   Padilha AF, Escriba DM, Morris EM, *et al* Precipitation in AISI 316L(N) during creep tests at 550 and 600C up to 10 years. J Nucl Mater 2007; 362(1): 132-8.

[59]   Nagae Y. A study on detection of creep damage before crack initiation in austenitic stainless steel. Mater Sci Eng A 2004; 387-89: 665-9.

[60]   Arioka K, Yamada T, Terachi T, *et al.* Intergranular stress corrosion cracking behavior of austenitic stainless steels in hydrogenated high-temperature water. Corrosion 2006; 62(1): 74-83.

[61]   Arioka K, Yamada T, Terachi T, *et al.* Influence of carbide precipitation and rolling direction on intergranular stress corrosion cracking of austenitic stainless steels in hydrogenated high-temperature water. Corrosion 2006; 62(7): 568-75.

[62]   Fullman RL. A thermodynamic model of the effect of composition on the susceptibility of austenitic stainless steels to intergranular stress corrosion cracking. Acta Metall 1982; 30(7): 1407-15.

[63]   Parvathavarthini N, Dayal RK. Influence of chemical composition, prior deformation and prolonged thermal aging on the sensitization characteristics of austenitic stainless steels. J Nucl Mater 2002; 305(2-3): 209-19.

[64]   Bruemmer SM. Composition-based correlations to predict sensitization resistance of austenitic stainless steels. Corrosion 1986; 42(1): 27-35.

[65]   Trillo EA, Murr LE. Effects of carbon content, deformation, and interfacial energetics on carbide precipitation and corrosion sensitization in 304 stainless steel. Acta Mater 1999; 47(1): 235-45.

[66]   Beltran R, Maldonada JG, Murr LE, *et al.* Effects of strain and grain size on carbide precipitation and corrosion sensitization behavior in 304 stainless steel. Acta Mater 1997; 45(10): 4351-60.

[67]   Almanza E, Murr LE. A comparison of sensitization kinetics in 304 and 316 stainless steels. J Mater Sci 2000; 35(13): 3181-8.

[68]   Singh R, Ravikumar B, Kumar A, *et al.* The effects of cold working on sensitization and intergranular corrosion behavior of AlSl 304 stainless steel. Metall Mater Trans A 2003; 34(11): 2441-7.

[69]   Oh YJ, Hong JW. Nitrogen effect on precipitation and sensitization in cold-worked Type 316L(N) stainless steels. J Nucl Mater 2000; 278(2-3): 242-50.

[70]   Advani AH, Murr LE, Atteridge DG, *et al.* Mechanism of deformation-induced grain-boundary chromium depletion (sensitization) development in type316 stainless steel. Metall Trans A 1991; 22(12): 2917-34.

[71]   Mao X, Zhao W. Preplastic strain effect on chromium carbides precipitation of type316 stainless steel during high temperature aging. J Mater Sci Lett 1992; 11(9): 585-7.

[72]   Mao X, Zhao W. Preplastic deformation and subsequent aging effect on low-temperature (-196C) tensile properties of type 316 stainless steel. J Mater Sci Lett 1992; 11(16): 1137-9.

[73]   Briant CL, Ritter AM. Effect of cold work on the sensitization of 304 stainless steel. Scr Metall 1979; 13(3): 177-81.

[74]   Murr LE, Staudhammer KP, Hecker S. Effect of strain state and strain rate on deformation-induced transformation in 304 stainless steel 2. microstructural study. Metall Trans A 1982; 13(4): 627-35.

[75]   Staudhammer KP, Murr LE. Hecker S. Nucleation and evolution of strain-induced martensitic (bcc) embryos and substructure in stainless steel – a transmission electron-microscopy study. Acta Metall Mater 1983; 31(2): 267-74.

[76]   Alvarez CJ, Almanza E, Murr LE. Evaluation of the sensitization process in 304 stainless steel strained 50% by cold-rolling. J Mater Sci 2005; 40(11): 2965-9.

[77]   Trillo EA, Beltran R, Maldonado JG, *et al.*Combined effects of deformation (strain and strain rate), grain size, and carbon content on carbide precipitation and corrosion sensitization in 304 stainless steel. Mater Charact 1995; 35(2): 99-112.

[78]   Advani AH, Murr LE, Matlock DJ, *et al.* Deformation-induced microstructure and martensite effects on transgranular carbide precipitation in type304 stainless steels. Acta Metall Mater 1993; 41(9): 2589-600.

[79]   Shaikh H, Sivaibharasi N, Sasi B, *et al.* Use of eddy current testing method in detection and evaluation of sensitisation and intergranular corrosion in austenitic stainless steels. Corrosion Sci 2006; 48(6): 1462-82.

[80]   Kain V, Prasad RC, De PK. Testing sensitization and predicting susceptibility to intergranular corrosion and intergranular stress corrosion cracking in austenitic stainless steels. Corrosion 2002; 58(1): 15-37.

[81]   Aydogdu GH, Aydinol MK. Determination of susceptibility to intergranular corrosion and electrochemical reactivation behaviour of AISI 316L type stainless steel. Corrosion Sci 2006; 48(11): 3565-83.

[82]   Tavares SSM, Fonseca MPC, Maia A, *et al.* Influence of the starting condition on the kinetics of sensitization and loss of toughness in an AISI 304 steel. J Mater Sci 2003; 38(17): 3527-33.

[83]   Kamada Y, Mikami T, Takahashi S, *et al.* Compositional dependence of magnetic properties on thermally sensitized austenitic stainless steels. J Magn Magn Mater 2007; 310(2): 2856-8.

[84]   Takaya S, Suzuki T, Matsumoto Y, *et al.* Estimation of stress corrosion cracking sensitivity of type 304 stainless steel by magnetic force microscope. J Nucl Mater 2004; 327(1): 19-26.

[85]   Huang YL, Kinsella B, Becker T. Sensitisation identification of stainless steel to intergranular stress corrosion cracking by atomic force microscopy. Mater Lett 2008; 62(12-13): 1863-6.

[86]   Szakalos P. Mechanisms and driving forces of metal dusting. Mater Corros 2003; 54(10): 752-62.

[87]   Grabke HJ. Metal dusting of low alloy and high alloy steels. Corrosion 1995; 51(9): 711-720.

[88]   Grabke HJ. Thermodynamics, mechanisms and kinetics of metal dusting. Mater Corros 1998; 49(5): 303-8.

[89]   Schneider A. Iron layer formation during cementite decomposition in carburising atmospheres. Corros Sci 2002; 44(10): 2353-65.

[90]   Schneider R, Pippel E, Woltersdorf J, *et al.* Microprocesses of metal dusting on nickel and Ni-base alloys. Steel Res 1997; 68(7): 326-32.

[91]   Zhang J, Young DJ. Kinetics and mechanisms of nickel metal dusting I. Kinetics and morphology. Corros Sci 2007; 49(3): 1496-512.

[92]   Schmid B, Grong O, Ødegard R. Coke formation in metal dusting experiments. Mater Corros 1999; 50(11): 647-53.

[93]   Lin C, Tsai W. Nano-sized carbon filament formation during metal dusting of stainless steel. Mater Chem Phys 2003; 82(3): 929-36.

[94]   Stevens KJ, Levi T, Minchington I, *et al.* Transmission electron microscopy of high pressure metal dusted 316 stainless steel. Mater Sci Eng A 2004; 385(1-2): 292-9.

[95]   Szakalos P, Pettersson R, Hertzman S. An active corrosion mechanism for metal dusting on 304L stainless steel. Corros Sci 2002; 44(10): 2253-70.

[96]   Sedriks AJ. Corrosion of Stainless Steels. John Wiley and Sons, New York. 1979.

[97]   Sedriks AJ. Effects of alloy composition and microstructure on the passivity of stainless steels. Corrosion 1986; 42(7): 377-89.

[98]   Lim YS, Kim JS, Ahn SJ, *et al.* The influences of microstructure and nitrogen alloying on pitting corrosion of type 316L and 20 wt.% Mn-substituted type 316L stainless steels. Corrosion Sci 2001; 43(1): 53-68.

[99]   Park JO, Suter T, Bohnl H. Role of manganese sulfide inclusions on pit initiation of super austenitic stainless steel's. Corrosion 2003; 59(1): 59-67.

[100]   Wranglen G. Pitting and sulphide inclusions in steel. Corrosion Sci 1974; 14(5): 331-49.

[101]   Castle JE, Ke R. Study by Auger spectroscopy of pit initiation at the site of inclusions in stainless steel. Corros Sci 1990; 30(4-5): 409-28.

[102]   Lott SE. Alkire RC. The role of inclusions on initiation of crevice corrosion of stainless steel 1. experimental studies. J Electrochem Soc 1989; 136(4): 973-9.

[103]   Eklund GS. Initiation of pitting at sulphide inclusion in stainless steel. J Electrochem Soc 1974; 121(4): 467-73.

[104]   Williams DE, Mohiuddin TF, Zhu YY. Elucidation of a trigger mechanism for pitting corrosion of stainless steels using submicron resolution scanning electrochemical and photoelectrochemical microscopy. J Electrochem Soc 1998; 145(8): 2664-72.

[105]   Newman RC. Beyond the kitchen sink. Nat 2002; 415(6873): 743-4.

[106]   Szummer A, Czachor MJ, Hofmann S. Discontinuity of the passivating film at non-metallic inclusions in stainless steels. Mater Chem Phys 1993; 34(2): 181-3.

[107]   Suter T, Bohni H. Microelectrodes for corrosion studies in microsystems. Electrochim Acta 2001; 47(1-2): 191-9.

[108]   Krawiec H, Vignal V, Heintz O, *et al.* Influence of the dissolution of MnS inclusions under free corrosion and potentiostatic conditions on the composition of passive films and the electrochemical behaviour of stainless steels. Electrochim Acta 2006; 51(16): 3235-43.

[109]   Wang W. Effect of MnS powder addition and sintering temperature on the corrosion resistance of sintered 303LSC stainless steels. Corros Sci 2003; 45(5): 957-66.

[110]   Williams DE, Zhu YY. Explanation for initiation of fitting corrosion of stainless steels at sulfide inclusions. J Electrochem Soc 2000; 147(5): 1763-1766.

[111]   Ryan MP, Williams DE, Chater RJ, *et al.* Why stainless steel corrodes. Nat 2002; 415(6873): 770-4.

[112]   Ryan MP, Williams DE, Chater RJ, *et al.* Metallurgy - Stainless-steel corrosion and MnS inclusions - Reply. Nat 2003; 424(6947): 390.

[113]   Meng Q, Frankel GS, Colijn HO, *et al.* High-resolution characterization of the region around manganese sulfide inclusions in stainless steel alloys. Corrosion 2004; 60(4): 346-55.

[114]   Meng Q, Frankel GS, Colijn HO, *et al.* Metallurgy - Stainless-steel corrosion and MnS inclusions. Nat 2003; 424(6947): 389-90.

[115]   Schmuki P, Hildebrand H, Friedrich A, *et al.* The composition of the boundary region of MnS inclusions in stainless steel and its relevance in triggering pitting corrosion. Corros Sci 2005; 47(5): 1239-50.

[116]   Ha HY, Park CJ, Kwon HS. Effects of misch metal on the formation of non-metallic inclusions and the associated resistance to pitting corrosion in 25% Cr duplex stainless steels. Scr Mater 2006; 55(11): 991-4.

[117]   Gupta CK, Krishnamurthy N. Extractive metallurgy of rare-earths. Int Mater Rev 1992; 37(5): 197-248.

[118]   Yang H, Yang K, Zhang B. Pitting corrosion resistance of La added 316L stainless steel in simulated body fluids. Mater Lett 2007; 61(4-5): 1154-7.

[119]   Krawiec H, Vignal V, Finot E, *et al.* Local electrochemical studies after heat treatment of stainless steel: Role of induced metallurgical and surface modifications on pitting triggering. Metall Mater Trans A 2004; 35(11): 3515-21.

[120]   Barbucci A, Cerisola G, Cabot PL. Effect of cold-working in the passive behavior of 304 stainless steel in sulfate media. J Electrochem Soc 2002; 149(12): B534-42.

[121]   Mudali UK, Shankar P, Ningshen S, *et al.* On the pitting corrosion resistance of nitrogen alloyed cold worked austenitic stainless steels. Corros Sci 2002; 44(10): 2183-9.

[122]   Gutman EM, Solovioff G, Eliezer D. The mechanochemical behavior of type 316L stainless steel. Corros Sci 1996; 38(7): 1141-5.

[123] Haanappel VAC, Stroosnijder MF. Influence of mechanical deformation on the corrosion behavior of AISI 304 stainless steel obtained from cooking utensils. Corrosion 2001; 57(6): 557-65.

[124] Peguet L, Malki B, Baroux B. Influence of cold working on the pitting corrosion resistance of stainless steels. Corros Sci 2007; 49(4): 1933-48.

[125] Salvago G, Fumagalli G, Sinigaglia D. The corrosion behavior of AISI304L stainless steel in 0.1M HCl at room temperature 2. the effect of cold working. Corros Sci 1983; 23(5): 515-23.

[126] Vignal V, Mary N, Valot C, *et al.* Influence of elastic deformation on initiation of pits on duplex stainless steels. Electrochem Solid State Lett 2004; 7(4): C39-42.

[127] Stefec R, Franz F. Study of pitting corrosion of cold-worked stainless steel. Corros Sci 1978; 18(2): 161-8.

[128] Semino CJ, Pedeferri P, Burstein GT, *et al.* Localized corrosion of resistant alloys in chloride solutions. Corros Sci 1979; 19(12): 1069-78.

[129] Merello R, Botana FJ, Botella J, *et al.* Influence of chemical composition on the pitting corrosion resistance of non-standard low-Ni high-Mn-N duplex stainless steels. Corros Sci 2003; 45(5): 909-22.

[130] Pettersson RFAJ. Application of the pitting resistance equivalent concept to some highly alloyed austenitic stainless steels. Corrosion 1998; 54(2): 162-8.

[131] Mesias LFG, Sykes JM, Tuck CDS. The effect of phase compositions on the pitting corrosion of 25 Cr duplex stainless steel in chloride solutions. Corros Sci 1996; 38(8): 1319-30.

[132] Uggowitzer PJ, Magdowski R, Spiedel MO. Nickel free high nitrogen austenitic steels. ISIJ Int 1996; 36(7): 901-8.

[133] Wang J, Uggowitzer PJ, Magdowski R, *et al.* Nickel-free duplex stainless steels. Scr Mater 1998; 40(1): 123-9.

[134] Pettersson RFAJ. Electrochemical investigation of the influence of nitrogen alloying on pitting corrosion of austenitic stainless steels. Corros Sci 1999; 41(8): 1639-64.

[135] Olsson COA.The influence of nitrogen and molydenum on passive films formed on the austenoferritic stainless steel 2205 studied by AES and XPS. Corros Sci 1995; 37(3): 467-79.

[136] Olefjord I, Wegrelius L. The influence of nitrogen on the passivation of stainless steels. Corros Sci 1996; 38(7): 1203-20.

[137] Hermas AA, Ogura K, Takagi S, *et al.* Effects of alloying additions on corrosion and passivation behaviors of type304 stainless steel. Corrosion 1995; 51(1): 3-10.

[138] Sourisseau T, Chauveau E, Baroux B. Mechanism of copper action on pitting phenomena observed on stainless steels in chloride media. Corros Sci 2005; 47(5): 1097-117.

[139] Pardo A, Merino MC, Carboneras M, *et al.* Pitting corrosion behaviour of austenitic stainless steels with Cu and Sn additions. Corros Sci 2007; 49(2): 510-25.

[140] Hamdy AS, Shenawy EE, Bitar TE. The corrosion behavior of niobium bearing cold deformed austenitic stainless steels in 3.5% NaCl solution. Mater Lett 2007; 61(13): 2827-32.

[141] Hermas AA, Allah IMH. Microstructure, corrosion and mechanical properties of 304 stainless steel containing copper, silicon and nitrogen. J Mater Sci 2001; 36(14): 3415-22.

[142] Rondelli G, Vicentini B, Cigada A. Influence of nitrogen and manganese on localized corrosion behavior of stainless steels in chloride environments. Mater Corros 1995; 46(11): 628-32.

[143] Hanninen H. Corrosion properties of HNS. Mater Sci Forum 1999; 318-320: 479-88.

[144] Park CJ, Kwon HS. Effects of aging at 475C on corrosion properties of tungsten-containing duplex stainless steels. Corros Sci 2002; 44(12): 2817-30.

[145] Perren RA, Suter TA, Uggowitzer PJ, *et al.* Corrosion resistance of super duplex stainless steels in chloride ion containing environments: investigations by means of a new microelectrochemical method I. Precipitation-free states. Corros Sci 2001; 43(4): 707-26.

[146] Perren RA, Suter TA, Solenthaler C, *et al.* Corrosion resistance of super duplex stainless steels in chloride ion containing environments: investigations by means of a new microelectrochemical method II. Influence of precipitates. Corros Sci 2001; 43(4): 727-45.

[147] Vannevik H, Nilsson JO, Frodigh J, *et al.* Effect of elemental partitioning on pitting resistance of high nitrogen duplex stainless steels. ISIJ Int 1996; 36(7): 807-12.

[148] Merello R, Botana FJ, Botella J, *et al.* Determination of the weaker phase in the pitting corrosion of non-standard low-Ni high-Mn-N duplex stainless steels. Mater Corros 2004; 55(2): 95-101.

[149] Ravichandran K, Sivakumar M, Narayanan TSNS, *et al.* Predicting the susceptibility of stainless steel materials to localized attack – a new approach. J Mater Sci Lett 1995; 14(5): 317-18.

[150] Rettig TW, Felsen MJ. Acoustic emission method of monitoring corrosion reactions. Corrosion 1976; 32(4): 121-6.

[151] Mansfeld F, Stocker PJ. Acoustic emission from corroding electrodes. Corrosion 1979; 35(12): 541-4.

[152] Darowicki K, Mirakowski A, Krakowiak S. Investigation of pitting corrosion of stainless steel by means of acoustic emission and potentiodynamic methods. Corrosion Sci 2003; 45(8): 1747-56.

[153] Fregonese M, Idrissi H, Mazille H, *et al.* Initiation and propagation steps in pitting corrosion of austenitic stainless steels: monitoring by acoustic emission. Corrosion Sci 2001; 43(4): 627-41.

[154] Cakir A, Tuncell S, Aydin A. AE response of 316L SS during SSR test under potentiostatic control. Corrosion Sci 1999; 41(6): 1175-83.

[155] Magaino S, Kawaguchi A, Hirata A, *et al.* Spectrum analysis of corrosion potential fluctuations for localized corrosion of type304 stainless steel. J Electrochem Soc 1987; 134(12): 2993-7.

[156] Jones RH, Friesel MA. Acoustic emission during pitting and transgranular crack initiation in type304 stainless steel. Corrosion 1992; 48(9): 751-8.

[157] Mazille H, Rothea R, Tronel C. An acoustic emission technique for monitoring pitting corrosion of austenitic stainless steels. Corrosion Sci 1995; 37(9): 1365-75.

[158] Kim YP, Fregonese M, Mazille H, *et al.* Ability of acoustic emission technique for detection and monitoring of crevice corrosion on 304L austenitic stainless steel. NDT&E Int 2003; 36(8): 553-62.

[159] Conde A, Williams DE. Crevice corrosion and pitting detection on 304 stainless steel using electrochemical noise. Mater Corrosion 1999; 50(10): 585-90.

[160] Habib K, Bouresli K. Detection of localized corrosion of stainless steels by optical interferrometry. Electro Chem Acta 1999; 44(25): 4635-41.

[161] Sreekumari KR, Nandakumar K, Takao K, *et al.* Silver containing stainless steel as a new outlook to abate bacterial adhesion and microbiologically influenced corrosion. ISIJ Int 2003; 43(11): 1799-806.

[162] Sreekumari KR, Nandakumar K, Kikuchi Y. Bacterial attachment to stainless steel welds: Significance of substratum microstructure. Biofouling 2001; 17(4): 303-16.

[163] Damborenea JJ, Cristobal AB, Arenas MA, *et al.* Selective dissolution of austenite in AISI 304 stainless steel by bacterial activity. Mater Lett 2007; 61(3): 821-3.

[164] Geesey GG, Gillis RJ, Avci A, *et al.* The influence of surface features on bacterial colonization and subsequent substratum chemical changes of 316L stainless steel. Corrosion Sci 1996; 38(1): 73-95.

[165] Sreekumari KR, Takao K, Ujiro T, *et al.* High nitrogen stainless steel as a preferred substratum for bacteria and other microfouling organisms. ISIJ Int 2004; 44(5): 858-64.

[166] Scotto V, Lai ME. The ennoblement of stainless steels in seawater: A likely explanation coming from the field. Corrosion Sci 1998; 40(6): 1007-18.

[167] Dickinson WH, Caccavo Jr F, Lewandowski Z. The ennoblement of stainless steel by manganic oxide biofouling. Corrosion Sci 1996; 38(8): 1407-22.

[168] George RP, Muraleedharan P, Sreekmuari KR. Influence of surface characteristics and microstructure on adhesion of bacterial cells onto a type 304 stainless steel. Biofouling 2003; 19(1): 1-8.

[169] Ibars JR, Moreno DA, Ranninger C. MIC of stainless steels – a technical review of the influence of microstructure. Int Biodeter Biodegrad 1992; 29(3-4): 343-55.

[170] Baena MI, Marquez MC, Matres V, *et al.* Bactericidal activity of copper and niobium-alloyed austenitic stainless steel. Curr Microbiol 2006; 53(6): 491-5.

[171] Hong IT, Koo CH. Antibacterial properties, corrosion resistance and mechanical properties of Cu-modified SUS 304 stainless steel. Mater Sci Eng A 2005; 393(1-2): 213-22.

[172] Peng X, Yan J, Zhou Y, *et al.* Effect of grain refinement on the resistance of 304 stainless steel to breakaway oxidation in wet air. Acta Mater 2005; 53(19): 5079-88.

[173] Basu SN, Yurek GY. Effect of alloy grain size and silicon content on the oxidation of austenitic Fe-Cr-Ni-Mn-Si alloys in pure O_2. Oxid Metal 1991; 36(3-4): 281-315.

[174] Harper MA, Rapp RA. Codeposited chromium and silicon diffusion coatings for Fe-base *via* pack cementation. Oxid Metal 1994; 42(3-4): 303-33.

[175] Wang W, Wu M. Effect of silicon content and aging time on density, hardness, toughness and corrosion resistance of sintered 303LSC–Si stainless steels. Mater Sci Eng A 2006; 425(1-2): 167-71.

[176] Wilde BE. Influence of silicon on the intergranular corrosion behavior of 18Cr-8Ni type stainless steels. Corrosion 1988; 44(10): 699-704.

[177] Kajimura H, Ogawa K, Nagano H. Effect of α/β phase ratio and N addition on the corrosion resistance of Si-bearing duplex stainless steel in nitric acid. ISIJ Int 1991; 31(2): 216-22.

[178] Grunling HW, Bauer R. The role of silicon in corrosion resistant high-temperature coatings. Thin Solid Films 1982; 95(1): 3-20.

[179] Lin D, Chang T, Liu GL. Effect of Si contents on the growth-behavior of sigma phase in SUS 309L stainless steels. Scr Mater 2003; 49(9): 855-860.

[180] Hosoi Y, Moteki S, Shimoide Y, *et al.* Effect of nitrogen, carbon and tungsten on high-temperature oxidation of 12%Cr-15%Mn austenitic steel. ISIJ Int 1996; 36(7): 834-9.

[181] Masuyama F, Hiromatsu K, Hasegawa Y. High temperature oxidation behavior of high nitrogen ferritic steels. ISIJ Int 1996; 36(7): 825-33.

[182] Nishimura R, Kudo K. Stress-corrosion cracking of AISI304 and AISI316 austenitic stainless steels in HCl and H_2SO_4 solutions – prediction of time-to-failure and criterion for assessment of SCC susceptibility. Corrosion Sci 1989; 45(4): 308-16.

[183] Nishimura R. SCC failure prediction of austenitic stainless steels in acid solutions – effect of PH, anion species, and concentration. Corrosion 1990; 46(4): 311-8.

[184] Nishimura R. Stress-corrosion cracking of type 430 ferritic stainless steel in chloride and sulfate solutions. Corrosion 1992; 48(11): 882-90.

[185] Nishimura R. Effect of specimen width on stress-corrosion cracking of type304 stainless steel. Corrosion 1993; 49(10): 796-801.

[186] Nishimura R. The effect of chloride ions on stress-corrosion cracking of type304 and type316 austenitic stainless steels in sulfuric acid solution. Corrosion Sci 1993; 34(11): 1859-68.

[187] Nishimura R. The effect of potential on stress-corrosion cracking of type316 and type310 austenitic stainless steels. Corrosion Sci 1993; 34(9): 1463-73.

[188] Nishimura R, Katim I, Maeda Y. Stress corrosion cracking of sensitized type 304 stainless steel in hydrochloric acid solution - Predicting time-to-failure and effect of sensitizing temperature. Corrosion 2001; 57(10): 853-62.

[189] Nishimura R, Musalam I, Maeda Y. The effect of sensitizing temperature on stress corrosion cracking of type 316 austenitic stainless steel in hydrochloric acid solution. Corrosion Sci 2002; 44(6): 1343-60.

[190] Nishimura R, Sulaiman A, Maeda Y. Stress corrosion cracking susceptibility of sensitized type 316 stainless steel in sulphuric acid solution. Corrosion Sci 2003; 45(2): 465-84.

[191] Nishimura R, Maeda Y. Stress corrosion cracking of sensitized type 316 austenitic stainless steel in hydrochloric acid solution - effect of sensitizing time. Corrosion Sci 2003; 45(8): 1847-62.

[192] Nishimura R. Characterization and perspective of stress corrosion cracking of austenitic stainless steels (type 304 and type 316) in acid solutions using constant load method. Corrosion Sci 2007; 49(1): 81-91.

[193] Hong S, Lee S, Byun T. Temperature effect on the low-cycle fatigue behavior of type 316L stainless steel: Cyclic non-stabilization and an invariable fatigue parameter. Mater Sci Eng A 2007; 457(1-2): 139-47.

[194] Hong S, Yoon S. Lee S. The effect of temperature on low-cycle fatigue behavior of prior cold worked 316L stainless steel. Int J Fatigue 2003; 25(9-11): 1293-300.

[195] Smaga M, Walther F, Eifler D. Deformation-induced martensitic transformation in metastable austenitic steels. Mater Sci Eng A 2008; 483(Sp.Iss.SI): 394-7.

[196] Zhang ZF, Wang ZG. Cyclic deformation behaviour of a copper bicrystal with common primary slip planes. Philos Mag A 2001; 81(2): 399-415.

[197] Kaneko Y, Fukui F, Hashimoto S. Electron channeling contrast imaging of dislocation structures in fatigued austenitic stainless steels. Mater Sci Eng A 2005; 400-401: 413-7.

[198] Hong Y, Qiao Y, Liu N, *et al.* Effect of grain size on collective damage of short cracks and fatigue life estimation for a stainless steel. Fatigue and Fracture Eng Mater Struct 1998; 21(11): 1317-25.

[199] Zhang GP, Takashima K, Shimojo M, *et al.* Fatigue behavior of microsized austenitic stainless steel specimens. Mater Lett 2003; 57(9-1): 1555-60.

[200] Zhang GP, Takashima K, Higo Y. Fatigue strength of small-scale type 304 stainless steel thin films. Mater Sci Eng A 2006; 426(1-2): 95-100.

[201] Mocarski S. Total approach to reliability of case-hardened parts. Metal Prog 1970; 98(3): 96-6.

[202] Chiu PK, Weng KL, Wang SH, *et al.* Low-cycle fatigue-induced martensitic transformation in SAF 2205 duplex stainless steel. Mater Sci Eng A 2005; 398(1-2): 349-59.

[203] Rajanna K, Pathiraj B, Kolster BH. X-ray fractography studies on austenitic stainless steels. Eng Frac Mech 1996; 54(1): 155.

[204] Pineau A, Swam LFV, Pelloux RM. Cyclic stress-strain curves of a stainless austenitic steel in Ms-Md range. Scr Matell 1973; 7(6): 657-60.

[305] Mei Z, Morris Jr JW. Influence of deformation-induced martensite on fatigue crack propagation in 304 type steels. Metall Trans A 1990; 21(12): 3137-52.

[306] Biswas S, Sivaprasad S, Narasaiah N, *et al.* Load history effect on FCGR behaviour of 304LN stainless steel. Int J Fatigue 2007; 29(4): 786-91.

[307] Miller A, Estrin Y, Hu XZ. Magnetic force microscopy of fatigue crack tip region in a 316L austenitic stainless steel. Scr Mater 2002; 47(7): 441-6.

[308] Topic M, Tait RB, Allen C. The fatigue behaviour of metastable (AISI-304) austenitic stainless steel wires. Int J Fatigue 2007; 29(4): 656-65.

[209] Oh KH, Jung CK, Yang YC, *et al.* Acoustic emission behavior during fatigue crack propagation in 304 stainless steel. Key Eng Mater 2004; 261-263: 1325-30.

[210] Masuda C, Ohta A, Nishijima S, *et al.* Fatigue striation in a wide-range of crack propagation rates up to 70 m/cycle in a ductile structural steel. J Mater Sci 1980; 15(7): 1663-70.

[211] Grinberg NM. Stage II fatigue crack growth. Int J Fatigue 1984; 6(4): 229-42.

[212] Birkbeck G, Inckle AE, Waldron GWJ. Aspects of stage II fatigue crack propagation in low carbon steel. J Mater Sci 1971; 6(4): 319-25.

[213] Moorthy V, Jayakumar T, Raj B. Influence of microstructure on acoustic emission behavior during stage 2 fatigue crack growth in solution annealed, thermally aged and weld specimens of AISI type 316 stainless steel. Mater Sci Eng A 1996; 212(2): 273-80.

[214] Backer FD, Schoss V, Maussner G. Investigations on the evaluation of the residual fatigue life-time in austenitic stainless steels. Nucl Eng Des 2001; 206(2-3): 201-19.

[215] Krupp U, Christ HJ, Lezuo P, *et al.* Influence of carbon concentration on martensitic transformation in metastable austenitic steels under cyclic loading conditions. Mater Sci Eng A 2001; 319-321(Sp.Iss.SI): 527-30.

[216] Oka M, Yakushiji T, Enokizono M. Fatigue dependence of residual magnetization in austenitic stainless steel plates. IEEE Trans Magn 2001; 37(4): 2045-8.

[217] Nakasone Y, Iwasaki Y, Shimizu T. Non-destructive detection of damage in an austenitic stainless steel SUS 304 by the use of martensitic transformation. Int J Appl Electromagn Mech 2001/2002; 15(1-4): 309-13.

[218] Park DG, Kim DW, Timofeev VP, *et al.* Detection of low cycle fatigue in type 316 stainless steel using HTS-SQUID. Key Eng Mater 2004; 270-3: 1224-8.

[219] Schaaff P, Holzwarth U. Nondestructive detection of fatigue damage in austenitic stainless steel by positron annihilation. J Mater Sci 2005; 40(23): 6157-68.

[220] Barbieri A, Ilzhofer SH, Ilzhofer A, *et al.* Nondestructive positron-lifetime measurements during fatigue of austenitic stainless steel using a mobile positron beam. Appl Phys Lett 2000; 77(12): 1911-3.

[221] Yamamoto Y, Brady MP, Lu ZP, *et al.* Creep-resistant, Al_2O_3-forming austenitic stainless steels. Sci 2007; 316(5823): 433-6.

[222] Yamamoto Y, Takeyama M, Lu ZP, *et al.* Alloying effects on creep and oxidation resistance of austenitic stainless steel alloys employing intermetallic precipitates. Intermet 2008; 16(3): 453-62.

[223] Yamamoto Y, Brady MP, Lu ZP, *et al.* Alumina-forming austenitic stainless steels strengthened by laves phase and MC carbide precipitates. Metall Mater Trans A 2007; 38(11): 2737-46.

[224] Keown SR, Pickering FB. In: Creep Strength in Steel and High-temperature Alloys, The Inst Metal Lond, 1974; pp.229-34.

[225] Masuyama F, In: Advanced Heat Resistant Steels for Power Generation (eds R.Viswanathan and R.Nutting), The Institute of Metals, London, 1998; pp.33-47.

[226] Sourmail T, Bhadeshia HKDH, Mackay DJC. Neural network model of creep strength of austenitic stainless steels. Mater Sci Technol 2002; 18(6): 655-63.

[227] Laha K, Kyono J, Sasaki T, *et al.* Effect of additions of Ti, B and Ce on microstructural stability, creep strength and creep damage in austenitic stainless steel. Mater Sci Technol 2005; 21(11): 1309-17.

[228] Kesternich W. Dislocation-controlled precipitation of TiC particles and their resistance to coarsening. Philos Mag A 1985; 52(4): 533-48.

[229] Rajaraman R, Gopalan P, Venkadesan S. Positron trapping at precipitates in titanium-modified stainless steels. Mater Lett 1995; 24(4): 243-5.

[230] Opila EJ. Volatility of common protective oxides in high-temperature water vapor: Current understanding and unanswered questions. Mater Sci Forums 2004; 461-4: 765-73.

[231] Brady MP, Yamamoto Y, Santella ML, *et al.* Effects of minor alloy additions and oxidation temperature on protective alumina scale formation in creep-resistant austenitic stainless steels. Scr Mater 2007; 57(12): 1117-20.

[232] Nunes FC, Dille J, Delplancke JL, *et al.* Yttrium addition to heat-resistant cast stainless steel. Scr Mater 2006; 54(9): 1553-6.

[233] Nunes FC, Almeida LH, Dille J, *et al.* Microstructural changes caused by yttrium addition to NbTi-modified centrifugally cast HP-type stainless steels. Mater Charact 2007; 58(2): 132-42.

[234] Laha K, Kyono J, Shinya N. An advanced creep cavitation resistance Cu-containing 18Cr-12Ni-Nb austenitic stainless steel. Scr Mater 2007; 56(10): 915-8.

[235] Yao XX. On the grain boundary hardening in a B-bearing 304 austenitic stainless steel. Mater Sci Eng A 1999; 271(1-2): 353-9.

[236] Laha K, Kyono J, Kishimoto S, *et al.* Beneficial effect of B segregation on creep cavitation in a type 347 austenitic stainless steel. Scr Mater 2005; 52(7): 675-8.

[237] Laha K, Kyono J, Sasaki T, *et al.* Improved creep strength and creep ductility of type 347 austenitic stainless steel through the self-healing effect of boron for creep cavitation. Metall Mater Trans A 2005; 36(2): 399-409.

[238] Tanaka H, Murata M, Abe F, *et al.* The effect of carbide distributions on long-term creep rupture strength of SUS321H and SUS347H stainless steels. Mater Sci Eng A 1997; 234-6: 1049-52.

[239] Bouchard PJ, Fiori F, Treimer W. Characterisation of creep cavitation damage in a stainless steel pressure vessel using small angle neutron scattering. Appl Phys A-Mater Sci Process 2002; 74: S1689-91.

[240] Bouchard PJ, Withers PJ, McDonald SA, *et al.* Quantification of creep cavitation damage around a crack in a stainless steel pressure vessel. Acta Mater 2004; 52(1): 23-34.

[241] Ohtani T. Acoustic damping characterization and microstructure evolution during high-temperature creep of an austenitic stainless steel. Metall Mater Trans A 2005; 36(11): 2967-71.

[242] Herenu S, Armas IA, Armas AF. The influence of dynamic strain aging on the low cycle fatigue of duplex stainless steel. Scr Mater 2001; 45(6): 739-45.

[243] Srinivasan VS, Sandhya R, KBS Rao, *et al.* Effects of temperature on the low-cycle fatigue behavior of nitrogen alloyed type316L stainless steel. Int J Fatigue 1991; 13(6): 471-8.

[244] Sandhya R, Rao BS, Mannan SL. The effect of temperature on the low cycle fatigue properties of a 15Cr-15Ni, Ti modified austenitic stainless steel. Scr Mater 1999; 41(9): 921-7.

[245] Wu J, Lin C. Effect of strain rate on high-temperature low-cycle fatigue of 17-4 PH stainless steels. Mater Sci Eng A 2005; 390(1-2): 291-8.

[246] Srinivasan VS, Valsan M, Sandhya R, *et al.* High temperature time-dependent low cycle fatigue behaviour of a type 316L(N) stainless steel. Int J Fatigue 1999; 21(1): 11-21.

[247] Hong S, Lee S. The tensile and low-cycle fatigue behavior of cold worked 316L stainless steel: influence of dynamic strain aging. Int J Fatigue 2004; 26(8): 899-910.

[248] Srinivasan VS, Sandhya R, Valsan M, *et al.* The influence of dynamic strain ageing on stress response and strain-life relationship in low cycle fatigue of 316L(N) stainless steel. Scr Mater 1997; 37(10): 1593-8.

[249] Hong S, Lee B. Mechanism of dynamic strain aging and characterization of its effect on the low-cycle fatigue behavior in type 316L stainless steel. J Nucl Mater 2005; 340(2-3): 307-14.

[250] Samuel KG, Ray SK, Sasikala G. Dynamic strain ageing in prior cold worked 15Cr-15Ni titanium modified stainless steel (Alloy D9). J Nucl Mater 2006; 355(1-3): 30-7.

[251] Peng K, Qian K, Chen W. Effect of dynamic strain aging on high temperature properties of austenitic stainless steel. Mater Sci Eng A 2004; 379(1-2): 372-7.

[252] Alain R, Violan P, Mendez J. Low cycle fatigue behavior in vacuum of a 316L type austenitic stainless steel between 20 and 600 C.1. fatigue resistance and cyclic behavior. Mater Sci Eng A 1997; 229(1-2): 87-94.

[253] Kim DW, Ryu W, Hong JH, *et al.* Effect of nitrogen on high temperature low cycle fatigue behaviors in type 316L stainless steel. J Nucl Mater 1998; 254(2-3): 226-33.

[254] Gupta C, Chakravartty JK, Wadekar SL, *et al.* Effect of serrated flow on deformation behaviour of AISI 403 stainless steel. Mater Sci Eng A 2000; 292(1): 49-55.

[255] Tjong SC, Zhu SM. Creep and low-cycle fatigue behavior of ferritic Fe-24Cr-4Al alloy in the dynamic strain aging regime: Effect of aluminum addition. Metall Mater Trans A 1997; 28(6): 1347-55.

[256] Tjong SC, Zhang JS. Low cycle fatigue behavior of Ferritic Fe-24Cr-4Al alloy at high temperatures. Scr Metall Mater 1995; 32(10): 1589-93.

[257] Venkadesan S, Phaniraj C, Sivaprasad PV, *et al.* Activation energy for serrated flow in a 15Cr-15Ni Ti-modified austenitic stainless steel. Acta Metall Mater 1992; 40(3): 569-80.

[258] Anglada M, Nasarre M, Planell JA. High-temperature mechanical twinning of 2 Fe-Cr-Mo-Ni ferritic stainless steels. Scr Metall 1987; 21(7): 931-6.

[259] Stewart GR, Jonas JJ. Static and Dynamic Strain Aging at High Temperatures in 304 Stainless Steel. ISIJ Int 2004; 44(7): 1263-72

[260] Samuel EI, Choudhary BK, Rao KBS. Influence of temperature and strain rate on tensile work hardening behaviour of type 316 LN austenitic stainless steel. Scr Mater 2002; 46(7): 507-12.

[261] Ilola R, Kemppainen M, Hanninen H. Dynamic strain aging of austenitic high nitrogen Cr-Ni and Cr-Mn steels. Mater Sci Forum 1999; 318-20: 407-12.

[262] Girones A, Llanes L, Anglada M, *et al.* Dynamic strain ageing effects on superduplex stainless steels at intermediate temperatures. Mater Sci Eng A 2004; 367(1-2): 322-8.

[263] Tsuzaki K, Hori T, Maki T, *et al.* Dynamic strain aging during fatigue deformation in type 304 austenitic stainless steel. Mater Sci Eng 1983; 61(3): 247-60.

[264] Samuel KG, Mannan SL, Rodriguez P. Serrated yielding in AISI316 stainless steel. Acta Metall 1988; 36(8): 2323-27.

CHAPTER 10

Novel and New Applications of Stainless Steels

Abstract: Stainless steels are used in many applications like heat exchangers, furnace liners, automotive components, solar panels, and kitchenware. This chapter attempts to cover new applications that have recently been reported. A few novel uses of duplex stainless steels are also covered, these are the use of duplex stainless steels for temperature sensing and magnetic drug targeting.

Keywords: Feroplug, Sigmaplug, magnetic recording, hot-spot indication, temperature indication, stent, radiopaque, drug targeting, PRESS, strength retention, stiffness retention.

1. TEMPERATURE AND HOT-SPOT INDICATION

Duplex stainless steels, whose ferromagnetic ferrite phase decomposes upon thermal ageing, have been utilised for temperature measurement below about 600°C and the technique is named Feroplugs [1, 2]. The Feroplug technique works on the fact that the amount of remnant ferrite in duplex stainless steels after extended ageing is temperature-dependent [1, 2]. And so one would be able to deduce the prior working temperature of a component, to which a Feroplug was attached, by measuring the remnant ferrite content of the Feroplug. Nevertheless, the ferrite phase decomposes almost completely into secondary austenite and the intermetallic sigma phase above 600°C and so there the Feroplug technique is rendered useless, because there is virtually no ferrite in the aged materials.

Recent research by one of the inventors of the Feroplug technique (Lai) has proposed to use the cryogenic magnetic transition of the sigma phase for temperature measurement and the new technique is named Sigmaplug [3, 4]. Preliminary results have shown that the Sigmaplug technique is employable for hot spot indication [5].

2. MAGNETIC RECORDING AND APPLICATIONS THAT UTILISE MAGNETISM

The transformation from the non-ferromagnetic γ to the ferromagnetic α' has been utilised for measurement of contact pressure in indentation, monitoring of stresses in sliding experiments and magnetic recording [6, 7]. The use of metastable austenitic stainless steels as magnetic recording materials is quite interesting. It has been demonstrated that by putting a piece of paper on a stainless steel backing sheet and then penning on the paper, traces of martensite will be induced on the backing sheet and these martensite traces more or less replicate the trajectories travelled by the pen tip (Fig. **1**). What was written on the paper can therefore be revealed later on by examining the martensitic regions. Therefore, metastable austenitic stainless steels may serve the purpose of signature authentication (Fig. **1**).

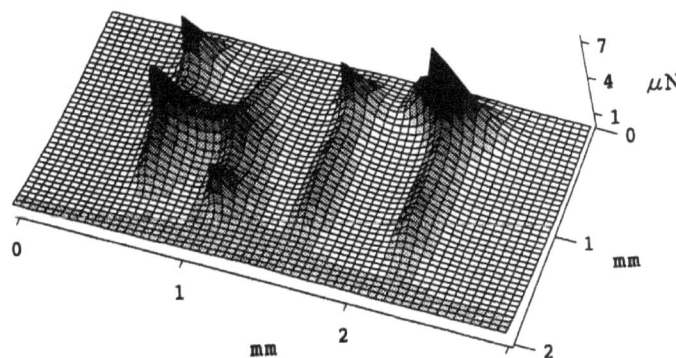

Figure 1: Magnetic force distribution on an AISI304 steel after three characters 'H', 'I' and 'T' were written on it [Ishigaki H, Konishi Y, Kondo K, *et al.* Possibility as a magnetic recording media of austenitic stainless steel using stress-induced phase transformation. J Magn Magn Mater 1999; 193(1-3): 466-9. With permission for reproduction from Elsevier].

While the applications just introduced take advantage of the ferromagnetism of deformation-induced martensite (α '), ferromagnetism is unwelcome under some circumstances. For example, even the presence of trace amounts of residual martensite in the experimental equipment for cryogenic physics is undesirable, in as much as artefacts will be produced. In this case, very stable austenitic stainless steels exhibiting cryogenic paramagnetism must be used. To ensure cryogenic paramagnetism, steel chemistry must be carefully designed such that the austenite matrix is stable under the loading conditions and working temperatures to which the steel is exposed. A specially made austenitic stainless steel possessing cryogenic paramagnetism and good low-temperature strength has been fabricated and used as the collars for the LHC dipole magnets at CERN [8].

3. TRANSPORTATION

Use of ferritic stainless steels in the automotive industry has had a long history and is pretty well known [9]. Equally useful for the automotive industry are the austenitic grades and the duplex grades [10]. While austenitic stainless steels may rival their competitors like aluminium in terms of properties like specific stiffness and specific strength [10], they stand out because of their crashworthiness (very absorbent of energy) under dynamic loading conditions [10]. The high-cost of nickel may sometimes render the use of austenitic stainless steels uneconomical. For this reason, new steels with high Mn and N levels have been made for use as automobile structural materials [11]. These steels may have mechanical and corrosion properties that rival the 304 steel [11].

Besides vehicles moving on the land, stainless steels have also found their niche in the making of those surfing the sea. The 15-5PH stainless steel has been confirmed to be a suitable candidate for use as the hydrofoil for high-speed passenger craft [12]. A study by Fujita *et al.* [12] has suggested the heat treating conditions that may inhibit the deformation of the 15-5PH steel associated with the re-solution treatment after welding. Superaustenitic stainless steels, with raised N and Mo levels, are also suitable for applications involving contact with sea water [13].

4. CONSTRUCTION

Stainless steels have long been used as roofing materials and facades by the construction industry. The Chrysler building in New York is an epitome. One of the main reasons for the use of stainless steels for construction purposes is that their corrosion resistance in atmospheric conditions is substantially higher compared with those of zinc and copper [14]. A recent interesting study by Wallinder *et al.* [15] has looked into the release rates of Ni and Cr of 304 and 316 upon exposure to atmospheric conditions. The influences of rain intensity and its pH value have been studied by these authors, too [16].

Using stainless steels for construction purposes has been proved to be cost-saving in the long haul. When corrosion of steel reinforcements of concrete structures occurs, the costs of repair can be prohibitively high. For steel reinforcements, galvanising may provide short-term protection only [16, 17]. Coating the reinforcements with epoxy may be an option, but this may degrade the adhesion between the reinforcements and the concrete [18]. To avoid this problem, use of stainless steels as concrete reinforcements may be a wise alternative [19]. It has to be noted that even though stainless steel reinforcements may cut cost in the long run, the high initial cost, which is mainly incurred by the use of Ni, may be prohibitive in some cases. Bautista *et al.* [20] have shown that the less expensive low-Ni stainless steels may be a feasible alternative to traditional stainless steels as concrete reinforcements.

Besides reinforcements, stainless steels have also been suggested for use as loading-sustaining structural components. A concrete bridge built by the Oregon Department of Transportation is a good illustration. The rebars of this bridge are made of duplex stainless steels and this bridge is designed to last 120 years with 'literally no maintenance' whatsoever, in spite of the salty atmosphere and seismic activity in Oregon [21].

A very good review article on the advantages of using stainless steels for construction purposes has been contributed by Gardner [22]. Some of the advantages of stainless steels include fire resistance and better retention of strength and stiffness at elevated temperatures compared with carbon steels [22-25] (refer to Figs. **1.1** and **1.2** in Chapter **1**). Therefore, stainless steel structures will be able to retain better structural integrity in a fire accident. Good plastic deformation behaviour and the ability to absorb energy also make stainless steels stand out as structural components [26].

5. MEDICAL USES

Medically, coronary stents may be made of stainless steels like 316L. But their high Fe contents may render them non-compatible with magnetic resonance imaging (MRI) and to be poor fluoroscopic materials [27]. In spite of these limitations and a myriad of materials that have been chosen as stent materials (like titanium), stainless steels are still favoured, as evidenced by the fact that seven out of the eight coronary stents approved by the US Food and Drug Administration are made of stainless steels [27]. Compared with other competitive materials like Ti, stainless steels may be better in formability, weldability and affordability.

To overcome the fluoroscopic radiopacity, recent attempts have been made to produce radiopaque stainless steel stents [28]. The radiopaque stents, known as Platinum-Enhanced Stainless Steel Stents (PRESS), are based on UNS S31673 whose enhanced fluoroscopic radiopacity is imparted by platinum [28].

Surface microstructuring using electrochemical etching of the stainless steel stents has been attempted by Stover *et al.* [29] to make drug-coated stents. To control magnetically the rate of drug release, Huang *et al.* [30] have attempted to coat the surfaces of stents made of SUS316 with a drug-laden magnetic-gelatin hydrogel.

Stents made of the stainless steel 430, which is ferromagnetic, have been utilised by Aviles *et al.* [31] for magnetic drug targeting. This ferromagnetic stent, when implanted near the target zone in the body and energised by an externally applied magnetic field, may assist retention of magnetic drug carrier particles (Fig. **2**), thereby enhancing drug delivery to the desired part of the body.

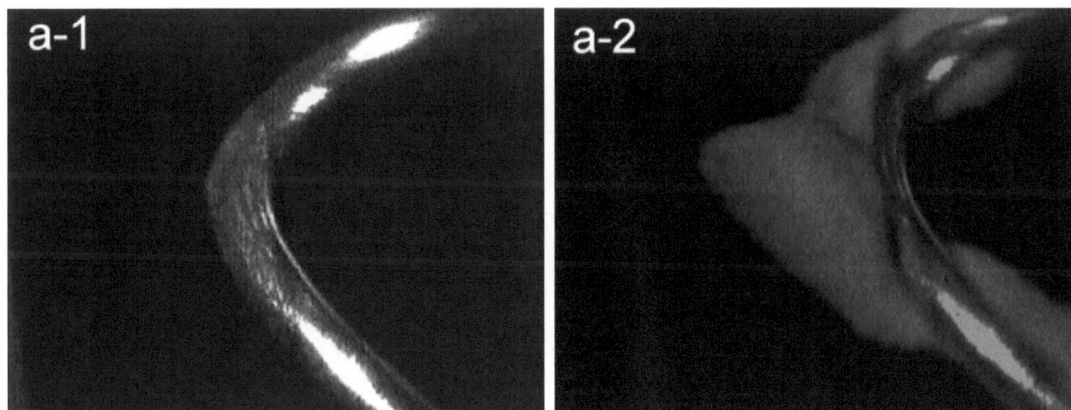

Figure 2: Drug particles captured by a 430 stainless steel stent after the application of a magnetic field. (**a-1**): before application of field. (**a-2**): after application of field. [Aviles MO, Chen H, Ebner ED, *et al. In vitro* study of ferromagnetic stents for implant assisted-magnetic drug targeting. J Mag Mag Mater 2007; 311(1): 306-11. With permission for reproduction from Elsevier].

Shih *et al.* [32], using a proprietary process, have produced an amorphous oxide layer in a 316L wire which offers outstanding corrosion resistance in Ringer's physiological solution at 37℃. These authors [33] have also found that the amorphous oxide layer may reduce thrombosis when the treated 316L wire is used as stents. Downsizing stainless steel stents may bring even more benefits such as reduction of restenosis [34].

Stainless steels are also used as retinal tacks [35]. A very striking recent study has found that stainless steel retinal tacks used for 21 years shows almost no toxicity.

6. CONTAINERS FOR CHEMICAL AND RADIOACTIVE SUBSTANCES

Stainless steels are used in the petroleum industry where they will be subjected to pitting and stress corrosion cracking in the presence of Cl⁻. Oberndorfer *et al.* [36] have recently simulated the effects of such environments by using a NaCl solution containing 3 to 100000ppm Cl⁻ whose temperatures were varied between 40 and 200℃. They have established a diagram that is usable for guaranteeing safe operations of a range of different stainless steels in such environments.

Besides chemicals, stainless steels are also used for storing radioactive wastes. It has recently been demonstrated that containers for medium-level radioactive waste made of AISI304L may last for 30 years [37].

In spite of the dangers of hydrogen embrittlement and rather serious alkaline attack, austenitic stainless steels 304 and 316 have been demonstrated to be suitable candidates as the containers of Ni/MN$_X$ batteries, which may be used as the power supply for electric vehicles [38].

7. MISCELLANEOUS USES

Titanium oxide (TiO_2) is utilisable in a variety of applications like photocatalytic purification and photovoltaic cells. Recently, Yang *et al.* [39] have shown that the photoactivity of the titanium oxide layer obtained by using a metal organic chemical vapour deposition on a stainless steel substrate is higher than the photoactivies of TiO_2 layers grown on substrates like Ti alloy, Al alloy, silicon and glass.

Other recent applications of stainless steels include force transducers possessing small errors associated with hysteresis [40], impellers for centrifugal compressors [41], electrode antenna for the observation of electric fields in shallow seas [42], and very sensitive (yet stable) and robust spray tips for sheathless microelectrospray ionisation (μESI) [43].

REFERENCES

[1] Lai JKL, Shek CH, Duggan BJ. The transformation characteristics of ferrite in a cast of duplex stainless steel and its applications in temperature measurement. Mater High Temp 1992; 10(1): 60-2.

[2] Lai JKL, Lo KH, Shek CH, *et al.* Magnetic and ageing behaviour of 7MoPLUS and the viability of monitoring ferrite decomposition using AC magnetic susceptibility. Mater Sci Eng A 2005; 406(1-2): 110-8.

[3] Lai JKL, Shek CH, Wong KW. A novel technique to detect hot spots in high temperature boilers. Sens Actuat A 2001; 95(1): 51-4.

[4] Lai JKL, Wong KW, Li DJ. Magnetic properties of Feroplug material at sub-ambient temperature Acta Mater 1996; 44(2): 567-71.

[5] Wong KW. Ph.D Thesis. Ageing characteristics and microstructural studies of duplex stainless steels. City Univ Hong Kong, Hong Kong (1996).

[6] Ishigaki H, Konishi Y, Kondo K, *et al.* Possibility as a magnetic recording media of austenitic stainless steel using stress-induced phase transformation. J Magn Magn Mater 1999; 193(1-3): 466-69.

[7] Sanda H, Takai M, Shigematsu H, *et al.* MeV ion beam induced localized enhancement of magnetization in stainless steel foils. Nucl Instrum Method Phys Res B 1991; 59: 778-80.

[8] Bertinelli F, Fudanoki F, Komori T, *et al.* Production of austenitic steel for the LHC superconducting dipole magnets. IEEE Trans Appl Supercond 2006; 16(2): 1773-6.

[9] Lula RA. Stainless Steel. Am Soc Metal. 1986.

[10] Cunat P. Stainless Steel Properties for Structural Automotive Applications. A paper presented on the occasion of the Metal Bulletin International Automotive Materials Conference, Cologne, June, 2000 (www.euro-inox.org/pdf/auto/StructuralAutomotiveApp_EN.pdf, document assessed on 02-April-2010).

[11] Toor I, Hyun PJ, Kwon HS. Development of high Mn-N duplex stainless steel for automobile structural components. Corros Sci 2008; 50(2) : 404-10.

[12] Fujita A, Kawano T, Sueoka H, *et al.* Heat Treatment of Weld Part in 15-5 PH Stainless Steel and Manufacturing of Actual Hydrofoil for a High Speed Passenger Craft. ISIJ Int 1998; 38(8): 866-74.

[13]   Latha G, Rajendran N, Rajeswari S. Evaluation of superaustenitic stainless steels as construction materials in sea water service systems. J Mater Sci Lett 1998; 17(6): 519-22.

[14]   Leygraf C, Graedel TE. Atmospheric Corrosion. John Wiley & Sons, New York, 2000.

[15]   Wallinder IO, Lu J, Bertling S, *et al.* Release rates of chromium and nickel from 304 and 316 stainless steel during urban atmospheric exposure - a combined field and laboratory study. Corrosion Sci 2002; 44(10): 2303-19.

[16]   Ramirez E, Gonzalez JA, Bautista A. The protective efficiency of galvanizing against corrosion of steel in mortar and in Ca(OH)(2) saturated solutions containing chlorides. Cem Concr Res 1996; 26(10): 1525-36.

[17]   Bautista A, Gonzalez JA. Analysis of the protective efficiency of galvanizing against corrosion of reinforcements embedded in chloride contaminated concrete. Cem Concr Res 1996; 26(2): 215-24.

[18]   Rasheeduzzafar F, Dakhil FH, Bader MA, *et al.* Performance of corrosion resisting steels in chloride-bearing concrete. ACI Mater J 1992; 89(5): 439-48.

[19]   Castro H, Rodriguez C, Belzunce FJ, *et al.* Mechanical properties and corrosion behaviour of stainless steel reinforcing bars. J Mater Process Technol 2003; 143-4(Sp Iss): 134-7.

[20]   Bautista A, Blanco G, Velasco F. Corrosion behaviour of low-nickel austenitic stainless steels reinforcements: A comparative study in simulated pore solutions. Cem Concr Res 2006; 36(10): 1922-30.

[21]   Carpenter Technology Corporation. New Bridge Using Stainless Steel Rebar to Last 120 Years in Corrosive Marine and Earthquake Environment. New materials International. (http://www.newmaterials.com/Customisation/News/General/General/New_Bridge_Using_Stainless_Steel_Rebar_to_Last_120_Years_in_Corrosive_Marine_and_Earthquake_Environment.asp) (assessed on Apr-2010).

[22]   Gardner L. The use of stainless steel in structures. Prog Struct Eng Mater 2005; 7(2): 45-55.

[23]   Gardner L, Ng KT. Temperature development in structural stainless steel sections exposed to fire. Fire Saf J 2006; 41(3): 185-203.

[24]   Gardner L, Baddoo NR. Fire testing and design of stainless steel structures. J Constr Steel Res 2006; 62(6): 532-43.

[25]   Sakumoto Y, Nakazato T, Matsuzaki A. High-temperature properties of stainless steel for building structures. J Struct Eng 1996; 122(4): 399-406.

[26]   Sarno LD, Elnashai AS, Nethercot DA. Seismic retrofitting of framed structures with stainless steel. J Constr Steel Res 2006; 62(1-2): 93-104.

[27]   Mani G, Feldman MD, Patel D, *et al.* Coronary stents: A materials perspective. Biomater 2007; 28(9): 1689-710.

[28]   Craig CH, Friend CM, Edwards MR, *et al.* Mechanical properties and microstructure of platinum enhanced radiopaque stainless steel (PERSS) alloys. J Alloy Compd 2003; 361(1-2): 187-99.

[29]   Stover M, Gluszko MR, Schratzenstaller T, *et al.* Microstructuring of stainless steel implants by electrochemical etching. J Mater Sci 2006; 41(17): 5569-75.

[30]   Huang L, Yang M. Behaviors of controlled drug release of magnetic-gelatin hydrogel coated stainless steel for drug-eluting-stents application. J Magn Magn Mater 2007; 310(2): 2874-6.

[31]   Aviles MO, Chen H, Ebner ED, *et al. In vitro* study of ferromagnetic stents for implant assisted-magnetic drug targeting. J Magn Magn Mater 2007; 311(1): 306-11.

[32]   Shih C, Shih C, Su Y, *et al.* Effect of surface oxide properties on corrosion resistance of 316L stainless steel for biomedical applications. Corrosion Sci 2004; 46(2): 427-41.

[33]   Shih C, Shih C, Su Y, *et al.* Characterization of the thrombogenic potential of surface oxides on stainless steel for implant purposes. Appl Surf Sci 2003; 219(3-4): 347-62.

[34]   Rittersma SZH, Winter RJ, Koch KT, *et al.* Impact of strut thickness on late luminal loss after coronary artery stent placement. Am J Cardiol 2004; 93(4): 477-80.

[35]   Javey G, Schwartz SG, Flynn HW, *et al.* Lack of Toxicity of Stainless Steel Retinal Tacks During 21 Years of Follow-Up. Ophthalmic Surg Laser Imaging 2009; 40(1): 75-6.

[36]   Oberndorfer M, Thayer K, Kastenbauer M. Application limits of stainless steels in the petroleum industry. Mater Corrosion 2004; 55(3): 174-80.

[37]   Vehovar L, Tandler M. Stainless steel containers for the storage of low and medium level radioactive waste. Nucl Eng Des 2001; 206(1): 21-33.

[38]   Chuang HJ, Chen SY, Chan SLI. Corrosion and hydrogen damage resistance of stainless steels in Ni MHx batteries. Corrosion Sci 1999; 41(7): 1347-58.

[39]   Yang J, Li Y, Wang F, *et al.* New application of stainless steel. J Iron Steel Res Int 2006; 13(1): 62-66.

[40] Aydemir B, Kaluc E, Fank S. Influence of heat treatment on hysteresis error of force transducers manufactured from 17-4PH stainless steel. Measurement 2006; 39(10): 892-900.

[41] Nowacki J. Weldability of 17-4 PH stainless steel in centrifugal compressor impeller applications. J Mater Process Technol 2004; 157-58 (Sp.Iss): 578-83.

[42] Nakamura T, Hirose C, Hirose R, *et al.* Observation of electric fields in the shallow sea using the stainless steel electrode antenna system. Phys Chem Earth 2006; 31(4-9): 352-5.

[43] Ishihama Y, Katayama H, Asakawa N, *et al.* Highly robust stainless steel tips as micro electro spray emitters. Rapid Commun Mass Spectrosc 2002; 16(10): 913-8.

CHAPTER 11

Improvement of Bulk and Surface Properties of Stainless Steels

Abstract: This chapter is on techniques that can be used to improve the properties of stainless steels. Some of these techniques improve surface properties, while some of them enhance the properties throughout the bulk of the material. Many of the techniques improve properties through grain refinement, whereas some rely on a phase transformation. Some of the techniques are specifically devised for a particular type of stainless steel (the technique that utilises reverse transformation of martensite is mainly for the austenitic type) or for steels in the liquid state.

Keywords: Surface modification, laser surface modification, grain refinement, carburisation, colossal carbon saturation, Kolstering, S-phase, nitridation, heavy plastic deformation, martensite, reversion transformation, intercritical annealing.

1. SURFACE MODIFICATION AND TREATMENT

Surface engineering is the branch of science that aims at improving the surface properties of materials and components. Compared with conventional techniques for property enhancement, surface enhancing techniques such as lasers are usually less energy intensive and may be deployed on-site. This chapter will go through the recent techniques for surface treatment of stainless steels.

1.1. Surface Grain Refinement

Very often, as the grain sizes decrease, mechanical and corrosion properties of metals become better. The improvement is even more noticeable if the grain sizes are reduced to the nanometre scale. It is to be noted even if only the surface region is nanostructured, the properties of the bulk material are also improved [1]. Fatigue life, for example, is prolonged because the surface nanocrystalline layer can impede dislocation movement [2]. Corrosion properties are enhanced since the copious amounts of grain boundaries in the nanostructured surface region enable fast diffusion of Cr [3].

Recent techniques for obtaining stainless steels with a nanostructured surface include ultrasonic peening (UP) [3], high-energy shot peening [4], ultrasonic shot peening(USP) [5], surface mechanical attrition (SMA) [2, 6, 7] and sandlblasting [8].

The nanostructuring mechanisms involved in these techniques have been investigated in detail [2, 5, 9]. In ultrasonic shot peening, the colossal amounts of dislocations and the formation of small shear bands are though to be the underlying mechanism [3]. On the other hand, the formation of dense dislocation walls (DDWs) and dislocation tangles (DTs) and their subsequent transformations to highly misoriented grain boundaries have been suggested to be responsible in the SMA of pure Fe. In AISI304, the subdivision of grains due to intersecting twins and the formation of nanosized martensite regions are regarded as the contributing factors [7]. In these surface nanostructuring processes, the top surface layer is subjected to very severe plastic deformation at very high strain rates. Recently, in studying the grain refinement process within adiabatic shear localisation, which also involves severe plastic deformation at high strain rates, it has been found that secondary twinning inside shear bands plays a critical role in the formation of the nanostructure in 316L [10].

In addition to plastic deformation, other techniques such as magnetron sputter deposition has been employed to fabricate nanostructured thin films of 330 austenitic stainless steel, too [11, 12]. The grains of these films are heavily twinned on [111] with a nanoscale (~4nm) spacing.

1.2. Laser Surface Modification

Lasers, and electron beams, which may bring about melting, phase transformation, *etc.*, in the surface regions of materials have been utilised for surface enhancement for a long time. Amongst them, lasers are arguably the most commonly used. Many types of lasers (excimer, Nd:YAG, CO2) have been utilised for surface modification of stainless steels. These modifications include surface melting [13-19], laser annealing [20-22], laser cladding [23, 24], laser alloying [25-31], laser peening [32], and laser transformation hardening [33, 34].

In laser surface melting (LSM), a thin surface layer is melted by the laser beam and then this shallow layer is cooled down very quickly by the bulk of the matrix. LSM has been reported to improve the intergranular corrosion resistance [13, 18, 35, 36] and pitting corrosion resistance [14-18, 37] of stainless steels, because carbides [38] and impurities like MnS [39] are diminished in size. The formation of Mn-Si compounds is also thought to contribute to pitting corrosion resistance because the quantity of Mn available for forming MnS is lowered [40]. It has also been reported that LSM may promote preferred orientation along the [200] direction and the formation of ferrite in the melted layer. These factors are thought to be beneficial to pitting corrosion resistance [14]. A recent study by Yang *et al.* [41] has demonstrated that laser surface melting and annealing may bring about a high density of low-Σ CSL boundaries in 304. These boundaries may enhance intergranular corrosion resistance substantially. When the surfaces of stainless steels are lightly melted by lasers, the extremely high cooling rate due to the bulk of the substrate may bring about a nanostructured surface layer, resulting in substantial surface enhancement [42].

Although LSM is in general viewed as a property-enhancing technique, some researchers have reported negatively on this method [43-45]. A study by Kwok *et al.* [16] reveals that LSM may degrade the cavitation erosion resistance of S30400 and S32760 stainless steels because of the formation of delta ferrite. The work by Parvathavarthini *et al.* [43] on laser surface treatment in a nitrogen atmosphere showed a reduction in the critical pitting potential because of the roughness and inhomogeneity of the melted surface. When nitrides are present after the LSM treatment in a nitrogen atmosphere, resistance to pitting is compromised [44]. Nevertheless, pitting is improved if N is in solid solution [17]. So, it is important to control the processing parameters (scan speed, intensity of the laser, etc) in such a way that the laser-treated surfaces are smooth and nitride-free.

The incorporation of intermetallics and carbides to stainless steels through laser-cladding can enhance tribological properties [23-29]. Laser cladding of stainless steels with ceramic powders like WC and SiC to form a surface metal matrix composite may improve wear resistance and corrosion-erosion resistance [25-29, 46-49].

Lasers may also be used to cleanse the surface of stainless steels that is contaminated with micro-organisms [50-52] and surface oxides [53]. When a laser beam is shone briefly on the stainless steel, these contaminants will be vaporised, leaving behind a shiny stainless steel surface.

Laser peening, which utilises a laser beam of very high intensity (~GW/cm2), is capable of generating very large compressive stresses and hence can shore up fatigue properties [32, 54] and resistance to stress corrosion cracking [54]. Laser peening can better the pitting corrosion resistance of stainless steels, too [32]. It is to be noted that compressive surface residual stresses may also be achieved by laser-annealing [20].

In addition to lasers, high-energy electron beams [55, 56], high-intensity pulsed ion beam (HIPIB) [57, 58], and saddle field neutral fast atom beam [59] have been used to enhance the surface properties of stainless steels [55, 56]. Typically, these high-energy sources are used to slightly melt the surface of stainless steels, thereby vaporising impurities like MnS. Surface melting also homogenises the composition of the surface region and smooth out any irregularities. Removal of compositional heterogeneities and surface roughness surely are beneficial to fatigue and corrosion properties.

1.3. Carburisation (Colossal Carbon Supersaturation)

In Chapter **8**, the generation of the S phase (which is obtained by introducing carbon (nitrogen) in stainless steels without causing precipitation) for surface enhancement has been introduced. This section will thus not delve on this phase. Instead, a few proprietary methods which are capable of introducing a colossal amount of carbon (nitrogen) into stainless steels without the precipitation of carbides are covered.

One common problem associated with conventional carburisation is the precipitation of carbides when the carbon content reaches a high level. Therefore, the possibility of introducing a high level of carbon to stainless steels without causing carbide formation has been the subject of a lot recent research. For duplex stainless steels, Jauhari *et al.* [60] have found that by superplastically deforming them during carburising, the surface tribological properties can be greatly improved compared with conventional carburisation, as the diffusion of carbon is facilitated [60]. While this may increase the surface carbon content somewhat, it never gets very high. To achieve substantial carbon contents in the surface regions, several proprietary techniques have been developed recently.

The Pointe treatment for austenitic stainless steels involves carburising in a gas mixture of CO and H_2 below 773K [60] and has been demonstrated to be a useful surface enhancement technique [60-62]. Another new, novel gas-phase carburising technique has been developed by the Swagelok Company for austenitic stainless steels. The carburisation process involves removal of the Cr_2O_3 surface oxide at 200°C in a dry HCl atmosphere and a subsequent carburisation treatment (CO and H2 gases) between 465°C and 475°C for 20-30h [63-66].

Removal of the surface oxide enables very efficient entry of C into the substrate. Between 465°C and 475°C, substitutional elements are not mobile, whereas interstitial elements are able to move around. As a result, a steel carburised in this way is said to be carburised under paraequilibrium [64, 65], which allows a colossal supersaturation of carbon (12at%, about 800 times the equilibrium solubility) in the surface region without the precipitation of carbides. The avoidance of carbide precipitate is attributed to the immobility of the substitutional elements. Beyond 12at%, M5C2 (the Hagg carbide) and M7C3 form, the former being the dominant one [63, 75]. It is to be noted that the Hagg carbide seldom forms in austenitic stainless steels. The colossal C-supersaturation gives rise to superb corrosion, fatigue and mechanical properties [63, 65-69].

Kolserising® is also a proprietary low-temperature surface carburising technique (developed by the Bodycote Metal Technology Group [70]). The Kolstering technique can bring about substantial C saturation in the surface of austenitic stainless steels (6-7wt%). The Kolserised surface, which is a mixture of the S phase and carbides (possibly the Hagg carbide) [71], exhibits a combination of high hardness and good ductility and enhanced corrosion resistance [70, 71].

In addition to the supersaturation of carbon, the supersaturation of nitrogen has also been used for surface property enhancement. Christiansen *et al.* [72] have devised a patented method for attaining colossal N-supersaturation (~38at%) in austenitic stainless steels. In this method, the surface chromium oxide is first removed to expedite the entry of N. And then a thin film of Ni is deposited on the stainless steel substrate, followed by a nitriding treatment in an ammonia-containing gas.

1.4. Other Surface Enhancing Techniques

Recently, many interesting, novel surface modification methods have been shown to be feasible for the enhancement of corrosion resistance. Saleh *et al.* [73] have found that by applying cyclic potentiodynamic polarisation to a 316LVM stainless steel between the potential of hydrogen and oxygen evolution, the passive surface film formed will possess very good resistance to general corrosion and pitting. Exposure to UV light during passivation [74] and electropolishing [75] have also been reported to enhance the corrosion resistance of the surface film. Shih *et al.* [76] have utilised a proprietary process to attain an amorphous oxide layer in a 316L wire. The amorphous oxide layer possesses outstanding corrosion resistance in Ringer's physiological solution at 37°C compared with the protective oxides that formed *via* thermal oxidising at 450°C for 15 minutes and electropolishing in a phosphoric acid-glycerin solution [76]. Fredj *et al.* [77, 78] have shown that by using liquid nitrogen as a coolant during grinding, a surface layer of lower roughness, with less defects, of higher work hardening and a residual stress that is less tensile will form. The resulting surface possesses better resistance to both fatigue [77] and pitting [78].

2. GRAIN REFINEMENT OF THE BULK

2.1. Grain Refinement in the Solid State

2.1.1. Grain Refinement of Austenitic Stainless Steels

Plastic deformation is quite often used to increase the strength of austenitic stainless steels. Nevertheless, strengthening in this way leads to poor formability and ductility. In order to overcome this problem, strengthening through grain refinement is more desirable. Besides strength, wear resistance [79], pitting resistance [80], and resistance to cavitation and cavitation-erosion [81, 82] can also be improved by reducing the grain size.

2.1.1.1. Reversion Transformation of Martensite

For metastable austenitic stainless steels, using plastic deformation to induce martensitic transformation and then reverting the martensite to austenite ($\alpha' \to \gamma$) through re-heating can lead to grain refinement [83-90], provided that some criteria are met [87]. These criteria are complete transformation to α' after prior cold deformation and a low temperature for the reversion transformation of $\alpha' \to \gamma$. The fulfilment of these criteria may be achieved by properly adjusting the composition [87].

As far as alloying elements are concerned, carbon deserves special attention. Both the final grain size and the mode of $\alpha' \to \gamma$ reversion are influenced by the level of carbon, this is because it affects both boundary mobility [91, 92] and the propensity to carbide precipitation (which may alter the chromium content of the matrix [83]). Gavard *et al.* [93] have found that when the carbon content drops down to about 15 μ g/g in plastically-deformed 18Cr-12Ni austenitic stainless steels, the recrystallised grain size will be smaller because of an enhanced martensitic transformation. But below 15 μ g/g, raised grain boundary mobility will result in enhanced grain growth. Therefore, the carbon content must be carefully controlled when grain refinement is attained through thermomechanical treatments.

An early study by Guy *et al.* [94] indicated that the starting and final temperatures of the $\alpha' \to \gamma$ reversion is affected by composition. This reversion may proceed either through a shear mechanism or a diffusion-controlled mechanism [86, 89], with the former encouraged by prior deformation [90] and a high Ni/Cr ratio [89]. In the shear mechanism, reverted austenite will inherit the high dislocation densities from the prior α'. These dislocations will rearrange themselves during reversion annealing, forming cells and subgrains in the reverted austenite phase and giving rise to a fine microstructure [89]. In the diffusion-controlled mechanism, reverted austenite forms through the nucleation of austenite embryos in the α'. These embryos then grow as annealing goes by. The growth of the reverted austenite is hindered by the tempered martensite particles at grain junctions [89].

The morphology of the prior deformation-induced martensite α' changes from laths to a dislocation-cell-type as the amount of deformation increases. For the lath-type martensite, the laths having the same crystallography will form lath blocks. In the diffusion-controlled mechanism, the reverted austenite forming from the martensite lath blocks will inherit the same lath-block morphology, with each lath block having several austenite laths. It is the sizes of the lath blocks, but not the sizes of the individual laths in these blocks, that have to be used in the Hall-Petch relation for strength prediction [86]. For the reverted austenite that forms from the dislocation-cell-type martensite, it forms through the recovery of the cell-type martensite. This reverted austenite is equiaxied with a variety of crystallographic orientations [86].

2.1.1.2. Heavy Plastic Deformation

Methods that may refine the grains throughout the bulk of metallic materials by utilising very severe plastic deformations have been devised in recent years. These methods include high-pressure, equal channel angular pressing [95], high-pressure torsion, multiple forging and accumulative roll-bending [96-103] and mechanical milling [104].

Hot deformation of metals having low or medium stacking fault energies (austenitic stainless steels, *e.g.*) is accompanied by discontinuous dynamic recrystallisation [105]. Since the evolution of grains is affected by this recrystallisation, it is therefore possible to utilise this phenomenon to achieve grain refinement [105]. However, for metals having high stacking fault energies (aluminium, *e.g.*), formation of new grains results from continuous dynamic recrystallisation [106, 107]. In continuous dynamic recrystallisation, new grains form through the gradual increase in misorientations amongst subgrains generated by plastic deformation. This strain-induced, continuous dynamic recrystallisation, which involves dislocation cells and dense dislocation walls and their eventual rearrangement into high-angle grain boundaries, is the underlying mechanism in the grain refinement process of an ECAPed (Equal Channel Angular Pressing) 304L steel at high temperatures [104].

The microstructural evolutions of metals during hot deformation have been well studied, but the microstructural evolutions during 'warm' deformation, *i.e.*, T≤0.5Tm (Tm : melting point) were less studied until recently. The microstructural evolutions of austenitic [101, 108, 109], ferritic [110, 111] and duplex stainless steels [112] during warm deformation (T≤0.5Tm), the underlying mechanisms for their subsequent grain refinements, and the microstructural changes in the deformed stainless steels during annealing have been examined in detail by Belyakov *et al.* [108-112].

For a 304 austenitic stainless steel (initial grain size ~25 μ m), Belyakov *et al.* [101, 108] imparted severe deformation to it by using multiple compressions at about 873K(0.5Tm), with a change in loading direction by 90° in each pass. In the early stage of deformation, high densities of dislocations form nearly homogeneously in parallel layers. These layers are crossed by dislocation walls with low-to-medium misorientations. At moderate levels of strain, more equiaxed subgrains having high densities of dislocations will appear, with an attendant increase in the misorientations among their boundaries. These subgrains (~0.3 μ m in size) become well developed with narrower walls and are distributed homogeneously in the material. The misorientations among these subgrains would eventually increase to such an extent that their boundaries are very similar to conventional grain boundaries. The noticeable features at high strain levels is that the dislocation densities in the grain interiors are reduced by the severe plastic deformation.

During deformation, the grains are broken up by dense dislocation walls (geometrically necessary boundaries, GNBs [102, 113, 114]) into domains/cell blocks, which possess different combinations of operative slip systems. The multiple compressions with successive rotations by 90° in the loading direction promote the evolution of crisscrossing GNBs and increase their misorientations rapidly. These dislocations are distributed inhomogeneously in the subgrain boundaries and they do not relax easily because of the relatively low deformation temperature (≤0.5Tm). Grain misorientations are increased due to the absorption of dislocations to grain boundaries and grain sliding. Severe elastic lattice distortions and a continuous build-up of internal stresses in the subgrains therefore result as deformation proceeds. At even higher strain levels, the high internal stresses may act as back stresses to hinder the movement of dislocations, leading to a decreased number of dislocations in the interiors of the fine grains and a high hardening of the material. The strain-induced, fine grains containing reduced amounts of dislocations are distributed homogeneously in the bulk of the material. The fine-grained microstructure with high-angle boundaries developed in this way is quite resistant to discontinuous (primary) recrystallisation and grain coarsening, although its strain energy is very high [108].

While primary recrystallisation occurs heterogeneously in materials having low-to-medium levels of deformation (those attainable in conventional deformation techniques), the recrystallisation in strain-induced, fine-grained austenitic stainless steels occurs homogeneously (the deformations are very high in these steels). This recrystallisation is called continuous recrystallisation [108], and the nuclei for continuous recrystallisation are the strain-induced fine grains themselves. Insofar as the nuclei are distributed homogeneously in the material, the continuous recrystallisation process occurs homogeneously in the material.

The continuous recrystallisation process are composed of three stages: recovery, transient recrystallisation and grain growth. During recovery, the high internal stresses are relaxed and many of the grains will then statically recrystallise at the end of the recovery process. Some of the grains (favourable grains) grow by consuming their neighbouring grains that contain higher densities of dislocations (these are unfavourable grains). However, this grain growth process is limited in extent and short-lived. This is because the number of favourable grains that may grow is high and they spread nearly homogeneously throughout the bulk of the material, there are thus very few regions that are particularly heterogeneous (unfavourable) such that they are vulnerable to being consumed by their favourable neighbours. Furthermore, high-angle boundaries are more resistant to primary (discontinuous) recrystallisation. As a consequence, the recrystallising grains will impinge on one another after just limited growth. This is called transient recrystallisation and it is difficult to discern clearly the recrystallised and un-recrystallised regions, in as much as the nucleation process for transient recrystallisation is homogeneous and takes place uniformly throughout the bulk of the material. Concurrent with transient recrystallisation is the gradual change in the misorientations of grain boundaries towards a random distribution. As annealing proceeds further, the material reaches a fully recrystallised state [108].

For ferritic stainless steels, it has been found that small and equiaxed grains are also achievable in ferritic stainless steels by using the same multiple compressions as described above [111]. Also, continuous recrystallisation has been observed to occur in strain-induced, fine-grained ferritic stainless steels [110]. For duplex stainless steels, the austenite and ferrite phases have different deformation kinetics [112]. The austenite phase will attain an ultrafine-grained microstructure faster than will the ferrite phase, because multiple twinning occurs readily in the former phase at the early stage of deformation [112].

Subdivision of the grain structure through twinning may also give rise to nanostructured materials. The extensive secondary twinning generated inside primary twins in an AISI316L stainless steel by high-strain-rate shear deformation (obtained through the Hopkinson bar test) has been found to generate nanosized equiaxed subgrains [115]. During shear deformation, some of the primary twins gradually rotate under the action of the shear force. When some of the secondary twin systems inside the rotating primary twins become aligned with the direction of the shear force, rotations of the primary twins stop and the subsequent deformation is accommodated through the activation of secondary twins. The result is that the initially straight primary twin boundaries become curved. And high densities of secondary twins exist inside the curved primary twins, forming a nanosized microstructure [115].

Transmutation of twin-matrix lamellae that are generated by heavy rolling deformation may also be utilised to obtain a fine-grained structure along the original grain boundaries [116]. Upon heavy rolling deformation, lamellae that are orientated along the rolling direction and comprising deformation twins and matrix will form. Slips along the [111] planes in the rolling direction and twinning are gradually suppressed as deformation proceeds, but slips along the [111] planes that are not parallel to the lamellae are activated. This process leads to accumulations of high numbers of dislocations at the twin-matrix boundaries. The dislocations are of the same sign for a certain boundary, but they are of opposite signs between two neighbouring boundaries. In the end, mutual annihilations of neighbouring boundaries of the opposite signs will take place locally, forming new fine grains as a consequence. The formation of these fine grains is further enhanced in regions along the original grain boundaries, because of the additional constraints imposed.

2.1.1.3. Other Methods of Refinement for Austenitic Stainless Steels

For high-N austenitic stainless steels, besides the $\alpha' \rightarrow \gamma$ reversion transformation, the reversion transformation of the eutectoid structure consisting of ferrite and Cr_2N to austenite may also be used to refine the grain size, because of the large amounts of boundaries in the eutectoid structure [117].

Conventional thermomechanical treatments like hot deformation in the austenite region followed by accelerated cooling may generate a fine-grained microstructure, too. A recent study by Ikeda *et al.* [118] has set out the details of the processing parameters (*e.g.*, cooling rate) for attaining a fine-grained microstructure in a 22Mn-13Cr-5Ni austenitic stainless steel. It has been found that fine grains can be obtained by decreasing the temperature for finish-rolling, the cooling rate and the final finish temperature.

2.2. Grain Refinement of Ferritic Stainless Steels

In this method, powders of yttria (Y_2O_3), iron and chromium are mechanically milled in an argon atmosphere. The mixture is then annealed in the presence of hydrogen. This processing results in an ultrafine ferritic microstructure (~20nm) with amorphous Y_2O_3 dissolved along the grain boundaries [119]. The amorphous grain boundaries can effectively inhibit grain growth.

2.3. Grain Refinement of Duplex Stainless Steels

Conventionally, a fine microstructure in a duplex stainless steel may be obtained by first solution-annealing it at high temperatures (at 1573K, e.g). This high-temperature annealing results in a supersaturated, fully ferritic microstructure. The steel having a fully ferritic microstructure is then subjected to heavy cold-rolling, followed by a subsequent re-annealing in the duplex ($\alpha+\gamma$) field. The choices of the solution-treating temperature and the subsequent re-annealing treatment are critical in order to obtain and maintain the microduplex microstructure. The initial as-rolled fully ferritic matrix may recover easily and form subgrains before γ precipitates out. If the re-annealing temperature is too high, then the subgrains recrystallise rapidly and a coarse ferritic matrix containing small γ grains at triple junctions will form. However, when the microduplex microstructure is achieved, its α matrix does not recrystallise easily [120]. Recently, Maki *et al.* [121] have been able to obtain a microduplex microstructure in as-hot-forged, coarse-grained duplex stainless steels without the initial high-temperature solution-treatment, adding to the convenience of this method.

The reversion transformation of the intermetallic sigma phase has also been employed for grain refinement of duplex stainless steels [122]. The finer the starting ($\gamma + \sigma$) microstructure, the smaller the grain size of the resulting ($\gamma + \alpha$) microstructure will be. And an ultrafine ($\gamma + \sigma$) microstructure is obtainable by using high-strain powder metallurgy [123].

2.4. Grain Refinement of Precipitation-hardening Stainless Steels

For the grain refinement of precipitation-hardening stainless steels, the use of intercritical annealing (annealing at temperatures just below the phase boundary between the γ and ($\alpha+\gamma$) regions) may be handy [124]. If the austenitising treatment is denoted as Q and intercritical annealing as L, then grain refinement may be achieved by using either (QLQL) or (LQLQ) [124].

The underlying mechanism for achieving grain refinement through intercritical annealing is the redistribution of alloying elements [124]. For example, when a L-treated ($\alpha+\gamma$) microstructure is austenitised (Q treatment), the α phase transforms into γ. This newly reverted γ differ from the original γ in composition. Because the redistribution of substitutional alloying elements between these two compositionally different γ phases is not efficient, the nomially single-phased γ microstructure is in fact 'duplex' (one alloy-rich γ plus one alloy-lean γ). Upon cooling, the alloy-lean γ will transform martensitically first and the alloy-rich γ acts to constrain this transformation. Hence, the alloy-lean martensite and the alloy-rich γ mutually deform and constraint each other. On further cooling, the alloy-rich, deformed γ will eventually undergo a martensitic transformation, under the constraint of the alloy-lean martensite that has formed already. Hence, the alloy-rich γ tends to transform into martensite variants that are compatible with the constraint imposed by the alloy-lean, prior martensite, resulting in a refined microstructure [124].

3. GRAIN REFINEMENT IN THE LIQUID STATE

Besides solid-state grain refinement, liquid-state grain refinement has also been attempted in recent years. Pulsed magnetic fields [125], pulsed current [126], and ultrasonic treatment [127] have all been demonstrated to be employable for refining the solidification microstructure of austenitic stainless steels. The on and off periods of pulsed magnetic fields affect the liquidus temperature and the degree of constitutional supercooling at the liquid/solid interface differently [128]. Electromagnetic force also affects the melt flow by causing fluctuations in composition and temperature in the vicinity of the liquid/solid interface. These factors then influence the rate of freezing at the interface, thereby engendering different growth modes. In the ultrasonic treatment approach, long dendrites in the melt are broken into small pieces because of vibration and induced radiation forces [127]. As a result, the grain structure is refined.

As far as weld metals are concerned, electromagnetic stirring [129], liquid metal chilling [130], pulsed direct current or pulsed alternate current [131] have all be shown to be effective in refining the microstructures of ferritic stainless steel weld metals.

REFERENCES

[1]   Chen XH, Lu J, Lu L, *et al.* Tensile properties of a nanocrystalline 316L austenitic stainless steel. Scr Mater 2005; 52(10): 1039-44.

[2]   Roland T, Retraint D, Lu K, *et al.* Fatigue life improvement through surface nanostructuring of stainless steel by means of surface mechanical attrition treatment. Scr Mater 2006; 54(11): 1949-54.

[3]   Mordyuk BN, Prokopenko GI, Vasylyev MA, *et al.* Effect of structure evolution induced by ultrasonic peening on the corrosion behavior of AISI-321 stainless steel. Mater Sci Eng A 2007; 458(1-2): 253-61.

[4]   Wang T, Yu J, Dong B. Surface nanocrystallization induced by shot peening and its effect on corrosion resistance of 1Cr18Ni9Ti stainless steel. Surf Coat Technol 2006; 200(16-17): 4777-81.

[5]   Liu G, Lu J, Lu K. Surface nanocrystallization of 316L stainless steel induced by ultrasonic shot peening. Mater Sci Eng A 2000; 286(1): 91-5.

[6]   Zhang HW, Hei ZK, Liu G, *et al.* Formation of nanostructured surface layer on AISI 304 stainless steel by means of surface mechanical attrition treatment. Acta Mater 2003; 51(7): 1871-81.

[7]   Lu K, Lu J. Nanostructured surface layer on metallic materials induced by surface mechanical attrition treatment. Mater Sci Eng A 2004; 375-7: 38-45.

[8]   Wang XY, Li DY. Mechanical and electrochemical behavior of nanocrystalline surface of 304 stainless steel. Electrochim Acta 2002; 47(24): 3939-47.

[9]   Tao NR, Wang ZB, Tong WP, *et al.* An investigation of surface nanocrystallization mechanism in Fe induced by surface mechanical attrition treatment. Acta Mater 2002; 50(18): 4603-16.

[10]   Xue Q, Liao XZ, Zhu YT, *et al.* Formation mechanisms of nanostructures in stainless steel during high-strain-rate severe plastic deformation. Mater Sci Eng A 2005; 410-1: 252-6.

[11]   Zhang X, Misra A, Wang H, *et al.* Nanoscale-twinning-induced strengthening in austenitic stainless steel thin films. App Phys Lett 2004; 84(7): 1096-8.

[12]   Zhang X, Misra A, Wang H, *et al.* Effects of deposition parameters on residual stresses, hardness and electrical resistivity of nanoscale twinned 330 stainless steel thin films. J App Phys 2005; 97(9): 094302-1.

[13]   Kumar S, Banerjee MK. Improvement of intergranular corrosion resistance of type 316 stainless steel by laser surface melting. Surf Eng 2001; 17(6): 483-9.

[14]   Chong PH, Liu Z, Wang YY, *et al.* Pitting corrosion behaviour of large area laser surface treated 304L stainless-steel. Thin Solid Films 2004; 453-4: 388-93.

[15]   Chong PH, Liu Z, Skeldon P, *et al.* Characterisation and corrosion performance of laser-melted 3Cr12 steel. Appl Surf Sci 2005; 247(1-4): 362-8.

[16]   Kwok CT, Man HC, Cheng FT. Cavitation erosion and pitting corrosion of laser surface melted stainless steels. Surf Coating Technol 1998; 99(3): 295-304.

[17]   Conde A, Garcia I, De Damborenea JJ. Fitting corrosion of 304 stainless steel after laser surface melting in argon and nitrogen atmospheres. Corros Sci 2001; 43(5): 817-28.

[18]   Akgun OV, Inal OT. Laser surface melting and alloying of type 304L stainless steel.2. Corrosion and wear resistance properties. J Mater Sci 1995; 30(23): 6105-12.

[19]   Conde A, Colaço R, Vilar R, *et al.* Corrosion behaviour of steels after laser surface melting. Mater Des 2000; 21(5): 441-5.

[20]   Tsay LW, Young MC, Chou FY, *et al.* The effect of residual thermal stresses on the fatigue crack growth of laser-annealed 304 stainless steels. Mater Chem Phys 2004; 88(2-3): 348-52.

[21]   Tsay LW, Liu YC, Lin DY, *et al.* The use of laser surface-annealed treatment to retard fatigue crack growth of austenitic stainless steel. Mater Sci Eng A 2004; 384(1-2): 177-83.

[22]   Shiue RK, Chang CT, Young MC. The effect of residual thermal stresses on the fatigue crack growth of laser-surface-annealed AISI 304 stainless steel: Part I: computer simulation. Mater Sci Eng A 2004; 364(1-2): 101-8.

[23]   Abenojar J, Velasco F, Torralba JM, *et al.* Reinforcing 316L stainless steel with intermetallic and carbide particles. Mater Sci Eng A 2002; 335(1-2): 1-5.

[24]   Cuppari MGDV, Souza RM, Sinatora A. Effect of hard second phase on cavitation erosion of Fe-Cr-Ni-C alloys. Wear 2005; 258(1-4): 596-603.

[25]   Tassin C, Laroudie F, Pons M, *et al.* Improvement of the wear resistance of 316L stainless steel by laser surface alloying. Surf Coat Technol 1996; 80(1-2): 207-10.

[26]   Zhang D, Zhang X. Laser cladding of stainless steel with Ni-Cr3C2 and Ni-WC for improving erosive-corrosive wear performance. Surf Coat Technol 2005; 190(2-3): 212-7.

[27]   Abbas G, Ghazanfar U. Two-body abrasive wear studies of laser produced stainless steel and stainless steel plus SiC composite clads. Wear 2005; 258(1-4): 258-64.

[28]   Cheng FT, Kwok CT, Man HC. Cavitation erosion resistance of stainless steel laser-clad with WC-reinforced MMC. Mater Lett 2002; 57(4): 969-74.

[29]   Lo KH, Cheng FT, Kwok CT, *et al.* Improvement of cavitation erosion resistance of AISI 316 stainless steel by laser surface alloying using fine WC powder. Surf Coat Technol 2003; 165(3): 258-67.

[30]   Kwok CT, Cheng FT, Man HC. Laser surface modification of UNS S31603 stainless steel using NiCrSiB alloy for enhancing cavitation erosion resistance. Surf Coat Technol 1998; 107(1): 31-40.

[31]   Majumdar JD, Manna I. Laser surface alloying of AISI 304-stainless steel with molybdenum for improvement in pitting and erosion–corrosion resistance. Mater Sci Eng A 1999; 267(1): 50-9.

[32]   Peyre P, Scherpereel X, Berthe L, *et al.* Surface modifications induced in 316L steel by laser peening and shot-peening. Influence on pitting corrosion resistance. Mater Sci Eng A 2000; 280(2): 294-302.

[33]   Lo KH, Cheng FT, Kwok CT, *et al.* Effects of laser treatments on cavitation erosion and corrosion of AISI 440C martensitic stainless steel. Mater Lett 2004; 58(1-2): 88-93.

[34]   Lo KH, Cheng FT, Man HC. Laser transformation hardening of AISI 440C martensitic stainless steel for higher cavitation erosion resistance. Surf Coat Technol 2003; 173(1): 96-104.

[35]   Mudali UK, Dayal RK. Improving intergranular corrosion resistance of sensitized type 316 austenitic stainless steel by laser surface melting. J Mater Eng Perform 1992; 1(3): 341-6.

[36]   Mudali UK, Dayal RK, Goswami GL. Laser surface melting for improving intergranular corrosion resistance of cold-worked and sensitised type 316 stainless steel. Anti-corros Method Mater 1998; 45(3): 181-181.

[37]   Bransden AS, Megaw JHPC, Balkwill PH, *et al.* Metal surface treatment and reduction in pitting corrosion of 304L stainless steel by excimer laser. J Mod Opt 1990; 37(4): 813-28.

[38]   Kumar S, Banerjee MK. Desensitization of type 316 stainless steel by laser surface melting. Anti-Corros Method Mater 2000; 47(1): 20-25.

[39]   Akgun OV, Inal OT. Laser surface melting and alloying of type 304L stainless steel Part II Corrosion and wear resistance properties. J Mater Sci 1995; 30(23): 6105-12.

[40]   Akgun OV, Urgen M, Cakir AF. The effect of heat treatment on corrosion behavior of laser surface melted 304L stainless steel. Mater Sci Eng A 1995; 203(1-2): 324-31.

[41]   Yang S, Wang ZJ, Kokawa H, *et al.* Grain boundary engineering of 304 austenitic stainless steel by laser surface melting and annealing. J Mater Sci 2007; 42(3): 847-53.

[42]   Yang J, Lian J, Dong Q, *et al.* Nano-structured films formed on the AISI 329 stainless steel by Nd-YAG pulsed laser irradiation. Appl Surf Sci 2004; 229(1-4): 2-8.

[43]   Parvathavarthini N, Dayal RK, Sivakumar R, *et al.* Pitting corrosion resistance of laser alloyed 304 stainless steel. Mater Sci Technol 1992; 8(12): 1070-4.

[44]   Yue TM, Yu JK, Man HC. The effect of excimer laser surface treatment on pitting corrosion resistance of 316LS stainless steel. Surf Coat Technol 2001; 137(1): 65-71.

[45]   Kwok CT, Man HC, Cheng FT. Cavitation erosion and pitting corrosion behaviour of laser surface-melted martensitic stainless steel UNSS42000. Surf Coat Technol 2000; 126(2-3): 238-55.

[46]   Yue TM, Hu QW, Mei Z, *et al.* Laser cladding of stainless steel on magnesium ZK60/SiC composite. Mater Lett 2001; 47(3): 165-70.

[47]   Cheng FT, Kwok CT, Man HC. Laser surfacing of S31603 stainless steel with engineering ceramics for cavitation erosion resistance. Surf Coat Technol 2001; 139(1): 14-24.

[48]   Zhang DW, Lei TC, Li FJ. Laser cladding of stainless steel with Ni-Cr3C2 for improved wear performance. Wear 2001; 251: 1372-6.

[49]   Lo KH, Cheng FT, Man HC. Cavitation erosion mechanism of S31600 stainless steel laser surface-modified with unclad WC. Mater Sci Eng A 2003; 357(1-2): 168-80.

[50]   Watson IA, Wang RK, Peden I, *et al.* Effect of laser and environmental parameters on reducing microbial contamination of stainless steel surfaces with Nd : YAG laser irradiation. J Appl Microbiol 2005; 99(4): 934-44.

[51]   Ward GD, Watson IA, Tull DESS, *et al.* Bactericidal action of high-power Nd : YAG laser light on Escherichia coli in saline suspension. J Appl Microbiol 2000; 89(3): 517-25.

[52]   Ward GD, Watson IA, Tull DESS, *et al.* Inactivation of bacteria and yeasts on agar surfaces with high power Nd:YAG laser light. Lett Appl Microbiol 1996; 23(3): 136-40.

[53]   Psyllaki P, Oltra R. Preliminary study on the laser cleaning of stainless steels after high temperature oxidation. Mater Sci Eng A 2000; 282(1-2): 145-52.

[54]   Sano Y, Obata M, Kubo T, *et al.* Retardation of crack initiation and growth in austenitic stainless steels by laser peening without protective coating. Mater Sci Eng A 2006; 417(1-2): 334-40.

[55]   Sano Y, Obata M, Kubo T, *et al.* Retardation of crack initiation and growth in austenitic stainless steels by laser peening without protective coating. Mater Sci Eng A 2006; 417(1-2): 334-40.

[66]   Lee J, Euh K, Oh JC, *et al.* Microstructure and hardness improvement of TiC/stainless steel surface composites fabricated by high-energy electron beam irradiation. Mater Sci Eng A 2002; 323(1-2): 251-9.

[67]   Zhang K, Zou J, Grosdidier T, *et al.* Improved pitting corrosion resistance of AISI 316L stainless steel treated by high current pulsed electron beam. Surf Coat Technol 2006; 201(3-4): 1393-400.

[68]   Wang X, Zhu XP, Lei MK, *et al.* Influence of high-intensity pulsed ion beam irradiation on the creep property of 316 L stainless steel. Nucl Instrum Method Phys Res B 2007; 259(2): 937-42.

[59]   Rahman M, Hashmi MSJ. Effect of treatment time on low temperature plasma nitriding of stainless steel by saddle field neutral fast atom beam source. Thin Solid Films 2006; 515(1): 231-8.

[60]   Jauhari I, Rozali S, Masdek NRN, *et al.* Surface properties and activation energy analysis for superplastic carburizing of duplex stainless steel. Mater Sci Eng A 2007; 466(1-2): 230-4.

[61]   Tokaji K, Kohyama K, Akita M. Fatigue behaviour and fracture mechanism of a 316 stainless steel hardened by carburizing. Int J Fatigue 2004; 26(5): 543-51.

[62]   Akita M, Tokaji K. Effect of carburizing on notch fatigue behaviour in AISI 316 austenitic stainless steel. Surf Coat Technol 2006; 200(20-21): 6073-8.

[63]   Aoki K, Kitano K. Surface hardening for austenitic stainless steels based on carbon solid solution. Surf Eng 2002; 18(6): 462-4.

[64]   Cao Y, Ernst F, Michal GM. Colossal carbon supersaturation in austenitic stainless steels carburized at low temperature. Acta Mater 2003; 51(14): 4171-81.

[65]   Michal GM, Ernst F, Kahn H, *et al.* Carbon supersaturation due to paraequilibrium carburization: Stainless steels with greatly improved mechanical properties. Acta Mater 2006; 54(6): 1597-606.

[66]   Ernst F, Cao Y, Michal GM. Carbides in low-temperature-carburized stainless steels. Acta Mater 2005; 52(6): 1469-77.

[67]   Michal GM, Ernst F, Heuer AH. Carbon paraequilibrium in austenitic stainless steel. Metall Mater Trans 2006; 37(6): 1819-24.

[68]   Qu J, Blau PJ, Jolly BC. Tribological properties of stainless steels treated by colossal carbon supersaturation. Wear 2007; 263(Sp.Iss.SI): 719-26.

[69]   Agarwal N, Kahn H, Avishai A, *et al.* Enhanced fatigue resistance in 316L austenitic stainless steel due to low-temperature paraequilibrium carburization. Acta Mater 2007; 55(16): 5572-80.

[70]   Kolserising®. CD ROM from Bodycote Metal Technology, August 2001.

[71]   Farrell K, Specht ED, Pang J, *et al.* Characterization of a carburized surface layer on an austenitic stainless steel. J Nucl Mater 2005; 343(1-3): 123-33.

[72]   Christiansen T, Somers MAJ. Controlled dissolution of colossal quantities of nitrogen in stainless steel. Metall Mater Trans A 2006; 37(3): 675-82.

[73]   Saleh ZB, Shahryari A, Omanovic S. Enhancement of corrosion resistance of a biomedical grade 316LVM stainless steel by potentiodynamic cyclic polarization. Thin Solid Films 2007; 515(11): 4727-37.

[74]   Hong T, Ogushi T, Nagumo M. Effect of chromium enrichment in the film formed by surface treatments on the corrosion resistance of type 430 stainless steel. Corros Sci 1996; 38(6): 881-8.

[75]   Lee SJ, Lai JJ. The effects of electropolishing (EP) process parameters on corrosion resistance of 316L stainless steel. J Mater Process Technol 2006; 140(1-3): 206-10.

[76]   Shih C, Shih C, Su Y, *et al.* Effect of surface oxide properties on corrosion resistance of 316L stainless steel for biomedical applications. Corros Sci 2004; 46(2): 427-41.

[77]   Fredj NB, Sidhom H. Effects of the cryogenic cooling on the fatigue strength of the AISI 304 stainless steel ground components. Cryogencris 2006; 46(6): 439-48.

[78]   Fredj NB, Sidhom H, Braham C. Ground surface improvement of the austenitic stainless steel AISI 304 using cryogenic cooling. Surf Coat Technol 2006; 200(16-17): 4846-60.

[79]   Valentini L, Schino AD, Kenny JM, *et al.* Influence of grain size and film composition on wear resistance of ultra fine grained AISI 304 stainless steel coated with amorphous carbon films. Wear 2002; 253(3-4): 458-64.

[80]   Schino AD, Kenny JM. Effect of grain size on the corrosion resistance of a high nitrogen-low nickel austenitic stainless steel. J Mater Sci Lett 2002; 21(24): 1969-71.

[81]   Bregliozzi G, Schino AD, Ahmed SIU, *et al.* Cavitation wear behaviour of austenitic stainless steels with different grain sizes. Wear 2005; 258(1-4): 503-10.

[82]   Bregliozzi G, Schino AD, Haefke H, *et al.* Cavitation erosion resistance of a high nitrogen austenitic stainless steel as a function of its grain size. J Mater Sci Lett 2003; 22(13): 981-3.

[83]   Yagodzinskyy Y, Pimenoff J, Tarasenko O, *et al.* Grain refinement processes for superplastic forming of AISI 304 and 304L austenitic stainless steels. Mater Sci Technol 2004; 20(7): 925-9.

[84]   Wang TS, Peng JG, Gao YW, *et al.* Microstructure of 1Cr18Ni9Ti stainless steel by cryogenic compression deformation and annealing. Mater Sci Eng A 2005; 407(1-2): 84-8.

[85]   Schino AD, Salvatori I, Kenny JM. Effects of martensite formation and austenite reversion on grain refining of AISI 304 stainless steel. J Mater Sci 2002; 37(21): 4561-5.

[86]   Takaki S, Tomimura K, Ueda S. Effect of pre-cold-working on diffusional reversion of deformation-induced martensite in metastable austenitic stainless steel. ISIJ Int 1994; 34(6): 522-7.

[87]   Tomimura K, Takaki S, Tanimoto S, *et al.* Optimal chemical composition in Fe-Cr-Ni alloys for ultra grain refining by reversion from deformation-unduced martensite. ISIJ Int 1991; 31(7): 721-7.

[88]   Ma Y, Jin J, Lee Y. A repetitive thermomechanical process to produce nano-crystalline in a metastable austenitic steel. Scr Mater 2005; 52(12): 1311-5.

[89]   Tomimura K, Takaki S, Tokunaga Y. Reversion mechanism from deformation-induced martensite to austenite in metastable austenitic stainless steels. ISIJ Int 1991; 31(12): 1431-7.

[90]   Zhang Y, Jing XT, Lou BZ, *et al.* Mechanism and reversible behavior of the $\alpha' \rightarrow \gamma$ transformation in 1Cu18Ni9Ti stainless steel. J Mater Sci 1999; 34(13): 3291-6.

[91]   El Wahabi M, Gavard L, Montheillet F, *et al.* Effect of initial grain size on dynamic recrystallization in high purity austenitic stainless steels. Acta Mater 2005; 53(17): 4605-12.

[92]   El Wahabi M, Gavard L, Cabrera JM, *et al.* EBSD study of purity effects during hot working in austenitic stainless steels. Mater Sci Eng A 2005; 393(1-2): 83-90.

[93]   Gavard L, Montheillet F, Coze JL. Recrystallization and grain growth in high purity austenitic stainless steels. Scr Mater 1998; 39(8): 1095-9.

[94]   Guy KB, Butler EP, West DRF. Reversion of bcc α' martensite in Fe-Cr-Ni austenitic stainless steels. Metal Sci 1983; 17(4): 167-76.

[95]   Huang CX, Yang G, Gao YL, *et al.* Influence of processing temperature on the microstructures and tensile properties of 304L stainless steel by ECAP. Mater Sci Eng A 2008; 485(1-2): 643-50.

[96]   Saito Y, Tsuji N, Utsunomiya H, *et al.* Ultra-fine grained bulk aluminum produced by accumulative roll-bonding (ARB) process. Scr Mater 1998; 39(9): 1221-27.

[97]   Vinogradov A, Patlan V, Suzuki Y, *et al.* Structure and properties of ultra-fine grain Cu-Cr-Zr alloy produced by equal-channel angular pressing. Acta Mater 2002; 50(7): 1639-51.

[98]   Sakai G, Nakamura K, Horita Z, *et al.* Developing high-pressure torsion for use with bulk samples. Mater Sci Eng A 2005; 406(1-2): 268-73.

[99]   Furukawa M, Horita Z, Nemoto M, *et al.* Microhardness measurements and the Hall-Petch relationship in an Al-Mg alloy with submicrometer grain size. Acta Mater 1996; 44(11): 4619-29.

[100] Mishin OV, Gottstein G. Microstructural aspects of rolling deformation in ultrafine-grained copper. Philos Mag A 1998; 78(2): 373-88.

[101] Belyakov A, Sakai T, Miura H, *et al.* Grain refinement under multiple warm deformation in 304 type austenitic stainless steel. ISIJ Int 1999; 39(6): 592-9.

[102] Belyakov A, Gao W, Miura H, *et al.* Strain-induced grain evolution in polycrystalline copper during warm deformation. Metall Mater Trans A 1998; 29(12): 2957-65.

[103] Richert M, Liu Q, Hansen N. Microstructural evolution over a large strain range in aluminium deformed by cyclic-extrusion-compression. Mater Sci Eng A 1999; 260(1-2): 275-83.

[104] Huang H, Ding J, McCormick PG. Microstructural evolution of 304 stainless steel during mechanical milling. Mater Sci Eng A 1996; 216(1-2): 178-84.

[105] Sakai T, Jonas JJ. Dynamic recrystallization – mechanical and microstructural considerations. Acta Metall 1984; 32(2): 189-209.

[106] Gourdet S, Montheillet F. An experimental study of the recrystallization mechanism during hot deformation of aluminium. Mater Sci Eng A 2000; 283(1-2): 274-88.

[107] Gourdet S, Knonpleva EV, McQueen HJ, *et al.* Recrystallization during hot deformation of aluminium. Mater Sci Forums 1996; 217: 441-6.

[108] Belyakov A, Sakai T, Miura H, *et al.* Continuous recrystallization in austenitic stainless steel after large strain deformation. Acta Mater 2002; 50(6): 1547-57.

[109] Belyakov A, Tsuzaki K, Miura H, *et al.* Effect of initial microstructures on grain refinement in a stainless steel by large strain deformation. Acta Mater 2003; 51(3): 847-61.

[110] Belyakov A, Kimura Y, Tsuzaki K. Recovery and recrystallization in ferritic stainless steel after large strain deformation. Mater Sci Eng A 2005; 403(1-2): 249-59.

[111] Belyakov A, Tsuzaki K, Kimura Y, *et al.* Comparative study on microstructure evolution upon unidirectional and multidirectional cold working in an Fe-15%Cr ferritic alloy. Mater Sci Eng A 2007; 456(1-2): 323-31.

[112] Belyakov A, Kimura Y, Tsuzaki K. Microstructure evolution in dual-phase stainless steel during severe deformation. Acta Mater 2006; 54(9): 2521-32.

[113] Wilsdorf DK, Hansen N. Geometrically necessary, incidental and subgrain boundaries. Scr Metallurgica et Mater 1991; 25(7): 1557-62.

[114] Hughes DA, Hansen N. High angle boundaries formed by grain subdivision mechanisms. Acta Mater 1997; 45(9): 3871-86.

[115] Xue Q, Liao XZ, Zhu YT, *et al.* Formation mechanisms of nanostructures in stainless steel during high-strain-rate severe plastic deformation. Mater Sci Eng A 2005; 410-411(Sp.Iss.SI): 252-6.

[116] Morikawa T, Higashida K, Sato T. Fine-grained structures developed along grain boundaries in a cold-rolled austenitic stainless steel. ISIJ Int 2002; 42(12): 1527-33.

[117] Nakada N, Hirakawa N, Tsuchiyama T, *et al.* Grain refinement of nickel-free high nitrogen austenitic stainless steel by reversion of eutectoid structure. Scr Mater 2007; 57(2): 153-6.

[118] Ikeda S, Tone S, Takashima S, *et al.* Effect of thermomechanical control process on strengthening of a 22Mn13Cr-5Ni austenitic stainless steel plate for cryogenic use. ISIJ Int 1990; 30(8): 600-7.

[119] Kimura Y, Takaki S, Suejima S, *et al.* Ultra grain refining and decomposition of oxide during super-heavy deformation in oxide dispersion ferritic stainless steel powder. ISIJ Int 1999; 39(2): 176-82.

[120] Huang X, Tsuzaki K, Maki T. Effect of initial structure on recrystallization of the α matrix in an (α+γ) microduplex stainless steel. Scr Metallurgica et Mater 1995; 33(7): 1087-92.

[121] Maki T, Furuhara T, Tsuzaki K. Microstructure development by thermomechanical processing in duplex stainless steel. ISIJ Int 2001; 41(6): 571-9.

[122] Jiang ZL, Chen XY, Huang H, *et al.* Grain refinement of Cr25Ni5Mo1.5 duplex stainless steel by heat treatment. Mater Sci Eng A 2003; 363(1-2): 263-7.

[123] Ameyama K. Low temperature recrystallization and formation of an ultra fine (gamma+sigma) microduplex structure in a SUS316L stainless steel. Scr Mater 1998; 38(3): 517-22.

[124] Guo Z, Sha W, Wilson EA, *et al.* Improving toughness of PH13-8 stainless steel through intercritical annealing. ISIJ Int 2003; 43(10): 1622-29.

[125] Li Q, Li H, Zhai Q. Structure evolution and solidification behavior of austenitic stainless steel in pulsed magnetic field. JIron Steel Res Int 2006; 13(5): 69-72.

[126] Fan JH, Chen Y, Li RX, *et al.* Effects of pulse current on solidification structure of austenitic stainless steel. J Iron Steel Res Int 2003; 11(6): 37-9.

[127] Liu Q, Zhang Y, Song Y, *et al.* Influence of ultrasonic vibration on mechanical properties and microstructure of 1Cr18Ni9Ti stainless steel. Mater Des 2007; 28(6): 1949-52.

[128] Song C, Li Q, Li H, *et al.* Effect of pulse magnetic field on microstructure of austenitic stainless steel during directional solidification. Mater Sci Eng A 2008; 485(1-2): 403-8.

[129] Villafuerte JC, Kerr HW. Electromagnetic stirring and grain refinement in stainless steel GTA welds. Weld J 1990; 69(1): 1s-13s.

[130] Villafuerte JC, Kerr HW, David SA. Mechanisms of equiaxed grain formation in ferritic stainless steel gas tungsten arc welds. Mater Sci Eng A 1995; 194(2): 187-91.

[131] Reddy GM, Mohandas T. Explorative studies on grain refinement of ferritic stainless steel welds. J Mater Sci Lett 2001; 20(8): 721-3.

CHAPTER 12

Colouration of Stainless Steels

Abstract: In quite a number of applications, stainless steels are chosen for their aesthetic appeal. Besides being shiny, the fact that many colours may be imparted to stainless steels adds to their popularity, especially in the construction industry. This chapter covers the basics of colouration of stainless steels and the various techniques for achieving different colours.

Keywords: Colouration, INCO method, electrochemical colouring, laser colouring, oxidation, coating, current pulse method, potential pulse method, colour tone, one-step colouration, two-step colouring.

1. INTRODUCTION

Quite a large tonnage of stainless steels is used for construction and architectural purposes. In many of these applications, aesthetics is a primary concern. The stainless-steel crown of the Chrysler building in New York must be shiny in order to be appealing. The famous 'Gateway to the West' arch on the bank of the Mississippi River in St.Louis, Missouri, which is highly reflective in daytime and is magnificently illuminated at night, is another example.

While these examples make use of silvery stainless steels, there is no reason to limit the colour choice to just silver. Like other metals, stainless steels may be plated with other metals like gold and copper alloys. Gold is soft and pricey and so the same colour is more economically achieved using the methods discussed below (like ceramic coating). The painting of stainless steels does not differ much from the painting of other metals, except for some process modifications that must be made to enhance paint adherence. Stainless steels are typically coil or resin paint coated. The paints on stainless steels usually last longer than those applied to other metals, insofar as the underlying stainless steel does not corrode easily, thereby minimising peeling. Nevertheless, paint does fade over time. However, the methods introduced in subsequent sections yield more long-lasting colour films on stainless steels.

Over the years, quite a number of methods, many of them proprietary and patented, have been proposed for imparting different colours to stainless steels. Different colours can be obtained because of interference of lights reflected at the film/air and film/steel interfaces, although a colour subtraction effect might also not be important [1]. Earwaker *et al.* [2] pointed out that the film colour would be a function of the processing time, but the film composition is unchanged with processing time. Different film thicknesses may be achieved by using chemical methods [2-5], electrolytic oxidation [6] and thermal oxidation in air [7]. The film thickness may be deliberately varied over the surface to produce a rainbow effect. Alternatively, uniform film thickness, and thus a uniform colour, may be obtained.

A variety of colours are obtainable on stainless steel surfaces, these are gold, bronze, purple, red, blue, green, black and charcoal. Thicker films produce darker colours. Usually, the colour will not fade upon exposure to sunlight. For instance, the electrochemically coloured roof of the Shakaden Temple in Japan has not undergone any colour change since its installation in 1975 [8]. Patterning of colours may be obtained by selective etching, polishing or engraving.

2. CHEMICAL OXIDATION AND THE INCO METHODS

Chemical colouring methods involve the use of a hot solution (70°C or higher) containing chromic and sulphuric acids (e.g. 2.5mol/l CrO_3 + 5.0mol/l H_2SO_4) to produce an oxide film on the stainless steel surface [2]. Formation of the film has been attributed to anodic sites (slip bands, twins and grain boundaries) where the steel dissolves into the solution, and cathodic sites where chromate (or dichromate) anions are reduced to trivalent chromium ions [4]. The cathodic reaction contributes to growth of the oxide film due to the hydrolysis reaction of dissolved metal ions with Cr^{3+} formed by the reduction of chromic acid. The potentials of the

Joseph Ki Leuk Lai, Kin Ho Lo and Chan Hung Shek

stainless steel specimens are continuously monitored using a platinum electrode. The colour tone of the film may be controlled *via* the potential difference with respect to the platinum electrode in the potential-time curve. The film produced in this way is usually porous and soft. A subsequent electrolytic hardening process is thus usually required to improve film properties. Reproducibility of colour tones using the purely chemical method can be difficult to achieve [9].

The two-step colouring method, INCO, developed by Evans has been very popular [4]. The INCO method involves a hot dipping process and a cathodic hardening treatment. The potential rise of the stainless steel specimen relative to a reference electrode in the hot dipping process may be controlled to obtain the desired colour tone [4]. The colour film formed on the steel is then hardened by a cathodic electrolytic process in a solution containing CrO_3 and H_2SO_4. Evans suggested the complete reaction sequence as follows:

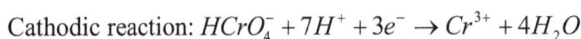

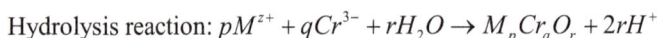

$$\text{Anodic reaction: } M \rightarrow M^{z+} + ze^-$$

$$\text{Cathodic reaction: } HCrO_4^- + 7H^+ + 3e^- \rightarrow Cr^{3+} + 4H_2O$$

$$\text{Hydrolysis reaction: } pM^{z+} + qCr^{3+} + rH_2O \rightarrow M_pCr_qO_r + 2rH^+$$

The films formed in this way on austenitic stainless steels have been found to be composed of two layer [10, 11]. The lower oxide layer is a uniform passive film, whereas the upper oxide layer is porous and enriched with Cr^{3+} [10].

A drawback of the INCO method lies in the colour change of the film during the cathodic hardening treatment. The original colour of the film after the hot dipping process is altered during cathodic hardening because the film continuously thickens. The hot dipping process must thus be properly controlled to take into account the subsequent colour change, which may not be an easy task. The time for initiation of colouring depends heavily on the temperature of the solution and the initial concentrations of H^+ and SO_4^{2-} [12].

3. ELECTROCHEMICAL COLOURATION METHOD

Electrochemical colouring methods utilising the potential pulse method [13] and the current pulse method [14, 15] have also been tried with success. The colour shade may be controlled by properly selecting the pulse frequency [16], the range of anodic current [16]/potential [17] and the electrolysis time [16-18]. The alternating current electrolysis method is a method in which anodic electrolysis (for film colouring) and cathodic electrolysis (for film hardening) are performed alternately. So, this method enables film colouring and hardening to be done simultaneously in a single solution and is essentially a one-step process. The colour tone may be controlled by properly adjusting the electrolysis frequency and time. The abrasion resistance of films produced in this way is better than those produced conventionally.

In the current pulse method, the current pulse may disperse nucleation sites of the coloured film and/or decrease the etching of the anodic resolution in the grain boundary area [15]. The positive current is associated with the anodic reaction, while the negative current is responsible for film hardening and so is associated with the cathodic reaction. The absolute ratio of anodic to cathodic charge density, and the corresponding final value of the cathodic potential are easy ways for colour control [14, 19]. Brown, blue, golden and purple films of reasonable mechanical properties have been successfully obtained [19]. The different colour tones and their reproducibility may be obtained through manipulation of the cathodic peak potential [20]. The thickness of the film, of submicrometre order, obtained using the square current pulse method is comparable to that achieved with the INCO method [14]. In addition to a square waveform, a current scan with a triangular waveform has also been used. The thickness of the film depends on the electrolysis time and the logarithm of the space between two triangular waves [21]. It is worth mentioning that the process is conducted at room temperature, rather than at T > 70°C as for the INCO process. Therefore, corrosion of the reaction vessel and the need for good ventilation to disperse pollutant gases are reduced.

Wang *et al.* [20] have compared the corrosion resistance of films produced by using the purely chemical method, the INCO method and the electrochemical method. These authors have concluded that the pitting potential and the induction time for pitting are the longest for the electrochemical method, while the purely chemical method fares the worst. The cathodic hardening process used in the electrochemical method and the INCO method restricts the diffusion of reactive species in the film, thereby enhancing its corrosion resistance.

Recently, a one-step process has been devised by Chen *et al.* [22] (Fig. **1**). The one-step process involves two circuits. The first circuit is made up of a digital voltmeter **(1)**, a stainless steel specimen **(7)**, and a reference platinum electrode **(5)** that is used to control the potential difference (and thus the colour of the steel) during colouration. The second circuit used for electrolytic hardening is composed of the stainless steel specimen **(7)** (as cathode), auxiliary electrodes **(6)** (two lead plates as anodes), an adjustable resistor **(2)**, a DC power supply **(3)** and a digital ammeter **(4)**. The colouring aqueous solution contains 490gl^{-1} H_2SO_4, 250 gl^{-1} CrO_3, 7.0 gl^{-1} $(NH_4)Mo_7O_{24}\cdot4H_2O$, 6.0 gl^{-1} $ZnSO_4$ and 4.0 gl^{-1} $MnSO_4$. In this solution, $(NH_4)Mo_7O_{24}\cdot4H_2O$ enhances wear resistance of the film, $ZnSO_4$ stabilises the colouring potential and $MnSO_4$ accelerates the colouration process. This one-step process has been shown to be more colour-reproducible. Also, films obtained using this method are essentially pore-free.

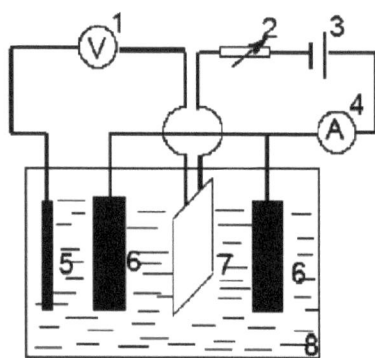

Figure 1: Schematic of the one-step colouration process proposed by Cheng *et al.* [Cheng Z, Xue Y, Tang Z, *et al.* A one-step process for chemical colouring on stainless steel. Surf Coat Technol 2008; 202(17): 4102-6. With permission for reproduction from Elsevier].

All of the afore-mentioned methods utilise chromic acid containing the highly toxic Cr(VI). In order to lessen the harmful effect on the environment, several research efforts have suggested the use of sulphuric acid [23, 24]. Fujimito *et al.* [23] have demonstrated that films of about 600nm may be achieved on the 304 steel by immersing it in deaerated H_2SO_4 (0.5mol L^{-1} and 5.0mol L^{-1}) at 50°C, followed by potential polarisation with a square wave for 40 minutes. On the other hand, Vasconcelos *et al.* [24] have shown that the use of a current scan (triangular wave) is also possible.

4. COLOURING USING COATING

Ceramic colouring/coating (also known as sputtering, plasma vapour deposition (CVD) or by the identification of materials and colour (like Ti-gold)) is a technique that may impart a uniform film of high-quality, mirror-like finish on stainless steel surfaces [25]. Films of ZrN, TiN, TiAlN and TiAlCN have been successfully coated on stainless steels *via* sputtering. Compared with electroplating and painting, ceramic colouring may produce very uniform films with high repeatability in colour tones. The latter technique may also produce very thin films such that the finish and texture of the metal sheets are not masked. Plus, the ceramic coatings are usually more durable and scratch-resistant than electrochemically-produced films, making them suitable for more aggressive use such as door hardware.

5. LASER COLOURING

Colouration of stainless steels has been demonstrated to be achievable *via* laser-colouration. Li *et al.* [26] obtained thick oxides on the 304 steel from multipasses of laser irradiation using a Q-switched, third harmonic Nd:YVO$_4$ laser. The oxide film is bi-layered, with Fe_2O_3 being the outer layer and Cr_2O_3 the inner one). The interface between the two layers is more even after multipasses of laser irradiation. Excimer laser (KrF) has also been used for this purpose by Zheng *et al.* [27]. These authors have propounded that during laser irradiation, oxygen molecules are absorbed into the solid surface, which then react with metal ions/atoms to form an ultrathin oxide layer. Electron tunnelling and diffusion of species through the growing oxide layer enable further reactions between absorbed oxygen molecules and metal ions/atoms.

REFERENCES

[1] Santoyo LDV, Bueno JJP, Ramirez AM, *et al.* Origin of interference colours on austenitic stainless steel. Inorg Mater 2005; 41(9): 955-60.

[2] Earwaker LG, Piddock V, Cole JM, *et al.* An investigation of films formed on stainless steel on immersion hot chromic/sulphuric acid. Nucl Instrum Method Phys Res B 1986; 15(1-6): 367-71.

[3] Evans TE, Hart AC, Skedgell AN. The nature of the film on coloured stainless steel. Trans Inst Metal Finish1973; 51: 108-12.

[4] Evans TE. Film formation on stainless steel in a solution containing chromic and sulphuric acids. Corros Sci 1977; 17(2): 105-24.

[5] Hideki I, Kastuski S. Japanese Patent 6 227 596. 1987.

[6] Evans TE, Sutton WH. British Patent 1 402 184. 1975.

[7] Sygeoda SA. Belgium Patent 869 885. 1979.

[8] Houska C. coloured Stainless Offers A Rainbow of Possibilities. Part 1. Newsletters of Sheet Metal and Air Conditioning Contractors' National Association (SMACNA). 12(4); August 24, 2005.

[9] Ogura K, Tsujigo M, Sakurai K, *et al.* Electrochemical colouration of stainless steel and the scanning tunneling microscopic study. J Electrochem Soc 1993; 140(5): 1311-5.

[10] Stoychev D, Stefanov P, Nicolova D, *et al.* Chemical composition and corrosion resistance of passive chromate films formed on stainless steels 316L and 1.4301. Mater Chem Phys 2002; 73(2-3): 252-8.

[11] Stefanov P, Stoychev D, Stoycheva M, *et al.* XPS and SEM studies of chromium oxide films chemically formed on stainless steel 316L. Mater Chem Phys 2000; 65(2): 212-5.

[12] Tan M. The mechanism of chemical colouring of stainless steels – I. Calculation of the colouring initiation time. Corros Sci 1992; 33(6): 873-8.

[13] Fujimoto S, Shibata T, Wada K. Formation process of coloured films on SUS304 stainless steel with the square wave potential pulse method. J Iron Steel Inst Japan 1991; 77(7): 1192-7.

[14] Lin CJ, Duh JG. Elemental redistribution in coloured films on SUS304 stainless steel produced by current pulse method. Surf Coat Technol 1996; 85(3): 175-83.

[15] Lin CJ, Duh JG. Fretting and Scratch wear characteristics of coloured films on stainless steel obtained by the current pulse method. Surf Coat Technol 1995; 73(1-2): 52-59.

[16] Sone Y, Wada K, Kurahashi H, Nakaim Y, Narutani T, Suzuki S. US Patent. 4 859 287. 1989.

[17] Ogura K, Sajurai K, Uehara S. Room temperature-colouration of stainless steel by alternating potential pulse method. J Electrochem Soc1994; 141(3): 648-51.

[18] Wang JH, Duh JG. Colour tone and chromaticity in a coloured film on stainless steel by alternating current electrolysis method. Surf Coat Technol 1995; 73(1-2): 46-51.

[19] Lin CJ, Duh JG. Mechanical characteristics of coloured film on stainless steel by the current pulse method. Thin Solid Films 1996; 287(1-2): 80-6.

[20] Wang JH, Duh JG, Shih HC. Corrosion characteristics of coloured films on stainless steel formed by chemical, INCO and a.c. processes. Surf Coat Technol 1996; 78(1-3): 248-54.

[21] Ogura K, Lou W, Nakayama M. colouration of stainless steel at room temperature by triangular current scan method. Electrochim Acta 1996; 41(18): 2849-53.

[22] Cheng Z, Xue Y, Tang Z, *et al.* A one-step process for chemical colouring on stainless steel. Surf Coat Technol 2008; 202(17): 4102-6.

[23] Fujimoto S, Shibata T, Wada K, *et al.* The electrochemical conditions for coloured film formation on type 304 stainless steel with square wave polarization. Corros Sci 1993; 35(1-4): 147-52.

[24] Vasconcelos KO, Bocchi N, Filho RCR, *et al.* An environmentally friendly and practical method for obtaining colour on stainless steel by interference. J Electrochem Soc 2005; 152(11): B491-4.

[25] Nakamoto K, Shiotani K, Makimoto M, *et al.* Development of coloured stainless steel sheets by ceramics coating. ISIJ Int 1993; 33(9): 968-75.

[26] Li ZL, Zheng HY, Teh KM, *et al.* Analysis of oxide formation induced by UV laser colouration of stainless steel. Appl Surf Sci 2009; 256(5): 1582-8.

[27] Zheng HY, Lim GC, Wang XC, *et al.* Process study for laser-induced surface colouration. J Laser Appl 2002; 14(4): 215-20.

CHAPTER 13

Powder Metallurgy of Stainless Steels

Abstract: This chapter is devoted to powder metallurgy (PM) of stainless steels. The main focus is on the making of PM parts and problems that are encountered during the process. The manufacture of stainless steel powders is excluded. Porous stainless steels have also been included in this chapter because powder metallurgy may be utilised for their fabrication, but other non-PM methods are also covered.

Keywords: Powder, powder metallurgy, stainless steel, lotus type, porous alloys, sintering, liquid phase sintering, supersolidus, cellular metals, directional pore, continuous casting.

1. INTRODUCTION

Work on powder metallurgy (PM) of stainless steels may be traced back to the 1930's when the Hardy Metallurgical Company tried to produce the 18-8 grade austenitic stainless steel by mixing elemental powders of iron, nickel and chromium in the right proportions. The test piece was sintered in purified dry hydrogen for 44h at 1300°C, an egregious uneconomical process. Since then, the costs for the production of PM parts have decreased substantially relative to the early days. As a result, the use of PM has been quite popular nowadays. In the year 2000, powder shipments grew nearly by 22% to 7157 short tonnes, following a 12% increase in the previous year in North America, thanks to the increased use of PM in the automotive industry [1].

Stainless steel parts fabricated by PM are used in a variety of fields. Some of these are automotive temperature control valves, anti-lock braking system sensors, exhaust flanges and filters. Stainless steel powders have very good flow and die fill characteristics and so do not present any unique problems in compacting. The austenitic grades are the most extensively used, but the ferritic grades are also gaining importance. The ferritic grades may reach a higher sintered density than the austenitic grades, especially at high temperatures, on grounds of their higher self-diffusivity [2]. In the as-atomised condition, austenitic stainless steel powders are very soft and so have good compressibility for PM processing. As-atomised powders of the martensitic grades, on the other hands, are usually in the hard martensitic state and so are not readily compressible. Consequently, tempering the powders below the austenitisation temperature or re-austenitising them at slightly above the austenitisation temperature with a subsequent slow cooling may be required.

Powders are most often made by either water atomisation or gas atomisation, but methods like thermal decomposition of a chemical compound, and electrodeposition are also in use. Gas-atomised powders have a low surface oxygen content, and so their sinterability is better than their water-atomised counterparts. Gas-atomised powders are also more spherical than water-atomised powders, and hence the former's mixture viscosity is lower during mixing and moulding. However, the less-spherical nature of water-atomised powders may improve shape retention during thermal processing. To mitigate surface oxidation of powders, even paraffin has been suggested as an atomising liquid [3].

Most commercially available stainless steel powders contain a small amount of silicon for the formation of a silicon-rich surface layer that may prolong shelf life. This silicon-rich layer, although microscopic in scale, may hinder mass transfer during sintering [4]. It is to be noted that when it is desired to produce good soft magnetic properties (in 410L, 434L, for example), it is important to ensure that a high degree of sintering is obtained.

2. PROBLEMS IN PM OF STAINLESS STEEELS

Compared with conventional methods, a main drawback of PM is that parts made using this method might be porous. Porosity not only detracts from mechanical properties, but it also adversely affects corrosion resistance as pores and crevices are convenient sites for the accumulation of corrosion products [5]. Adding more alloying elements like Ni and Mo may improve corrosion performance, but high alloying levels add to cost and impair compressibility. A recent research shows that mixing small, spherical gas-atomised powders to bigger, irregular water-atomised powders may reduce porosity, because both the good compressibility of gas-atomised powders and the good sinterability of water-atomised powders are taken advantage of [5-7].

There are many problems that may occur during sintering. Formation of carbides [8] and nitrides [9, 10] during sintering is one of these problems. The sintering temperature for types 420 and 440 martensitic stainless steel powders must be carefully chosen within a narrow sintering window in order to reduce porosity and generate a finely distributed, spheroidal carbide in the sintered microstructure [8]. High-nitrogen sintered parts containing nitrides may be reheated at a low temperature (say at 1150°C) to dissolve the nitrides, followed by rapid cooling to avoid nitride reprecipitation [10]. Besides precipitation, a thermal gradient within the compact may also be problematic. To circumvent this problem, it has been shown that sintering with a multimode microwave furnace may be handy. Microwaves interact directly with the powder particles and so rapid volumetric heating of the compact becomes possible, whereas in conventional furnaces, parts are heated from the outside in [11].

Quite often, the corrosion and mechanical properties of PM stainless steels is lower than that of their wrought counterparts having similar compositions [12] mainly because of the presence of (interconnected) porosity [13]. Pore morphology affects corrosion properties significantly [14], because open, interconnected pores allow the free flow of the corrodent [15]. Electrolyte stagnation in pores may also lead to development of hydrogen concentration cells between the pore surfaces [9].

It is thought that pore morphology, rather than the pores *per se*, may be responsible for corrosion attack [16]. Residues left over after sintering may detract from oxidation and corrosion resistance. The pore morphology also affects local plastic behaviour, which in turn influences mechanical properties [17]. The sintering atmosphere also plays a role in corrosion behaviour. For example, while sintering in a nitrogen atmosphere may improve mechanical [18] and corrosion properties [19], nitrogen is beneficial only if it remains in solution. When nitrides and carbonitrides form, corrosion resistance is reduced [20, 21]. The distribution of nitrides also affects the extent of corrosion. In this respect, sintering in a vacuum or other inert gases may be desirable [20]. As a matter of fact, the sintering atmosphere affects sintered density, size and shape of pores, and grain size [22]. Therefore, the sintering atmosphere is a very critical factor in PM as it directly bears on the final mechanical and corrosion properties.

Besides the sintering atmosphere, cleanliness of the press area is also important. Just a few leftover iron particles embedded in the sintered part may be sufficient to initiate localised corrosion. It is also crucial that any pressing lubricant (carbon-containing stearates, waxes, zinc, lithium stearates, *etc.*) be thoroughly burned off in a presintering treatment, or else it may cause carburisation during sintering, thereby compromising corrosion resistance and machinability. In general, lubricants containing a metallic component have to be dealt with at a higher temperature than with purely organic ones. The final finishing processes of the sintered part may also affect corrosion resistance. The influences of several commonly used finishing processes on the corrosion resistance of PM stainless steel parts have been assessed, with tumbling found to be the most deleterious [23].

In spite of the afore-mentioned problems, PM does have its upsides. For instance, it allows near-net-shape fabrication of complicated components [24, 25]. The use of PM, coupled with novel techniques like selective laser sintering [26, 27], direct laser sintering [28] and softlithograprahy [29], has been utilised to produce components of intricate shapes. Fabrication of defect-free micro-components with high precision and surface quality has been shown to be possible with PM [30]. Furthermore, porosity, when produced in a controlled manner, may be useful as sintered metallic foams. These foams may be used for sound attenuation, abradable seals and flow control devices. A microstructure with a very fine and uniform porosity may be attained by ball-milling stainless steel powders with ceramic powders, with a subsequent sintering treatment [31].

3. SINTERING ATMOSPHERE AND SUPERSOLIDUS LIQUID PHASE SINTERING

Commercially, dissociated ammonia ($75\%H_2 + 25\%N_2$) is frequently used as a sintering atmosphere. When nitrogen is in solution, it may bring about considerable strengthening. Nonetheless, nitrogen has a high affinity for chromium and so the presence of N in dissociated ammonia may lead to the precipitation of nitrides, thereby reducing corrosion resistance. Therefore, cooling rates should be high enough to avoid the formation of these phases. Dissociated ammonia may substitute for pure hydrogen for many applications when reaction with nitrogen is not a problem. Other commonly used atmospheres are hydrogen and vacuum. Hydrogen is an active atmosphere and it reduces surface oxide. The removal of oxide results in a greater degree of bonding. When hydrogen is used, Sands *et al.* [32] showed that the sintering response of 316L would be very dependent on moisture content. These authorsfound that when water vapour content was increased, the cooling rate must be increased in order to ensure good corrosion resistance. Sintering in a vacuum produces properties that are quite similar to those attained by using hydrogen, as it also involves dissolution of the surface oxide.

PM is often used to fabricate particulate-dispersoid reinforced stainless steels, with the aim of achieving better properties. To this end, various additives, mostly fine oxides, carbides and intermetallics, have been tried. Y_2O_3 has been found to produce more homogeneous porosity in 316L compared with just using 316L alone [33]. Al_2O_3 has been shown to yield better mechanical properties [34] and a fine grain size [35]. SiC is believed to give rise to a higher density in the sintered state [36]. Intermetallics like Cr_2Al, TiAl and Cr_2Ti can improve the wear resistance and corrosion behaviour, as they can react with the matrix [37] and so may generate a better bonding than do oxides and carbides [38]. These reinforcing particles, in addition to being mixed to the starting powders, may be generated in-situ by properly adjusting the composition [39].

While addition of these reinforcing particulates may improve mechanical and tribological properties, their presence renders densification more difficult. To tackle this problem, recent research has attempted a liquid phase sintering process called supersolidus liquid phase sintering (SLPS) [40], which is more suitable for pre-alloyed powders than ordinary liquid phase sintering. In supersolidus liquid phase sintering, the addition of Y_2O_3 has been shown to enhance sintered density for both austenitic (316L) and ferritic (434L) stainless steels [41]. However, the corrosion resistance of 434L seems to be reduced [42]. Use of Y_2O_3 in SLPS reduces the corrosion resistance of 434L [42], but the intermetallics Ni_3Al and Fe_3Al have the opposite effect [43].

4. PM OF STAINLESS STEELS

Precipitation-strengthening, attained by proper composition design, has been utilised to produce PM precipitation-hardening stainless steels whose mechanical properties rival those displayed by the conventional 17-4PH precipitation-hardening and austenitic 304L steels. These steels may be duplex (ferritic-martensitic [44] and ferritic-austenitic [45]) and single-phased [46].

Besides single-phased stainless steels, it is also possible to make duplex stainless steels by using PM. In fact, the making of duplex stainless steels by PM may be more advantageous than by conventional techniques. For example, hot-cracking in forging and microsegregation in casting may be aggravated by high contents of Cr, Mo and N, whose presence in new duplex stainless steels are rather substantial and common. These problems may be alleviated (or eliminated) by the use of PM and HIPing. PM HIPed duplex stainless steels are usually fine-grained, homogeneous and isotropic, even for components of large wall thicknesses [47]. HIP stainless steels are also amenable to ultrasonic testing because of their fine metallurgical structures and isotropy [48].

PM duplex stainless steels have been shown to possess good resistance to stress corrosion cracking [49], fatigue strength [50] and toughness [51].

Sintered duplex stainless steels may be fabricated by 1. using pre-alloyed powders whose compositions produce a duplex microstructure [52], 2. mixing of ferritic or austenitic powder with elemental powders (Mo, Cr, Ni, *e.g.*) [53], or by mixing austenitic and ferritic stainless steel powders in proper proportions

[54, 55]. Usually, the sintered density of a ferritic stainless steel is higher than that of an austenitic stainless steel, because of the former's higher Fe self-diffusivity in the bcc lattice. However, Pucas *et al.* [56] have discovered that duplex stainless steels made by mixing the 'right' proportion of ferritic powder and austenitic powder may reach a higher sintered density than does a ferritic stainless steel made using pure ferritic powder. Diffusional fluxes created by compositional gradients of the mixed powders are believed to be responsible for the increased sintered density.

Compressibility may be reduced and densification increased by using boron [57, 58] or nickel boride [58] as an additive. Boron promotes the formation of a liquid phase which can activate sintering [56]. Compared with boron, nickel boride has been found to give rise to better elongation properties and higher strength [57]. As regards the determination of the 'right' proportions of powders for achieving the desired microstructures, the Schaeffler diagram may be handy. Although this diagram was developed for weld metals, its adoption for PM has been shown to be valid [58, 59].

5. PM OF HIGH-NITROGEN STAINLESS STEEL

The making of high-nitrogen stainless steels using PM is also advantageous [60]. The diffusivity of nitrogen in Fe-Cr austenite is very low [61, 62] and this means a very long treatment time may be required in order to homogenise nitrogen in steels. This makes PM of high-nitrogen stainless steels advantageous. The short diffusion distances in steel powders allow minimal segregation and easy homogenisation of N. The use of HIP after a sintering-nitriding treatment minimises machining. This is desirable because machinability of highly-alloyed, high-N stainless steels is usually poor.

The solubility of nitrogen may be more predictable in PM [60] and a very high nitrogen content may be introduced to the starting powders [63]. Mechanical alloying is a very good method for producing high-N stainless steel powder [64, 65]. Stainless steel powders having a nitrogen content of up to 2.47wt% has been produced using this method [65], and high-N PM stainless steels with ultrafine grains has been made recently [66]. The solubility issue is particularly important to the ferritic grades, because nitrogen has a low solubility in ferrite. Nitrogen introduced to the molten steel by high-pressure melting (higher than that used for the austenitic grades) may come out of solution when the liquid metal solidifies, resulting in foaming of the melt. PM thus is valuable in the fabrication of N-containing ferritic stainless steels [67, 68].

Although PM of high-N stainless steels possesses many advantages, the ease of contamination of powder surfaces may be a headache. Contamination due to impurities, especially oxygen, may render nitriding and sintering difficult because of the presence of surface oxide.

In making high-N stainless steels using PM, sintering and nitriding may be carried out simultaneously [69]. A merit of this sintering-nitriding process is that the infusion of nitrogen into the steel is facilitated by open, interconnected pores. Increasing the temperature leads to more efficient sintering and the interconnectivity of pores is thus destroyed faster. However, the collapse of pore interconnectivity hinders nitrogen infusion into the steel. Usually, nitriding of loosely packed powders is easier because of a greater exposure of the powder surfaces to the nitriding atmosphere. However, since solid-solution hardening is expedited, compressibility may be reduced and the handling of green compacts can be more difficult.

Nitriding of loosely packed powders also leads to their agglomeration of powders. To overcome this undesired phenomenon, Feichtinger designed a rotating furnace that constantly stirs the powders, thereby preventing them from agglomerating [70]. Nitriding of steel powders using a fluidised bed, as suggested by Virta *et al.* [71], may be used to avoid agglomeration, too.

6. POROUS STAINLESS STEELS

6.1. Methods for Producing Non-directional Pores

This section is on porous stainless steels, which may be fabricated by using PM and non-PM techniques (Fig. 1). Some of these techniques that have been utilised for fabricating porous stainless steels will be covered.

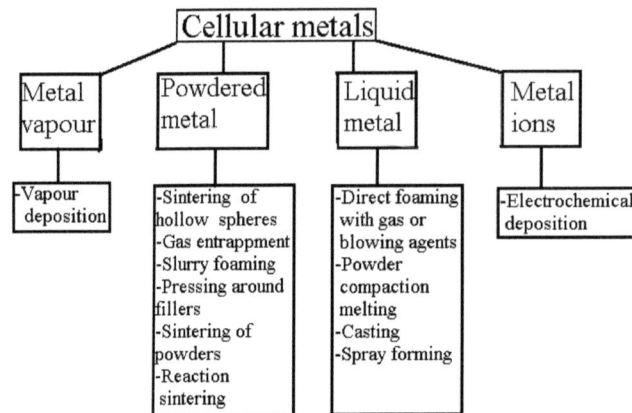

Figure 1: Some of the methods for fabricating cellular metals [Drawn with reference to Banhart J. Manufacture, characterisation and application of cellular metals and metal foams. Prog Mater Sci 2001; 46(6): 559-632].

Porous materials and metal foams have a number of very useful properties, such as high stiffness with low specific weight, and high fluid permeability. A large number of applications involve the use of porous metals, including filtration and separation, heat exchangers, control devices for fluid flows, noise control, and water purifying devices.

A variety of methods have been devised for fabricating porous metals (in the liquid state or solid state), as delineated in Fig. 1 [72]. Most of these methods have been employed for making porous aluminium, copper, and their alloys, whereas reports on the production of porous stainless steels are less numerous [72]. One of the hindrances to making porous stainless steels is their high melting points. For instance, foaming agents like TiH_2 or other metals (Pb, Al, Zn) decompose far below the required temperatures for making porous stainless steels.

In the spray forming route, molten metal is continuously atomised. A spray of metal droplets are collected on a substrate where they grow and form a deposit of a pre-set shape (sheet, tube, *e.g.*) The final properties of the deposit may be tailored by incorporating powders of oxides, carbides, or pure metals into the spray. Metal-matrix composites are obtained if oxides and carbides are used. When substances that will decompose on contact with the spray are used, pores will be generated in the deposit because of the release of gaseous decomposition products. Porous steel has been fabricated in this manner [72].

In the powder compact melting technique (invented at the Fraunhofer Institute, Germany), the starting materials are metal powders and foaming takes place in the liquid state. The process begins with mixing metal powders and blowing agents. Subsequently, the mixture is compacted by conventional means into a semi-finished product, which is then heated near the melting point of the matrix material. The uniformly distributed blowing agent will decompose during heating. Steel, whose melting point does not differ too much from that of stainless steel, has been fabricated in this way by using $MgCO_3$ and $SrCO_3$ as the foaming agents [73, 74].

Investment casting may also be used to produce porous stainless steels. In investment casting, a polymer foam having an open-pore structure is used as the starting material. A slurry of heat-resistant material (*e.g.*, mullite, phenolic resin) is then filled into the open cells of the polymer foam. After curing, the polymer foam is removed by a thermal treatment, leaving behind a structure of interconnected, open voids. At this stage, molten metal may be cast into the open voids, which are a replica of the original polymer foam structure. The final step involves removal of the mould material (Fig. 2). This technique may be used to make highly ordered porous metals (known as Lattice Block Materials, LBMs) having predictable, reproducible mechanical properties that are close to their theoretical optimum limits [72, 75]. Pre-fabricated injection-moulded polymer elements are bonded layer by layer together. The resulting structure is then used for investment casting.

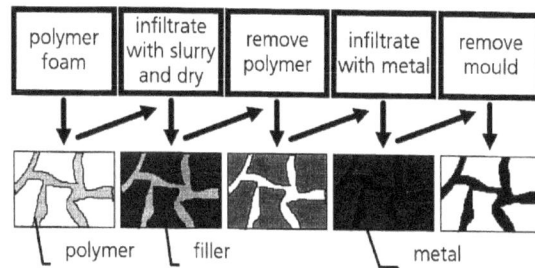

Figure 2: Schematic representation of the investment casting process [Banhart J. Manufacture, characterisation and application of cellular metals and metal foams. Prog Mater Sci 2001; 46(6): 559-632. With permission for reproduction from Elsevier].

The open cell structure as described in the investment casting route is also suitable for use as a precursor in making porous metals (*e.g.*, pure iron) by electro-deposition and vapour deposition. In electro-deposition, the starting polymer foam is first immersed into a slurry that is electrically conductive (*e.g.*, graphite), such that the polymer foam surface is covered by a conductive coating. Subsequently, metal is electrically deposited onto the coated polymer foam. The polymer mould is then removed by thermal treatment. Porous iron has been made in this way. In vapour deposition, metal vapour is allowed to condense on the precursor in a vacuum chamber. The thickness of the metal coating depends on the vapour density and deposition time.

Conventional sintering of stainless steel powders is certainly useable for producing porous products [76]. And products having a nanocrystalline structure are also producible using mechanical alloying and PM [77]. In the space-holder method, stainless steel powder is mixed with powders of space holder (*e.g.*, carbamide) and a polymeric moulding binder. The mixture is then compacted into a semi-finished product. A subsequent catalytic and/or thermal treatment is conducted to remove the space holder (*e.g.* carbamide [78]), followed by a conventional sintering process. The degree of porosity, pore morphology and size depend mostly on the size and shape of the space holder. Micro-sized porous parts having a near-net-shape may be fabricated by combining this space-holder method with powder injection moulding [79]. Recently, Bakan invented a novel technique in which carbamide is completely removed by water leaching at room temperature [80]. In this way, the release of harmful gases commonly associated with the thermal treatment and the anisotropic thermal expansion of the organic space holder material are eliminated. This water-leaching technique has been employed for producing porous 316L [80] and 17-4PH [81] stainless steels.

Highly porous stainless steels may be fabricated by bonding together hollow stainless steel spheres through sintering, too. Both open and closed pore structures having ordered or disordered arrangement of cells may be achieved (Fig. **3**). The stainless steel hollow spheres may be made by combined chemical and electrical deposition onto polymer spheres, or by coating polymer spheres with a metal powder suspension. The coated spheres are then sintered such that a dense metal shell is obtained [72]. Since the pore size distribution is not random, the properties of porous metals fabricated in this way are more predictable than those of other metal foams.

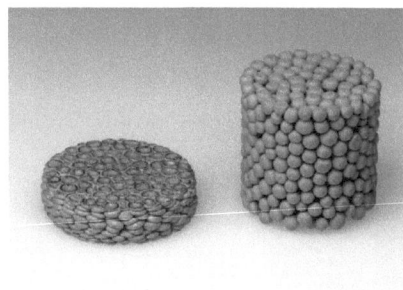

Figure 3: Porous stainless steel fabricated by sintering of hollow stainless steel spheres [Banhart J. Manufacture, characterisation and application of cellular metals and metal foams. Prog Mater Sci 2001; 46(6): 559-632. With permission for reproduction from Elsevier].

6.2. Methods for Producing Directional Pores

While the methods introduced in the above section produce pores that are largely non-directional, this section is on methods that can be used to fabricated stainless steels (and other metals like copper, magnesium, titanium, nickel and their alloys) having directional pores.

Some metals form a eutectic system with certain gases. When such a metal is melted in an atmosphere of the right gas under high pressure (up to 50atm), a homogeneous molten metal charged with the gas will form. As the temperature drops, the solubility of the gas in the molten metal decreases and so the system will undergo a eutectic reaction, resulting in a solid phase and a gas phase. The pores formed during the process are elongated and aligned parallel to the solidification direction. Hence, pore directions may be easily controlled as shown in Fig. **4**. Hydrogen may be used for stainless steel, iron, carbon steel, nickel, aluminium, copper, whereas oxygen may be used for gold and silver [82].

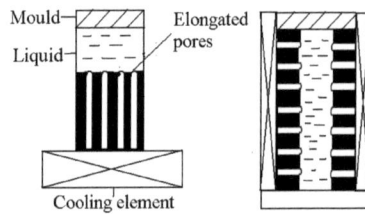

Figure 4: Different arrangement of moulds for obtaining different pore configurations [Drawn with reference to Banhart J. Manufacture, characterisation and application of cellular metals and metal foams. Prog Mater Sci 2001; 46(6): 559-632].

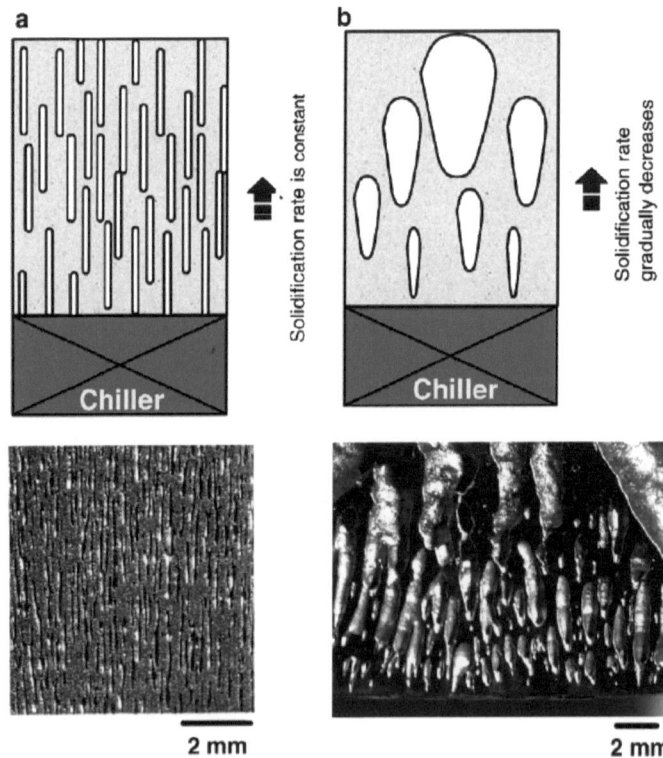

Figure 5: Pore structures in (a) copper and (b) stainless steel obtained by using the mould casting technique in a gas atmosphere [Nakajima H. Fabrication, properties and application of porous metals with directional pores. Prog Mater Sci 2007; 52(7): 1091-173. With permission for reproduction from Elsevier].

Porous copper and magnesium may be satisfactorily produced by using mould casting in a gas atmosphere (Fig. **5(a)**) as they possess high thermal conductivity. Solidification proceeds at an almost constant rate throughout the ingot, resulting in a uniform pore size and porosity. However, for metals having low thermal conductivity such as stainless steels, heat is less efficiently dissipated to the cooling plate. Therefore, the upper molten metal cools down more slowly and the pores are coarser (Fig. **5(b)**).

To overcome this problem, Nakajima *et al.* [82, 83] have devised the continuous zone melting technique (Fig. **6**). The induction coil melts part of the rod-shaped raw material which is moved at a constant speed. Since the set-up is placed in a high-pressure chamber filled with hydrogen (or other gases for other metals), hydrogen will be absorbed into the molten part up to the equilibrium solubility point. When the molten part leaves the induction coil, solidification takes place in its lower part and thus gas is ejected from the melt, resulting in elongated pores. The pore size may be controlled by varying the speed of the rod.

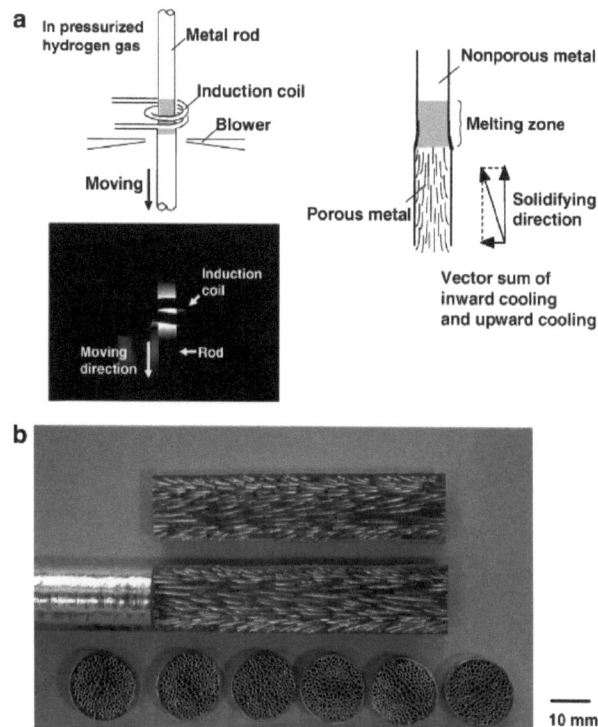

Figure 6: Continuous casting for making stainless steel of directional pores **(a)** schematic representation of the set-up **(b)** optical micrograph of porous stainless steel produced by using this method [Nakajima H. Fabrication, properties and application of porous metals with directional pores. Prog Mater Sci 2007; 52(7): 1091-173. With permission for reproduction from Elsevier].

Recently, Shen *et al.* [84] have also shown that directional pores of micrometre scales may be produced by selective laser melting with H_3BO_3 and KBF_4 as additives. H_3BO_3 is mainly used to produce gas in the melt pool, while KBF_4 is used for reaction with Cr_2O_3 to prevent the 'balling' effect of this oxide.

REFERENCES

[1]   Capus JM. Stainless steel. Adv Mater Process 2000; 158: p.57.

[2]   Baran MC, Shaw BA. Corrosion modes in P/M ferritic stainless steels. Int J Powder Metall 2000; 36(4): 57-68.

[3]   Dowson G. Introduction to powder metallurgy: the process and its products. Euro Powder Metal Assoc (EPMA). Shrewsbury, UK. 1992.

[4]   Lall C. Fundamentals of high temperature sintering: application to stainless steels and soft magnetic alloys. Intl J Powder Metall 1991; 27(4): 315-29.

[5] Moral C, Bautista A, Velasco F. Aqueous corrosion behaviour of sintered stainless steels manufactured from mixes of gas atomized and water atomized powders. Corros Sci 2009; 51(8): 1651-7.

[6] Bautista A, Moral C, Blanco G, *et al.* Mechanical and oxidation properties of high density sintered duplex stainless steels obtained from a mix of water and gas atomised powders. Powder Metall 2006; 49(3): 265-73.

[7] Bautista A, Moral C, Velasco F, *et al.* Density-improved powder metallurgical ferritic stainless steels for high-temperature applications. J Mater Process Technol 2007; 189(1-3): 344-51.

[8] Newell MA, Davies, HA, Messer PF, *et al.* Metal injection moulding of scissors using hardenable stainless steel powders. Powder Metall 2005; 48(3): 227-30.

[9] Iglesias FAC, Roman JMR, Prieto JMR, *et al.* Effect of nitrogen n sintered duplex stainless steels. Powder Metall 2003; 46(1): 39-42.

[10] Hwang KS, Hsueh YW. Post-sintering thermal treatment of nitrogen containing pressed and sintered and PIM stainless steels. Powder Metall 2007; 50(2): 165-71.

[11] Panda SS, Singh V, Upadhyaya A, *et al.* Sintering response of austenitic (316L) and ferritic (434L) stainless steel consolidated in conventional and microwave furnaces. Scr Mater 2006; 54(12): 2179-83.

[12] Otero E, Pardo A, Utrilla MV, *et al.* Corrosion behaviour of AISI 304l and 316L stainless steels prepared by powder metallurgy in the presence of sulphuric and phosphoric acid. Corros Sci 1998; 40(8): 1421-34.

[13] Sorbal W, Ristow W, Azambuja D, *et al.* Potentiodynamic tests and electrochemical impedance spectroscopy of injection molded 316L steel in NaCl solution. Corros Sci 43; 2001; 1019-30.

[14] Bautista A, Velasco F, Guzman S, *et al.* Corrosion behaviour of powder metallurgical stainless steels after two years of exposure in atmosphere. Corrosion Eng Sci Technol 2006; 41(4): 284-90.

[15] Seah KHW, Thampuran R, Teoh SH. The influence of pore morphology on corrosion. Corros Sci 1998; 40(4-5): 547-56.

[16] Peled P, Harush H, Itzhak D. The effect of Ni addition on the corrosion behaviour of sintered stainless steel in H$_2$SO$_4$. Corros Sci 1988; 28(4): 327-9.

[17] Lindstedt U, Karlsson B, Masini R. Influence of porosity on deformation and fatigue behavior of P/M austenitic stainless steel. Int J Powder Metall 1997; 33(8): 49-61.

[18] Mariappan R, Kumaran S, Rao TS. Effect of sintering atmosphere on structure and properties of austeno-ferritic stainless steels. Mater Sci Eng A 2009; 517(1-2): 328-33.

[19] Mudali UK, Shankar P, Ningshen S, *et al.* On the pitting corrosion resistance of nitrogen alloyed cold worked austenitic stainless steels. Corros Sci 2002; 44(10): 2183-98.

[20] Martin F, Garcia C, Tiedra PD, *et al.* Sensitization of powder metallurgy type 430L stainless steel sintered in nitrogen-hydrogen atmosphere. Corrosion 2008; 64(1): 70-82.

[21] Abdel-Karim R, Elmahallawi I, El-Menshawy K. Microstructure and corrosion properties of nitrogen stainless steel 316L produced by hipping. Powder Metall 2004: 47(10): 43-8.

[22] Li S, Huang B, Li D, *et al.* Influences of sintering atmospheres on densification process of injection moulded gas atomised 316L stainless steel. Powder Metall 2003; 46(3): 241-5.

[23] Molins A, Bas JA, Planas J. P/M stainless steel: types and their characteristics and applications. Adv Powder Metall Part Mater 1992; 5(3): 345-57.

[24] Jeon YC, Kim KT. Near-net-shape forming of 316L stainless steel powder under hot isostatic pressing. Int J Mech Sci 1999; 41(7): 815-30.

[25] Besson J, Abouaf M. Behavior of cylindrical HIP containers. Int J Solid Struct 1991; 28(6): 691-702.

[26] Liang LZ, Hui LJ, Sheng SY, *et al.* Characteristics of complicated AISI316L automobile components manufactured by powder/metallurgy. J Mech Sci Technol 2009; 23(7): 1924-31.

[27] Dewidar MM, Khalil KA, Lim JK. Processing and mechanical properties of porous 316L stainless steel for biomedical applications. Trans Nonferr Metal Soc China 2007; 17(3): 468-73.

[28] Gu D, Shentis Y. Processing conditions and microstructural features of porous 316L stainless steel components by DMLS. Appl Surf Sci 2008; 255(5): 1880-7.

[29] Imbaby M, Jiand K, Chang I. Fabrication of 316-L stainless steel micro parts by softlithography and powder metallurgy. Mater Lett 2008; 62(26): 4213-6.

[30] Imbaby M, Jiang K, Chang I. Net shape fabrication of stainless-steel micromachine components from metallic powder. Int J Micromech Microeng 2008; 18(11): 1-7.

[31] Park K, Lee TS. Fabrication and characteristics of porous AISI 304L stainless steels by ceramic powder additions. Mater Sci Technol 2004; 20(6): 711-4.

[32]   Sands RL, Bidmead GF, Oliver DA. The corrosion resistance of sintered austenitic stainless steel. Modern Developments in Powder Metallurgy (Hausner HH, ed). Vol.2, 1966. Plenum Press. NY. p.73.

[33]   Lal S. Sintering of 316L austenitic stainless steel-Y_2O_3 particulate composites. Ph.D Thesis. Indian Institute of Technology. Kanpur, India, 1988.

[34]   Mukherjee SK, Upadhyaya GS. Sintering of 434L ferritic stainless steel containing Al_2O_3 particles. Int J Powder Metall 1983; 19(4): 289-95.

[35]   Ng PG, Clarke E, Khoo CA, *et al.* Microstructural evolution during aging of novel superferritics stainless steel produced by the HIP process. Mater Sci Technol 2006; 22(7): 852-8.

[36]   Patankar SN, Tan MJ. Role of reinforcement in sintering of SiC/316L stainless steel composite. Powder Metall 2000; 43(4): 350-2.

[37]   Abenojar J, Velasco F, Bautista A, *et al.* Atmosphere influence in sintering process of stainless steels matrix composites reinforced with hard particles. Compos Sci Technol 2003; 63(10): 69-73.

[38]   Velasco F, Lima WM, Anton N, *et al.* Tribol Int 2003; 36(7): 547-51.

[39]   Navas EMR, Anton N, Gordo E, *et al.* Wear behavior of a ferritic stainless steel with carbides manufactured through powder metallurgy. J Mater Eng Perform 2001; 10(4): 479-83.

[40]   German RM. Supersolidus liquid-phase sintering of prealloyed powders. Metall Mater Trans A 1997; 28(7): 1553-67.

[41]   Jain J, Kar AM, Upadhyaya A. Effect of YAG addition on sintering of P/M 316L and 434L stainless steels. Mater Lett 2004; 58(14): 2037-40.

[42]   Shankar J, Upadhyaya A, Balasubramaniam R, *et al.* Electromechanical behaviour of sintered yttria reinforced ferritic stainless steels. Powder Metall 2003; 46(1): 25-9.

[43]   Balaji S, Joshi G, Upadhyaya A. Corrosion behavior of sintered aluminide-reinforced ferritic stainless steel. Scr Mater 2007; 56(2): 149-51.

[44]   Schade CT, Murphy TF, Lawley A, *et al.* Development of a dual-phase precipitation-hardening PM stainless steels. Int J Powder Metall 2009; 45(1): 38-46.

[45]   Smuk O, Nenonen P, Hanninen H, *et al.* Microstructures of a powder metallurgy-hot isostatically pressed superduplex stainless steel forming in industrial heat treatments. Metall Mater Trans A 2004; 35(7): 2103-9.

[46]   Schade CT, Stears PD, Lawley A, *et al.* Precipitation-hardening PM stainless steels. Int J Powder Metall 2007; 43(4): 51-9.

[47]   Hamalainen E, Laitinen A, Hanninen H, *et al.* "Properties of Nitrogen-Alloyed P/M Duplex Stainless Steels". In: Proceedings of Conference Duplex Stainless Steels 94. Glasgow. TWI.

[48]   Dellis Ch, Marois GL, Gentzbittel JM, *et al.* Properties of HIPed stainless steel powder. J Nucl Mater 1996; 233-237: 183-7.

[49]   Laitinen A, Hanninen H. Chloride-induced stress corrosion cracking of powder metallurgy duplex stainless steels. Corrosion 1996; 52(4): 295-306.

[50]   Laitinen A, Hanninen H. Corrosion fatigue resistance of powder metallurgy and forged duplex stainless steels. Mater Sci Technol 1996; 12(12): 1064-70.

[51]   Hamalainen E, Laitinen A, Hanninen H, *et al.* Mechanical properties of powder metallurgy duplex stainless steels. Mater Sci Technol 1997; 13(2): 103-9.

[52]   Ruiz-Prieto JM, Moriera WM, Torralba JM, *et al.* Powder metallurgical duplex austenitic-ferritic stainless steels from prealloyed and mixed powders. Powder Metall 1994; 37(1): 57-60.

[53]   Rosso M, Grande MA. In: Proceedings of the 2000 Powder Metallurgy World Congress Kyoto, vol. 2 (2000), p. 1017.

[54]   Munez CJ, Utrilla MV, Urena A. Effect of temperature on sintered austeno-ferritic stainless steel microstructure. J Alloy Compd 2008; 463(1-2): 552-8.

[55]   Datta P, Upadhyaya GS. Sintered duplex stainless steels from premixes of 316L and 434L powders. Mater Chem Phys 2001; 67(1-3): 234-42.

[56]   Puscas TM, Molinari A, Kazior J, *et al.* Sintering transofrmations in muixtures of austenitic and ferritic stainless steel powders. Powder Metall 2001; 44(1): 48-52.

[57]   Ornato D. High density sintering of duplex stainless steels. Powdcr Metall 2002; 45(4): 290-3.

[58]   Bakan HI, Heaney D, German RM. Effect of nickel boride and boron additions on sintering characteristics of injection moulded 316L powder using water soluble binder system. Powder Metall 2001; 44(3): 235-42.

[59]   Dobrzanski LA, Brytan Z, Grande MA, *et al.* Corrosion behavior of vacuum sintered duplex stainless steels. J Mater Process Technol 2007; 191(1-3): 161-4.

[60]   Bianncaniello FS, Jiggetts RD, Ricker RE, *et al.* Powder metallurgy high nitrogen stainless steel. Mater Sci Forum1999; 318-20: 649-54.

[61]   Tschiptschin AP. THERMEC2000. J Mater Process Technol 2001; 117(Sp.Iss.SI): 3-17.

[62]   Heger D, Duong TV. Nitrogen diffusion in highly alloyed Cr-steels. Defect Diffu Forum1997; 143-7: 443-8.

[63]   Romu JJ, Tervo JJ, Hanninen HE, *et al.* Development of properties of P/M austenitic stainless steels by nitrogen infusion. ISIJ Int 1996; 36(7): 938-46.

[64]   Haghir T, Abbasi MH, Golozar MA, *et al.* Investigation of α to γ transformation in the production of a nanostructured high-nitrogen austenitic stainless steel powder *via* mechanical alloying. Mater Sci and Eng A 2009; 507(1-2): 144-8.

[65]   Cisneros MM, Lopez HF, Manncha H, *et al.* Development of austenitic naostructures in high-nitrogen steel powders processed by mechanical alloying. Metall Mater Trans A 2002; 33(7): 2139-44.

[66]   Ishibashi R, Hiroyuki D, Aono Y. Development of ultrafine grained austenitic stainless steels by high strain powder metallurgy process. Mater Sci Forum 2003; 426: 4251-7.

[67]   Garcia C, Martin F, Tiedra PD, *et al.* Pitting corrosion behaviour of PM austenitic stainless steels sintered in nitrogen-hydrogen atmosphere. Corros Sci 2007; 49(4): 1718-36.

[68]   Garcia C, Martin F, Tiedra PD, *et al.* Improvement in mechanical properties and corrosion resistance of powder metallurgy type 430L stainless steel sintered in nitrogen-hydrogen atmosphere. Corrosion 2009; 65(6): 404-14.

[69]   A.P.Tschiptschin. Powder metallurgy aspects of high nitrogen steels. In: High Nitrogen Steels and Stainless Steels – Manufacturing, Properties and Applications (U.K.Mudali and B.Raj, eds). Alpha Sci. 2004. pp.94-112.

[70]   Feichtinger H. In: Stein, Witulski (Eds) Proceedings of the 2nd.International Conference on High Nitrogen Steels - I-INS90, Aachen, Germany, 1990, p. 298.

[71]   Virta J, Hannula SP. Nitrogen Alloying of Stainless Steel Powder using Ammonia Gas in the Fluidised Bed. In: Proceedings of the 5th International Conference on High Nitrogen Steels, Espoo-Finland, May 24-6, 1998, Stockholm-Sweden, May 27-8, 1998. p.655.

[72]   Banhart J. Manufacture, characterisation and application of cellular metals and metal foams. Prog Mater Sci 2001; 46(6): 559-632.

[73]   Park C, Nutt SR. PM synthesis and properties of steel foams. Mater Sci Eng A 2000; 288(1): 111-8.

[74]   Park C, Nutt SR. Effects of process parameters on steel foam synthesis. Mater Sci Eng A 2001; 297(1-2): 62-8.

[75]   Renault ML, Gia,ei AF, Thompson MS, *et al.* Porous and cellular materials for structural applications. In: MRS Symposium Proceedings (Schwartz DS, Shih DS, Evans AG Wadley HNG, EDS). Vol. 521, 1998. p.109.

[76]   Gradzka-Dahlke M, Dabrowski JR, Dabrowski B. Characteristic of the porous 316 stainless steel for the friction element of prosthetic joint. Wear 2007; 263(7-12): 1023-9.

[77]   Salahinejad E, Amini R, Marasi M, *et al.* Microstructure and wear behavior of a prous nanocrystalline nickel-free austenitic stainless steel developed by powder metallurgy. Mater Des 2010; 31(4): 2259-63.

[78]   Bram M, Stiller C, Buchkremer, *et al.* High-porosity titanium, stainless steel, and superalloy parts. Adv Eng Mater 2000; 2(4): 196-9.

[79]   Gulsoy HO, German RM. Production of micro-porous austenitic stainless steel by powder injection molding. Scr Mater 2008; 58(4): 295-8.

[80]   Bakan HI. A novel water leaching and sintering process for manufacturing highly porous stainless steel. Scr Mater 2006; 55(2): 203-6.

[81]   Gulsoy HO, German RM. Sintered foams from precipitation hardened stainless steel powder. Powder Metall 2008; 51(4): 350-3.

[82]   Nakajima H. Fabrication, properties and application of porous metals with directional pores. Prog Mater Sci 2007; 52(7): 1091-173.

[83]   Ikeda T, Aoki T, Nakajima H. Fabrication of lotus-type porous stainless steel by continuous zone melting technique and mechanical property. Metall Mater Trans A 2005; 36(1): 77-86.

[84]   Shen YF, Gu DD, Wu P. Development of porous 316L stainless steel with controllable microcellular features using selective laser melting. Mater Sci Technol 2008; 24(12): 1501-5.

Index

Austenitic conditioning, 64, 65

Austenitic stainless steel, 3-6, 8, 11, 17, 19, 23, 24, 26-31, 33, 35, 36, 52, 53, 57, 58, 66, 69, 72, 74, 75, 77, 78-83, 96, 98, 100-103, 111-116, 118-122, 133, 134, 136, 139, 141-145, 152, 156, 159

Austenitisation, 41, 43-47, 49, 64, 67, 74, 156

Carbon steel, 3, 5, 52, 135, 162

Carburisation, 72, 76, 102, 139-141, 157

Cellular metals, 156, 160-162

Chi phase, 17, 90, 96-98, 100

Colossal carbon saturation, 139

Conditioning treatment, 64, 65

Constitution diagram, 4, 8, 9, 17, 41, 44, 45, 52, 72, 74, 75

Continuous casting, 156, 163

Coloration, 151-153

Cr-depletion theory, 8, 18, 30, 32, 113

Creep, 58, 69, 96, 99-101, 111, 120, 121

Crevice corrosion, 52, 72, 76, 82, 83, 117

Cryogenic magnetic transition, 99, 133

Curie, 13, 34, 92, 99

Deformation-induced martensite, 6, 23-25, 27, 29, 35, 114, 116, 118, 119, 134, 142

Desensitisation, 23, 30, 31, 32, 35, 113, 114

Directional pore, 156, 159, 162

Drug targeting, 133, 135

Ductile-to-brittle transition, 8, 19, 72, 73, 80

Duplex stainless steel. 4, 6, 12-17, 52-61, 66, 76, 77-80, 82, 91-95, 97-99, 103, 112, 117, 122, 133, 134, 141, 143-145, 158, 159

Dynamic strain ageing, 79, 111, 121

Electrochemical colouring, 151, 152

Fatigue, 41, 45, 48, 49, 59, 66, 72, 78-80, 100, 111, 118-122, 139-141, 158

Ferritic stainless steel, 4, 8-12, 16-19, 24, 55, 56, 58, 72, 74, 99, 101, 115, 116, 118, 134, 144, 146, 158, 159

Feroplug, 57, 133

Ferrite, factor, 52, 55

Ferrite potential, 52, 55

G phase, 3, 6, 8, 14, 15, 68, 90, 92, 93, 102, 103

Galvanic corrosion, 91

Grain boundary engineering, 23, 36

Grain refinement, 8, 9, 11, 20, 27, 72, 80, 118, 139, 142-145

Healing, 3, 18, 23, 30, 32, 48

High-Ni stainless steel, 72, 75

High nitrogen stainless steel, 3, 72, 75, 76, 159

High-temperature embrittlement, 8, 17

High-temperature oxidation, 111, 117, 118

Hydrogen embrittlement, 6, 41, 78, 80, 111, 112, 136

Hyperduplex stainless steel, 52

INCO, 151-153

Intercritical annealing, 139, 145

Intergranular carbide, 8, 18, 19, 113

Intergranular corrosion, 18, 19, 30-32, 36, 140

J phase, 3, 6, 90, 93, 94

Kolstering, 139, 141

Laser colouring, 151, 153

Laser surface modification, 139

Laves phase, 17, 95, 101

Lean duplex stainless steel, 52, 53

Martensitic stainless steel, 4, 41-45, 47-49, 66, 67, 77, 103, 157

Martensitic transformation, 9, 17, 23-30, 35, 41, 44, 49, 64-66, 72, 118, 119, 142, 145

Mechanical stabilisation, 23, 29, 46, 67

Metastable austenitic stainless steel, 6, 23, 24, 27, 114, 118, 133, 142

Metal dusting, 111, 114, 115

Microbiologically-induced corrosion, 111, 117

Microduplex, 52, 58-60, 66, 145

Modified Schaeffler diagram, 55, 56, 72, 74, 75

Neel. 34. 98, 100

Ni-containing stainless steel, 72, 73, 76

Ni-free stainless steel, 72, 74, 75

Nitridation, 139

Overageing, 12, 64, 66-68, 72, 95

One-step coloration, 151, 153

Persistent slip bands, 118, 119

Phase diagram, 9, 10, 23, 24, 44, 93

Phase prediction, 52, 55, 56, 72, 75

Pitting, 3, 8, 12, 18, 41, 45, 52, 72, 76, 77, 80, 82, 83, 101, 111, 115-117, 136, 140, 141, 142, 152, 153

Powder metallurgy, 73, 76, 145, 156, 157, 159, 161, 163

Precipitation-hardening stainless steel, 4, 64-67, 69, 112, 145, 158

PREN, 52, 53, 77, 116, 117

Pt-alloyed stainless steel, 6

R phase, 94

R' phase, 95

Retained austenite, 29, 41, 42, 44-46, 48, 65, 66

Reversion transformation, 27, 139, 144, 145

S phase, 69, 90, 95, 96, 98, 139, 140, 141

Schaeffler diagram, 44, 45, 55, 56, 72, 74, 75, 159

Semiaustenitic, 64, 65, 69

Sensitisation, 8, 11, 17, 18, 23, 30-36, 47, 48, 83, 101, 111, 113, 114

Severe plastic deformation, 3, 6, 29, 139, 146, 142, 143

Sigma phase, 8, 11, 14, 17, 32, 47, 52, 57, 58, 90, 94,